Galileo's Instruments of Credit

Galileo's Instruments of Credit

Telescopes, Images, Secrecy

MARIO BIAGIOLI

The University of Chicago Press

CHICAGO AND LONDON

The University of Chicago Press, Chicago 60637
The University of Chicago Press, Ltd., London

Paperback edition 2007
Printed in the United States of America

15 14 13 12 11 10 09 08 07 2 3 4 5

ISBN-13: 978-0-226-04561-0 (cloth)
ISBN-13: 978-0-226-04562-7 (paperback)
ISBN-10: 0-226-04561-7 (cloth)
ISBN-10: 0-226-04562-5 (paperback)

Library of Congress Cataloging-in-Publication Data

Biagioli, Mario.
Galileo's instruments of credit : telescopes, images, secrecy /
Mario Biagioli.
p. cm.
Includes bibliographical references and index.
ISBN 0-226-04561-7 (cloth : alk. paper)
1. Research—Moral and ethical aspects. 2. Galilei, Galileo,
1564–1642—Criticism and interpretation. 3. Scientific
apparatus and instruments—History. 4. Discoveries
in science. I. Title.
Q180.55.M67B53 2006
520'.92—dc22
2005014562

⊗ The paper used in this publication meets the minimum
requirements of the American National Standard for
Information Sciences—Permanence of Paper for
Printed Library Materials, ANSI Z39.48-1992.

To Gabriel and Luka

Contents

Illustrations

Abbreviations

GA Maurice Finocchiaro, ed., *The Galileo Affair* (Berkeley: University of California Press, 1989)

GC Mario Biagioli, *Galileo, Courtier* (Chicago: University of Chicago Press, 1993)

GO Galileo Galilei, *Le opere di Galileo Galilei,* ed. Antonio Favaro, 20 vols. (Florence: Barbera, 1890–1909)

KGW Johannes Kepler, *Gesammelte Werke,* ed. Max Caspar and Franz Hammer, 20 vols. (Munich: Beck'sche Verlagsbuchhandlung, 1937–)

LE Antoni van Leeuwenhoek, *The Collected Letters of Antoni van Leeuwenhoek,* ed. G. van Rijnberk, A. Schierbeek, J. J. Swart, J. Heniger, and L. C. Palm, 12 vols. (Amsterdam: Sweets & Zeitlinger, 1939–89)

LGD Galileo Galilei, "Letter to the Grand Duchess Christina," in Maurice Finocchiaro (ed.), *The Galileo Affair* (Berkeley: University of California Press, 1989), 87–118.

OL Henry Oldenburg, *The Correspondence of Henry Oldenburg,* ed. Rupert Hall and Marie Boas Hall, 13 vols. (Madison: University of Wisconsin Press; London: Mansell; London: Taylor and Francis, 1965–86)

SN Galileo Galilei, *Sidereus nuncius: or the Sidereal Messenger,* trans. with introduction, conclusion, and notes by Albert van Helden (Chicago: University of Chicago Press, 1989)

INTRODUCTION

From Brass Instruments to Textual Supplements

IN six short years Galileo went from being a somewhat obscure mathematics professor who ran a student boarding house in Padua, to becoming a courtly star in Florence in the wake of his telescopic discoveries, to receiving a dangerous censure from the Inquisition for his support of Copernicanism. Galileo's tactics shifted as rapidly and dramatically as his circumstances. The fast pace of these changes (often measurable in weeks or months rather than years) required him to respond swiftly to the opportunities and risks posed by unforeseen inventions and discoveries, and to his opponents' interventions.[1]

The changing predicaments of Galileo as an author are the main focus of this book. I look at the various ways in which he disseminated his knowledge claims and instruments, but also at his secretive practices. I trace how the very meaning of credit changed as Galileo traversed different reward systems, ending with the controversy with the Inquisition, where his claims became associated not with credit but with responsibility—criminal responsibility. Galileo's "instruments of credit" refers not only to instruments like the geometrical compass he sold and gave lessons on in Padua or the telescope through which he made so many discoveries, but also to the tech-

1. Galileo's opportunistic tactics were first discussed in Paul Feyerabend, *Against Method: Outline of an Anarchistic Theory of Knowledge* (London: Verso, 1978). While Feyerabend saw Galileo's opportunism as a methodological and perhaps mental trait (one he took to be emblematic of science in general), I treat it more as a response to highly mutable, specific working conditions.

niques he used to maximize the credit he could receive from readers, students, employers, and patrons. I mean, for instance, the apparatus to make printable pictures out of telescopic observations, his systematic withholding of instrument-making techniques to establish a monopoly over telescopic astronomy, the bootstrapping techniques for constructing an authoritative persona, the new kinds of pictorial narratives he developed to gain assent for his discoveries, and the astute grafting of his theological arguments onto those of the theologians themselves.

While obviously connected to the claims they helped to establish, these techniques were not reducible to them any more than the telescope could be conflated with the satellites of Jupiter. Galileo's instruments of credit were as innovative as the discoveries they helped constitute and disseminate. Their deployment did not reflect long-term commitments like, say, the establishment of a community of instrument users or of like-minded natural philosophers. If Galileo had a plan at this time, it was mostly about Galileo. The set of conventions, values, and styles he adopted and rearranged while crafting his instruments of credit did not amount to a "form of life" but to bricolages of bits and pieces of practices he was familiar with and could use at a specific time, in relation to a specific problem or opportunity. He did not have literary technologies, only literary tactics—secrecy being one of them. My analysis of Galileo's instruments of credit is as local as the circumstances of their deployment: I wish to convey the quick pace of these moves and the pressure under which they were played rather than seek an overarching pattern that may (or may not) comprehend them.

The diversity of the economies (institutional, cultural, and legal) in which Galileo's work was rewarded or censured easily matches (and probably exceeds) the variety of its topics and genres. These economies framed his work as well as the various modalities of its presentation, diffusion, crediting, or condemnation. He started out as a mathematics teacher (rewarded for what we would now call "work for hire"), moved up into the courtly culture of wonder and alleged disinterestedness (but also into the economy of print and authorship), and ended his life condemned to house arrest by the Inquisition for having trespassed into a restricted discursive regime (and the jurisdiction of the institution in charge of it). But if that was the end of Galileo's career and life, it was by no means the death of Galileo-the-author. The condemnation ended up canonizing him as an Author, turning a punishment into posthumous credit. Although he could not enjoy it, the cultural capital that accrued around his image fostered what is aptly called the "Galileo Industry"—yet another economy inhabited by his work.

The twists and turns that I trace in Galileo's career are not instances of what we call professional mobility. He was not a science Ph.D. crossing over from academia into the private sector, or a physician moving back and forth between the worlds of clinical care and medical research. As Galileo was constructing new instruments and claims, he was also constructing the economy in which his claims could be credited. The "Medicean Stars"—what we call the satellites of Jupiter—were not discoveries in the modern sense of the term. Galileo construed them as a kind of object that, while displaying some of the features of our notion of scientific discovery, also participated in the economies of artworks and monuments. And as he was crafting such a new kind of product, he was also fashioning himself into a role—the philosopher and mathematician of the Medici grand duke—that had not quite existed before. Product, producer, and market were shaped simultaneously.

Galileo's economies were new but grew from grafts. At each phase of his career, he made room for his claims and activities within economies that had been previously set up for something else. At Padua he developed a thriving instrument-making activity by attaching it to his conventional academic post; at the Medici court in Florence he crafted a new persona for himself by borrowing from the roles and profiles of court artists and literati; and during the controversy with the Inquisition he cast himself as a theologian by grafting his defense of astronomy and Copernicanism on the discourse of his censors. In time, the things he circulated in his composite economies became increasingly less material (moving from brass instruments to textual devices), but their logic remained that of the graft.

ON THE VERGE OF PRINT

Galileo's entrance into an economy of print was neither precocious nor glamorous. By the time he published his first book—a slim sixty-copy edition of an instruction manual of his "geometrical and military compass"—he was already forty-two. By the same age, Kepler had already published most of the major contributions to astronomy and optics (four key books and several shorter texts) for which he is famous today.[2] What is more sur-

2. The *Mysterium cosmographicum,* the *Astronomia nova,* the *Ad Vitellionem paralipomena,* and the *Dioptrice,* as well as shorter volumes like the *Conversatio cum nuncio sidereo,* the *Narratio, De fundamentis astrologiae,* the *Tertius interveniens, De nive sexangula, De solis deliquio,* the *Phaenomenon singulare,* and the *Bericht vom Kometen.*

prising is that, left to his own devices, Galileo might have further delayed the pleasure of seeing his name in print. In the preface to the 1606 *Operations of the Geometrical and Military Compass* he explained his decision to print as a response to hearing that someone was "getting ready to appropriate" his multipurpose calculating instrument. Up to that point he had been perfectly content to use manuscript copies of the compass' instruction manual for his students, but now he wanted to enlist "the press as a witness" to his priority claims.[3] He printed to control, not to communicate. Galileo's preoccupation with credit and control over his work is evident in the text itself: while the *Operations* taught how to use the compass, it did not include any illustration or information about how to build it. Ten months later he entered a bitter dispute with a Paduan student, Baldassare Capra, whom he accused of plagiarizing both his book and his instrument.[4]

Galileo's life and career changed forever in 1610 with the publication of the *Sidereus nuncius*—a description of his telescopic discoveries. But if the philosophical caliber of his claims had grown exponentially in the few years since the *Operations,* the circumstances of their printing were comparable. Galileo told a friend that the *Nuncius* was "written for the most part as the earlier sections were being printed," fearing that "someone else might find the same things and precede me [to print]."[5] Like the 1606 *Operations,* the 1610 *Nuncius* did not provide information to build the instrument that was so central to the claims made in the book. Concerns with credit and priority stayed with him for the rest of his life. As late as 1632 he remarked: "Many pride themselves of having many authorities to support their claims, but I would rather have been the first and only one to make those claims."[6]

If it is easy to make sense of Galileo's view of print as a means to build

3. Galileo Galilei, *Operazioni del compasso geometrico e militare* (Padua: Marinelli, 1606), in *GO,* vol. II, p. 370.

4. An analysis of this dispute is in Mario Biagioli, "Galilei vs. Capra: Of Instruments and Intellectual Property," *History of Science* forthcoming. As suggested by the title, Galileo's *Defense against the Lies and Appropriations of Baldessar Capra* was not a contribution to mathematics or natural philosophy but a presentation of his grievances against Capra and a detailed history of the legal case that followed (Galileo Galilei, *Difesa di Galileo Galilei nobile fiorentino, lettore delle matematiche nello studio di Padova, contro le calunnie & imposture di Baldessar Capra milanese* [Venice: Baglioni, 1607], in *GO,* vol. II, pp. 515–601).

5. *GO,* vol. X, p. 300.

6. *GO,* vol. VII, p. 540.

or defend priority claims, it is more difficult to understand why he began to use it only, and reluctantly, in 1606. Given how prolific an author Galileo proved to be in the second half of his career, and the fact that he had been doing original work in mathematics since at least 1587, his absence from print until 1606 is all the more surprising.[7] Much of that behavior is probably traceable to the practices and expectations of the professional economy to which he belonged at that time. Given the small size of their audience, mathematicians published relatively little and often communicated their discoveries and theorems to other specialists through letters or personal visits.[8] Galileo's contacts with Clavius, Guidobaldo, and Sarpi were of this kind. Furthermore, his 1594 patent for a water pump as well as his duties at the Venetian Arsenal point to Galileo's participation in the community of inventors and engineers who tended to privilege trade secrets over publications.[9] Finally, his position as mathematics professor at the university provided him with a steady, if modest, income without the need to publish. Newton's minimal list of printed publications prior to the *Principia* indicates that Galileo was not the only mathematics professor to be unmoved by the charm of print.[10] Kepler's remarkable publication track reflects, in contrast, the unstable economy of patronage at the imperial court in Prague. Books allowed Kepler to maintain the kind of visibility needed to justify the title and stipend of Imperial Mathematician—resources that did not depend on teaching duties—but also to supplement his irregularly paid salary with reliable bonuses from book dedications.[11] For a university

7. Some early modern philosophers disliked printing—a form of publication they saw as driven by commercial interests—and continued to distribute their work through the "purer" medium of manuscript, but Galileo's patent for a horse-driven water pump indicates that he did not object to making a profit from the products of his intellect (*GO*, vol. XIX, pp. 126–29).

8. See also Paul Wittich's similar attitude about printing as discussed in Owen Gingerich and Robert Westman, *The Wittich Connection: Conflict and Priority in Late Sixteenth-Century Cosmology* (Philadelphia: American Philosophical Society, 1988).

9. For a comprehensive discussion of early engineers' publication practices see Pamela Long, *Openness, Secrecy, Authorship: Technical Arts and the Culture of Knowledge from Antiquity to the Renaissance* (Baltimore: Johns Hopkins University Press, 2001).

10. For a discussion of Newton's complex relationship to print see Rob Iliffe, "Butter for Parsnips: Authorship, Audience, and Incomprehensibility," in Mario Biagioli and Peter Galison (eds.), *Scientific Authorship* (New York: Routledge, 2003), pp. 33–66.

11. James Voelkel, "Publish or Perish: Legal Contingencies and the Publication of Kepler's *Astronomia Nova*," *Science in Context* 12 (1999): 33–59, p. 39 n. 3.

mathematician like Galileo, the production of pedagogical texts (like the *Operations*) would have been seen as a plus, but was not part of his job description.

Still, it is puzzling that Galileo did not publish on topics that were both cosmologically significant and of wide popular appeal like the lodestone (on which he experimented extensively at Padua) or on the nova of 1604.[12] In the case of the nova, he did toy with the idea of publishing a book, but eventually chose to deliver only a few lectures at the university.[13] Did he fear that the natural philosophers among his colleagues would object at seeing a mathematician trespass on their disciplinary turf in print?[14] Or did he think that a few large lectures in front of a quasi-captive audience would reach more people (and enhance his local fame) more effectively than a printed text? But when more astronomical wonders like his own telescopic discoveries emerged six years later, he reversed course and published a book on the topic, delivering lectures on them only as the book was being printed. In that case he seemed unwilling to publicize his findings before protecting his priority through the printed text.[15]

Galileo's shift from lectures to books to disseminate his observations may tell us a lot about his evolving perceptions of the nature, location, and

12. Written in Paduan dialect, and satirical in tone, a small book on the new star of 1604 was published in Padua and Verona by a pseudonymous Cecco di Ronchitti da Bruzene (reprinted in *GO*, vol. II, pp. 309–34). There have been reasonable speculations, but no conclusive evidence, that Galileo had a role in this publication. Concerning the lodestone, Galileo experimented with Sagredo's, acted as broker in its sale to the grand duke (*GC*, pp. 120–26) and described some of his work (and especially the wider implications of magnetism for Copernican cosmology) in the 1632 *Dialogo sopra i due massimi sistemi del mondo, tolemaico e copernicano* (*GO*, vol. VII, pp. 425–39).

13. We have only fragments of these lectures, reproduced in *GO*, vol. II, pp. 277–84. If he did not publish on the topic it was not because of lack of interest in it, as he did go back to that event almost thirty years later, in the first day of his *Dialogue on the Two Chief World Systems*.

14. If indeed Galileo was the author or coauthor of Cecco di Ronchitti's 1604 dialogue on the new star, the fact that it was published under a pseudonym and in Paduan dialect suggests that Galileo might have tried to avoid a direct engagement with the natural philosophers at the university by writing in a nonacademic genre.

15. On May 7, 1610, Galileo wrote Vinta that "I gave three public lectures on the subject of the Medicean Planets and my other observations" (*GO*, vol. X, pp. 348–49). The *Nuncius* was published in March. Galileo mentioned making public demonstrations of the Medicean Stars in Venice and Padua in a March 19 letter (*GO*, vol. X, p. 301) but even that announcement came about one week after the publication of the book (*GO*, vol. X, p. 288).

size of the markets for his work. It is important, then, to get a sense of Galileo's everyday activities in Padua during the first half of his career, before he became the star of early seventeenth-century astronomy and began to publish books at a fast clip.

FROM BOARDINGHOUSE TO COURT

Galileo's *Operations* could be bought only at his house.[16] Together with the small size of the printing, that arrangement confirms his concern with controlling the book's readership.[17] He knew most of the *Operations*' readers personally, as they either frequented his house or rented rooms there. They were the students who studied fortification, mechanics, and other applied mathematics with Galileo outside of the normal university curriculum. Together with the *Operations,* several students bought another pricier homemade item, the "geometrical and military compass." It was produced by Marcantonio Mazzoleni, a former employee of the Venetian Arsenal who had joined Galileo's extended household with his wife and daughter in 1599.[18] His main income came from selling Galileo compasses and other instruments, which Galileo then resold at a modest markup.

According to the figures in Galileo's accounts from 1600–1, he sold about 1,000 lire's worth of instruments in a year.[19] While far from trivial, this sum was just a fraction of what he charged for room and board.[20] In 1604, for instance, the "Hotel Galileo" took in about 7,500 lire from twenty paying guests.[21] Private teaching was also very remunerative, earning him 3,312 lire in 1603 and 2,600 in 1604. Even when we set aside the earnings from instruments, Galileo more than tripled his annual university salary (1,600 lire until 1606) with private teaching and boardinghouse alone.[22] Galileo, then, was not an instrument maker who provided in-

16. *GO,* vol. II, p. 365.

17. In the preface, Galileo confirms this preoccupation by stating that he printed "only sixty copies" to be given out with the instrument on an individual basis (*GO,* vol. II, p. 370).

18. *GO,* vol. XIX, p. 131.

19. *GO,* vol. XIX, pp. 133–47. The prices varied slightly, perhaps on account of the size and decoration of the instrument. Silver compasses were much more expensive than brass ones.

20. *GO,* vol. XIX, pp. 159–66.

21. *GO,* vol. XIX, pp. 159–66.

22. As he was negotiating a position at the Florentine court in 1610, Galileo told the Medici that he could more than triple his university salary with private lessons and with the

struction on how to use the products he made, but rather one who sold instruments to increase revenues from his other household activities.[23] A sector brought him twice its price from lessons on how to use it, not to mention other classes students might take with him after having mastered the instrument. Tycho's nephew—Otto Brahe—was quick to learn and paid only 106 lire for instrument and instruction, but slower students paid up to 180 lire for the classes alone.[24] In the end, Galileo's university salary represented a relatively small portion of his overall income, though it was his university position that made his other, more remunerative activities possible.

Galileo could not prevent the copying of his instruments and written instructions once they left Padua in the hands of his students, but he squeezed much credit out of them before their departure. It is difficult, however, to categorize what kind of credit this was. In science and science studies we find an opposition between credit as money and credit as reputation. The former is usually offered to instrument makers, engineers, or laboratory assistants while the latter is tied to authorship and professional authority, accrues around the scientist's name, and resists quantification.[25] But the notion of credit instantiated by Galileo's accusations of plagiarism against Baldassare Capra does not readily fit either of those categories. Galileo pursued Capra—swiftly and relentlessly—for the piracy of an instrument and

---

revenue from his boarding house. *GO*, vol. X, p. 350. It is interesting that Galileo did not mention his income from the sale of instruments. He was probably trying to avoid being perceived by the Medici as something of a shopkeeper. On Galileo's residences in Padua, see Antonio Favaro, "Delle case abitate da Galileo Galilei in Padova," in *Galileo Galilei a Padova* (Padua: Antenore, 1968), vol. I, pp. 57–95. Favaro argues that Galileo might have run more than one boarding house.

23. James Bennett, "Shopping for Instruments in Paris and London," in Pamela Smith and Paula Findlen (eds.), *Merchants and Marvels* (New York: Routledge, 2002), pp. 370–95.

24. *GO*, vol. XIX, p. 150.

25. For instance, definitions of scientific authorship in modern biomedicine still draw the line between those who qualify for authorship credit and should have their names listed on a publication's byline (usually because of their contributions to the conceptual and textual dimensions of the publications) and those who deserve only acknowledgments (usually those who have provided data, instruments, reagents, labor, etc.) (International Committee of Medical Journal Editors [ICMJE], "Uniform Requirements for Manuscripts Submitted to Biomedical Journals," *JAMA* 227 [1997]: 928). It is understood that their stipends are the appropriate compensation for these allegedly nonauthorial contributions. A line of demarcation between original ideas and "mere" labor underpins such a taxonomy of credit, one that can be traced back to the early modern period.

its instruction manual (the kind of grievance one would expect to see associated with claims of monetary damages). And yet he stated over and over that Capra's actions had hurt his honor, not his purse.[26] Galileo's reaction was more than an attempt to cast himself above the monetary interests that allegedly drove artisans and instrument makers. Capra, in fact, did not simply appropriate Galileo's book, but he also cast into doubt his original authorship of the instrument. By intimating that Galileo was greedy enough to print a book about an instrument he may have not invented without even including information on how to build it, and then to sell his copies of the instrument "at the highest cost," Capra tried to cast him as a price-gouging artisan, not as a proper academician.[27] These accusations endangered much more than Galileo's revenues from the sales of a short instruction manual. By tarnishing his image as an academic, they also tarnished the integrity of the entire business he had grafted on it. We could say that "Galileo" was not just the name of an individual, but that it functioned a bit as a "brand name" attached to his whole web of operations. In this case, honor *was* money.[28]

The audience for Capra's attack was largely limited to Padua, the only place where his veiled references to Galileo could have been easily deciphered. It is unlikely that Galileo would have pursued Capra had he been a foreign author printing in, say, Amsterdam. Like the economy of credit

26. See, among other passages, the opening of Galileo's *Difesa contro le calunnie & imposture di Baldessar Capra* in *GO*, vol. II, pp. 517–20.

27. A praising letter from Giovanni Antonio Petrarolo appended to Capra's book mentions a "real German inventor" of the compass (*GO*, vol. II, p. 433), while Capra, in his preface to the reader, says (in an implicit but quite transparent reference to Galileo) that "whereas others contend that they have invented the proportional compass and provide it at the highest cost, in the common interest I have decided to explain its structure and use as clearly as I could" (*GO*, vol. II, pp. 435–36).

28. With profuse apologies for the anachronism, I find it heuristically useful to think of what Galileo referred to as his honor as a category that, when viewed through contemporary lenses, would fall somewhere between trademark and personality rights. The latter is the right of famous persons to control the use of their image and name because that image and name are 'products' they have established through their labor and creativity. They are deemed to have authored their persona; thus, they are allowed to control it the way one would control his/her copyrighted work. On this legal doctrine see Justin Hughes, "The Personality Interest of Artists and Inventors in Intellectual Property," *Cardozo Arts and Entertainment Law Journal* 16 (1998): 81–181; and Jane Gaines, "Reincarnation as the Ring on Liz Taylor's Finger: Andy Warhol and the Right of Publicity," in Austin Sarat and Thomas Kearns (eds.), *Identities, Politics, and Rights* (Ann Arbor: University of Michigan Press, 1997), pp. 131–48.

and authority Galileo later developed around the Medici court, the credibility and profitability of his "operation" in Padua was a local affair that hinged on his university appointment and the networks branching off from that post.

In July 1610, only four months after the publication of the *Nuncius,* Galileo took up his new position of philosopher and mathematician of the grand duke of Tuscany. Different aspects of Galileo's move from Padua back to Florence have been singled out as emblematic of that transition: from university to court; from a tolerant republic to a pope-dependent absolutist state; from mathematician to philosopher; and from a modest salary and heavy teaching load to a generous paycheck and much free time. He also went, in a matter of months, from making money selling compasses and other instruments to refusing to sell his much more coveted telescopes. More precisely, he went from selling compasses to anyone who wanted them to giving telescopes as gifts to a few people of his choosing (mostly princes and cardinals).[29] This last shift suggests that the transition from Padua to Florence was not just between two jobs, two cities, or two titles, but between two different systems of exchange that attached credit and credibility to almost opposite practices. Galileo's strikingly different publication patterns before and after 1610 indicate that printing was one of the key aspects of the economy he entered at about the time he moved to Florence.

There was not much need for print in his Paduan professional context. Most of what Galileo had to offer fell either into the category of artifacts (instruments) or into that of services (instruction and lodging). From Galileo's teaching to Mazzoleni's instruments up to the services provided by people who cooked and cleaned at the "Hotel Galileo," what was provided was labor-intensive and inherently local. This work, however, had nonlocal effects. Most of Galileo's clients and students returned home, spreading his instruments and name in the process. We could say that in this period Galileo did not print books but imprinted students. In the second half of his career, by contrast, he published books but did not have actual students—only fans and disciples.

In Padua, Galileo's intellectual and economic interests (as well as his duties at the Venetian Arsenal) brought him in close contact with the world of inventors, instrument makers, and military engineers.[30] This is well known,

29. *GC,* pp. 42–44.

30. For instance, the *Operations* were written in Italian, "so that the book, coming into the hands of persons better informed in military matters than in the Latin language, can

but little attention has been given to the fact that in that period *Galileo's own mode of production and reward was artisanal.* Everything he received credit for in Padua was labor-intensive, produced and delivered locally—and in person. Any additional credit or income meant additional labor. Nick Wilding has commented that there was a tangible ceiling to the size of Galileo's Paduan economy: the more compasses he sold, the more he had to teach how to use them.[31] He quickly ran out of time. Galileo's desire to move to court may have reflected a dissatisfaction with the *kind of economy* he was in, not just with his university teaching duties, which, in fact, boiled down to a meager thirty hours of lecturing a year.[32]

Seen in this light, printed texts like the *Operations* were not meant to expand Galileo's credit the way a novel might earn royalties for a modern author, but rather to protect his finite, labor-intensive system of production and reward. According to Galileo the *Operations* was to function only as a memory aid to help recall the viva voce instructions the reader/student was to receive from the master himself.[33] While printed books are associated with mobility and multiple contexts of reading, the *Operations* was a supplement to the classroom experience. As such, it lost much of its utility (for both the reader and Galileo) outside of that context.

When he arrived in Florence riding the wave of the *Nuncius,* he also entered the "age of mechanical reproduction" of his work, leaving the craft economy he had practiced in Padua. While he still had to provide highly customized, in-person services (the occasional philosophical entertainment at court, the rare instruction of the princes, or the replies to patrons' questions), Galileo pursued a kind of credit he had not sought while at Padua. Now it was books that went to their unknown readers, not the students

be understood easily by them" (*GO*, vol. II, p. 371, as translated in Galileo Galilei, *Operations of the Geometric and Military Compass,* trans. Stillman Drake [Washington: Smithsonian Institution Press, 1978], p. 41).

31. Nick Wilding, "Writing the Book of Nature: Natural Philosophy and Communication in Early Modern Europe" (Ph.D. diss., European University Institute, 2000), pp. 21–22. Of course, this problem was not inherent in the instrument itself, but in the modality of its production. Industrially produced instruments one can buy off the shelf operate in an economy that is very different from that of Galileo's compass.

32. "L'obligo mio non mi tiene legato più di 60 mezz'ore dell'anno." *GO,* vol. X, p. 350.

33. "These are matters that do not permit themselves to be described with ease and clarity unless one has first heard them orally and seen them in the act of being carried out" (*GO,* vol. II, p. 370, [translated in Galilei, *Operations,* 41]).

who went to his home to buy his book. The credit he now received from books did not require extra authorial labor, only a larger print run. A comparison between the sixty-copy print run of the 1606 *Operations* and the two-thousand-copy run of the 1613 *Sunspots Letters* speaks for itself.

With the expansion of Galileo's market came a dramatic change in the function of his books. The *Nuncius* is the first book in which Galileo had to convince anybody of anything. The *Operations* did not have to convince the reader of the quality of the compass, or that it could perform the operations described. Most likely, by the time the student came to reading the book, he had already bought the instrument and signed up for classes. The *Operations* was an ordered compilation of *ex professo* statements about the use of the sector. It was handed to readers who, by assuming the role of the student, accepted the authority of the author-teacher even before the reading or the teaching began. And they accepted his authority not because he was the famous Galileo, but because he was the only professor of mathematics at the university. The *Operations* treated the student-reader as a nonissue, exactly the way it treated the performance and value of Galileo's instrument.

The students did not necessarily buy the compass because it was better or more original than other similar devices one could buy in Germany, France, or England. They bought it because they were there, Galileo was there, and he was their teacher. They trusted Galileo the way the residents of a certain neighborhood trust the local baker. They bought his goods because it would have been much less convenient (or in some cases materially impossible) to shop elsewhere.

In contrast, the credit his later books could generate for Galileo was unconnected to the sale of goods or services. The *Nuncius* could produce credit for Galileo only by convincing the reader of the new, controversial claims it presented. At once, Galileo stepped outside of the traditional boundaries of mathematics and outside the context of his pedagogical authority. In this new scenario credit became inextricably tied to credibility. The reader became an issue too. He was no longer a local customer, but a remote and typically unknown person who could contribute to Galileo's symbolic capital only if he were made to accept, or at least not oppose, his book's claims.

If Galileo's books operated very differently in Padua and Florence, it does not mean that from 1610 he worked in a modern book market where his credit came from book sales. Like his Paduan students who may have paid for Galileo's instruments and classes because he was the university pro-

fessor of mathematics, his later readers granted Galileo's books a certain amount of credibility because of his title of philosopher and mathematician of the grand duke of Tuscany. In both cases, Galileo's market depended on the identity and location of his employer. "To depend," however, meant different things in Padua and Florence.

The credit that Galileo earned within his clearly defined Paduan turf hinged on the perceived quality of his services, not on his priority claims over instruments or natural-philosophical discoveries. In Florence he could not count on a similarly well-delineated job description and sphere of credibility because his role at court was very much a work in progress. But if the authority bestowed by the unprecedented title of "philosopher and mathematician of the grand duke" was yet untested, it surely expanded his capacity to play on a larger, nonlocal stage.[34] Galileo's title traveled with the Medici name through Medici channels.[35] Furthermore, the fact that he could now call himself a philosopher gave him key resources to gain credit for claims of a kind that was essentially external to his Paduan economy. Galileo's credit in Padua depended primarily on the utility of his instruments, teaching, or patents. In Florence, on the other hand, we see that his credit stemmed mostly from being first at making new claims—claims whose immediate utility was much less of an issue.[36] Novelty, priority, and utility seemed to trade roles in Galileo's economy as he moved from Padua to Florence, from mathematics to natural philosophy.

34. It is not clear what kind of authority about what kind of knowledge Galileo could claim after 1610. His title of "philosopher and mathematician" would have been an oxymoron (or an anathema) to a traditional philosopher. Furthermore, such a title was issued by a prince, not by a professional corporate body like a university. As such, it had no formal power outside of the Medici jurisdiction. The definition of the power and boundaries of the authority bestowed by Galileo's title is probably an unanswerable question. But there is no question that it was a key resource for him to play. Given the changing scenarios he confronted, playing was the only thing he could do. If in Padua Galileo published little and taught a few well-defined topics over and over, in Florence he published numerous books on topics belonging to or bringing together different disciplines: observational astronomy, natural philosophy, cosmology, mathematical physics, and theology. The variety of the audiences further complicated the boundaries of Galileo's authority.

35. I refer to the mobility-enabling resources Galileo had in Florence but not in Padua: diplomatic networks, weighty letters of introduction, Medici name recognition, prestige confered by the court, etc.

36. For instance, the Medicean Stars gained much importance as astronomical clocks that could have helped to solve the longitude problem, but that application was perfectly irrelevant to why the Medici decided to reward Galileo for their discovery in 1610.

EPISODES

Galileo's Instruments of Credit focuses less on long-term continuities (such as the structuring of his career by court culture and patronage) and more on key episodes that provide windows on the various ways in which Galileo tried to gain different kinds of credit in the different economies he traversed. After a detailed discussion of the *Nuncius* and the events of 1610, the book follows Galileo for the first few years of his new career and ends with a discussion of the 1616 clash with the theologians over Copernican astronomy. This dispute played a key role in Galileo's career, as it framed much of his later work as well as his condemnation in 1633. It also entered him into yet another regime of discourse structured by rules quite different from those he had encountered either in the university or the court. The dispute with the theologians was not played out in the so-called republic of letters but in the chambers of the Inquisition's tribunal according to protocols that bore little or no resemblance to those of the debates about the compass, the *Nuncius,* or the discovery of sunspots. The space of this book is delineated by two transitions between two different economies and discursive regimes: one between Padua and Florence and the other between Florence and Rome.

Chapter 1 is about the constructive role played by distance in Galileo's move from Padua to Florence. It analyzes how limited perceptions across distance (not just the close-range personal interactions typical of both Galileo's Paduan setting and the Medici court) were crucial in constructing his authority at the time he needed it the most, that is, when he published the *Sidereus nuncius.* Stepping out of his local sphere of credibility with a book whose claims far exceeded the confines of his disciplinary authority (and without having yet secured the title of "philosopher and mathematician of the grand duke"), Galileo was in an authority limbo for a few short but remarkably tense months, until the confirmation of his court position that July. This chapter argues that the limited information that people outside Padua or Florence could obtain about Galileo, his position, and his telescope was not an obstacle to the acceptance of his early discoveries, but actually facilitated it. Limited information helped Galileo bridge the authority gap as he was crossing from Padua to Florence.

Most contemporary science studies treats distance as an obstacle that must be conquered in order for knowledge to be transformed from private to public, for inscriptions to move from the periphery toward centers of calculation and back out again. Here, by contrast, I argue that distance and

limited information were a condition of possibility for the construction of the personal credibility of Galileo as well as of his claims. Believing that the role of distance-based partial perception of a practitioner's status and authority is not limited to Galileo's career, I have included a second example, drawn from the early years of the Royal Society, both as an expansion of and a heuristic counterpoint to the analysis of the evidence from Galileo's career. I hope that Galileo specialists will not object too sternly to this brief detour to a foreign and rainy land. But in case they do, they can easily skip the trip and go directly to chapter 2.

There, they (and everyone else) will find a revisionist reading of the debate over Galileo's discoveries of 1610 through a detailed discussion of the writing and printing of the *Sidereus nuncius,* its argumentative strategies, and especially the introduction of visual narratives—movie-style sequences of pictures representing successive appearances and positions of the Moon and the satellites of Jupiter. I read the *Nuncius* as a text that participates, with unease, in the two economies Galileo was moving through at that time. I trace the *Nuncius'* tensions between secrecy and transparency and between pedagogico-authoritative and argumentative narratives to such a hybrid predicament.

The aspect of the text that perhaps best encapsulates these tensions is the transition from a regime of credit that prized useful instruments and inventions to one that privileged discoveries that were wondrous rather than materially useful. The status of the instrument changed in the process: from the compass as a calculating device, we move to the telescope as a producer of stunning discoveries—what Bennett would call a movement from mathematical to philosophical instruments.[37] If the Paduan economy rewarded Galileo for the geometrical compass, the Florentine court rewarded him mostly for the Medicean Stars—for having put the Medici name on a heavenly billboard.

The emphasis on the *Nuncius* as a point of singularity in Galileo's transition between two economies forces the rewriting of received narratives about the reception of his discoveries. As he was moving to a court (and print) economy, Galileo's tactics were still much informed by his Paduan background. Disclosure did not come naturally to him. The transition from seeing readers as colleagues he should convince rather than potential plagiarists he should control was equally difficult. I argue that the *Nuncius'*

37. James Bennett, "The Mechanics' Philosophy and the Mechanical Philosophy," *History of Science* 24 (1986): 1–28.

literary structure and pictorial narratives were part of a balancing act between communication and secrecy, between the desire to have his discoveries accepted and that of slowing down potential replicators so that they would not become his competitors. Placed in this context, the *Nuncius* and the very limited information Galileo gave concerning the building of telescopes appear as a monopoly-seeking strategy—the translation of his old Paduan habits onto a much larger scale.

The discovery of sunspots and the ensuing dispute between Galileo and the Jesuit mathematician Christoph Scheiner in 1612–13 is the topic of chapter 3. As with the *Nuncius,* the circumstances of the writing and publishing of Galileo's letters on sunspots were tightly related to issues of priority of discoveries that questioned traditional Aristotelean cosmology. This time, however, Galileo's priority was seriously contested. While he claimed to have been the first discoverer of the sunspots, Johannes Fabricius and Scheiner preceded him to print by more than a year. Galileo's attempt to regain priority was tied to claiming that the sunspots were not what Scheiner took them to be—perhaps a way to erode his opponent's discovery credit by implying that, no matter who saw what when, the Jesuit could not understand what he had observed. In a move he was to repeat a few years later in the dispute on comets, Galileo tried to attach credit not just to priority over the reporting of phenomena but to the proper interpretation of the evidence, even though such interpretation might lead him to say (as he claimed both about sunspots and about comets) that it was not possible to say what they were.[38] In the sunspots dispute, then, the source of credit became more dematerialized, moving from an object to an argument about that new object.

Such an argument hinged on the patterns of the sunspots' movements. As in the *Nuncius,* Galileo tackled the problem through pictorial narratives, this time longer and tighter ones. A detailed analysis of the pictorial apparatus is a central aspect of the chapter and engages ongoing discussions over imaging techniques in science studies—immutable mobiles, techniques of virtual witnessing, disciplinary styles of visual representation, etc. It does so by making a case for the difference between the functioning of visual sequences like those used by Galileo and the individual, snapshot-like images one typically finds in early modern anatomy and natural history. I argue that while single images tend to be part of taxonomical projects aimed at identifying the objects depicted, Galileo-style visual

38. *GC*, pp. 267–88.

narratives engage with the temporal dimension of the phenomena to argue for the existence (but not the essence) of the objects they track in a quasi-cinematic fashion. They are tools for nonessentialistic epistemologies. And they are about new emerging objects, not already stabilized ones.

During the debate on sunspots, Galileo moved further away from notions of credit attached to accessible, saleable objects like the geometrical compass or even proprietary instruments like the telescope. In the *Nuncius* he had already managed to gain credit for the discovery of new objects (the Medicean Stars) that, while visible, were not accessible and could not be sold (only dedicated). With the sunspots, he took a next step: he tried to get credit for objects he could simply locate and prove to be nonartifactual, but was unable to say what they were.

The discussion of Galileo's pictorial sequences, then, dovetails with arguments about credit as well as about realism and instrumentalism or nominalism in early modern astronomy. It is also part and parcel of an analysis of Galileo's full use of the technologies of printing, his unprecedented push toward mechanically produced and reproduced images, and the use of prints as a means to calibrate observations conducted and instruments owned by other practitioners. If the 1606 *Operations* represented Galileo's most reluctant engagement with printing technology and economy, his 1613 *Sunspots Letters* were probably the most extensive and successful.

A profoundly different scenario confronted Galileo just a few years later when a controversy erupted over the relationship between astronomy and theology. This time he did not have to seek credit, convince someone of his claims, or print and spread them far and wide to reach the audiences he was trying to develop. In 1615–16, Galileo found himself in the unenviable position of having to defend himself from theologians who took his claims at face value and saw them as dangerous. This time he knew perfectly well who the decisive readers of his text were, though to call them an "audience" would be to misrepresent their relationship to Galileo and his work. Unlike his previous interlocutors, the theologians had the authority not only to condemn his claims (or deny him credit for them) but to punish him for having uttered them. This striking shift from credit to responsibility (one that still took place within the author function as construed by print economy) is the subject of chapter 4.[39]

39. Foucault has argued that the modern author function emerged first in relation to book censorship (i.e., authorship as responsibility) and only later added the credit element with the development of copyright centered on the name of the author. I agree with Foucault's

I look at Galileo's predicament in 1616 as the direct result of his previous successes. While this chapter is mostly about what the theologians saw as Galileo's responsibility, such a responsibility was the other side of the credit he had managed to build during the previous five years. The reversal from credit to responsibility or from convincing to defending is mirrored in the argumentative structure and publication tactics of the texts he produced during this debate, especially the "Letter to the Grand Duchess Christina."

First of all, Galileo's texts on the relationship between astronomy and theology were not printed but written and circulated as letters to friends and patrons. These addressees, however, were not meant to be the sole readers of these texts. The "Letter to the Grand Duchess Christina" was actually directed at the person whose finger was on the Inquisition's trigger: Cardinal Bellarmine. It did not need to be printed because credit from other readers mattered little in this case. Moreover, a wide diffusion of Galileo's arguments could have triggered precisely the kind of discussions the Inquisition wished to avoid. This may look like a return to the Paduan days when he used to circulate manuscript copies of his class notes to his students, but this time his readers were not students who looked up to him as an authoritative teacher but theologians who looked down on him as a mere astronomer daring to trespass into the realm of theology.

Second, the disciplinary authority of theology over astronomy and of the theologians over Galileo as a person was such that he could not appear to try to impose his rules of discourse on them. He decided instead to articulate his arguments by grafting them onto the theologians' discourse (not unlike the way he had previously grafted his credit claims first on the conventions of artisanal culture and then on the culture of the court). While the theologians worried a great deal about the dangers posed by the printing press, their economy of truth still hinged on a book they deemed special: the Scripture. Unlike other books, Scripture did not function as a means for authorial credit but was cast as a truthful inscription of the word

analysis, especially in terms of the inherently Janus-face nature of authorship, and the inseparability of credit and responsibility. Michel Foucault, "What Is an Author?" in Donald Bouchard (ed.), *Language, Counter-Memory, Practice: Selected Essays and Interviews* (Ithaca: Cornell University Press, 1977), pp. 113–38. The inseparability of credit and responsibility is common in current discussions about scientific authorship among scientists, editors, and science administrators. Authorship is often presented as a coin, and credit and responsibility as its two sides (Drummond Rennie et al., "When Authorship Fails," *JAMA* 278 [1997]: 580).

of God—truth itself. Unable or unwilling to challenge the theologians' account of the truthfulness of the Scripture, Galileo tried to work within their regime of truth by arguing that the heliocentric structure of the world was written in a special book, the book of nature, that, although distinct from Scripture, shared its status as a divine text.

While this debate was not played out in the economy of print, the book took center stage anyway. It did so not as an object but as the dematerialized topos of the "book of nature" that structured the argument of these texts, functioning as a key ingredient of Galileo's defense. The image of the book of nature was put forward as part of an argument to avoid the condemnation of Copernicanism, that is, the censoring of Copernican books. It was a "book" aimed at saving real books.

The outcome of this controversy was not a happy one for Galileo and the Copernicans. Found to contradict scriptural passages, Copernicus' *De revolutionibus* and other texts supportive of heliocentrism were placed on the Index of Prohibited Books, and Galileo was warned not to pursue Copernican cosmology. The trial of 1633 followed from and referred back to these decisions. In fact, the debate of 1616 framed much of the rules of the game Galileo was (supposed) to follow until the end of his career and life, while the discoveries he produced up to that time provided most of the resources for the defense of Copernicanism he was to articulate in the *Dialogue* of 1632. Without suggesting in any way that what happened after 1616 was a mere filling in of the blanks, the conditions of possibility of Galileo's later career were already put in place by that date. And as the "Letter to the Grand Duchess" marks Galileo's entrance into yet another economy, it also brings *Galileo's Instruments of Credit* to a close.

CHAPTER ONE

Financing the Aura

Distance and the Construction of Scientific Authority

A FEW scientists have been enshrouded by an aura of greatness, genius, and perhaps even sacredness.[1] The substantial dismemberment of Galileo's body carried out almost a century after his death by fans eager to have some relic of the Florentine martyr of science is an example of such cultic, auratic perception of scientists.[2] The fascination with Einstein's brain may be another.[3]

This chapter takes the effects of the aura very seriously but looks at its genealogy from a distinctly mundane point of view. I do not connect the aura to a mythical evocation of the wholeness of a long-gone era (as Walter Benjamin does) but treat it as the mappable effect of negotiations carried out over distance and the delays produced by such a distance—negotiations in which each party has partial (and partially updated) information about the other's position, claims, resources, and authority. I treat the construction of the aura of scientific authority as the result of an investment

1. The locus classicus on the aura is Walter Benjamin, "The Work of Art in the Age of Mechanical Reproduction," in *Illuminations* (New York: Schocken, 1969), pp. 217–52.

2. Paolo Galluzzi, "The Sepulchers of Galileo: The 'Living' Remains of a Hero of Science," in Peter Machamer, *Cambridge Companion to Galileo* (Cambridge: Cambridge University Press, 1998), pp. 417–48. Galileo was not alone in receiving this kind of treatment. Tycho Brahe's body was disinterred and subjected to autopsy in 1901, three hundred years after his death (Victor Thoren, *The Lord of Uraniborg* [Cambridge: Cambridge University Press, 1990], p. 470).

3. Roland Barthes, "The Brain of Einstein," in *Mythologies* (New York: Noonday Press, 1991), pp. 68–71.

process—the lending of one's credit to a practitioner as a result *both* of the things one might *know* and of those one might *not know* about that person or his/her claims. The aura, therefore, is not just a result (the a posteriori recognition of one's work), but a resource for producing that work in the first place, as well as for securing its acceptance from patrons and fellow-practitioners (the way the financial backing of a project is a necessary step toward its possible, but by no means necessary, success).

The aura I am talking about is based on information that is necessarily partial due to the distance between those who are working at producing knowledge claims and those who may or may not decide to take the risk of investing in such claims—accepting the dedication of a discovery that could turn out to be an artifact, lending one's name to a claim by endorsing it, or spending time and money trying to replicate it, etc. At times I draw an analogy between the construction of scientific authority and investment decisions where the effects of distance are chronic and unavoidable. The scenarios I encounter along the way, however, are not the "perfect markets" idealized by neoclassical economics, but only specific actual markets in which information is inevitably limited and unevenly distributed rather than ubiquitous, free, and complete.[4]

Together with the relationship between distance and the production of value, I am equally interested in the capacity of distance to make possible the deferral of that very value. The effects of distance are ultimately temporal, not only in the banal sense that it took a certain number of days for the mail to go from Venice to Florence in 1610, but in the more interesting sense that distance and the time lag resulting from it made possible the delaying of the delivery of the knowledge one had invested in. Perhaps one could say that distance makes authoritative knowledge possible by enabling its deferral, making it appear acceptable that "the check is (always) in the mail," that is, that knowledge cannot be delivered fully stabilized here and now.[5]

4. Joseph Stieglitz, "The Contributions of Information to Twentieth Century Economics," *Quarterly Journal of Economics* 114 (2000): 1441–78, provides a comprehensive review of the literature that has questioned, in different ways, the role neoclassical economics attributes to information. On the paradoxes underpinning such roles see also James Boyle, "Information Economics," *Shamans, Software, and Spleens* (Cambridge: Harvard University Press, 2003), pp. 35–50.

5. I am thinking of the deferral-producing effect of distance in terms of *differance*. See Jacques Derrida, "Differance," in *Margins of Philosophy* (Chicago: University of Chicago Press, 1982), pp. 1–27, and also, in the same volume, "Signature Event Context," pp. 307–

Distance occupies a central but negative role in recent interpretive models in science studies and the history of science.[6] Since the demise of beliefs in the potential universality of science and in the transferability of its methods across geographical and cultural boundaries, science studies—especially the literature informed by the sociology of scientific knowledge (SSK)—has been looking at the less grand and more laborious processes through which local knowledge is rendered public by making it travel outwards from its site of production. With a parallel shift of focus from mental processes to bodily practices-in-space, the constitution of scientific knowledge has often been equated to the geographical diffusion and acceptance of those claims.[7]

It seems that the traditional philosophical distinction between the context of discovery and the context of justification has been now recast as a sociological distinction between local and nonlocal knowledge.[8] While the spatial turn in science studies and history of science has demonstrated its remarkable heuristic value, I believe we can revise that approach by considering distance neither as a problem nor as a resource, but rather as part of the conditions of possibility of knowledge. This is not a completely new proposal. Fleck, Collins, and MacKenzie have already pointed to different aspects of the relationship between authority and distance in technoscience.[9] With the exception of Fleck, however, discussions of the impor-

30, esp. p. 315. Colin Milburn, "Nanotechnology in the Age of Posthuman Engineering: Science Fiction as Science," *Configurations* 10 (2002): 261–95, and Michael Fortun, "Mediated Speculations in the Genomics Future Markets," *New Genetics and Society* 20 (2001): 139–56, have discussed the "forward-looking" nature of scientific claims in ways that are congruous with the deferring effects of *differance*.

6. The literature on the methodological features and the several varieties of SSK is extensive. Steven Shapin, "Here and Everywhere: Sociology of Scientific Knowledge," *Annual Review of Sociology* 21 (1995): 289–321, offers a concise review of the field and its debates.

7. "The localist thrust of recent SSK has generated one of the central problems for future work. If, as empirical research securely establishes, science is a local product, how does it travel with what seems to be unique efficiency?" Shapin, "Here and Everywhere," p. 307.

8. The cognitive movement from discovery to justification has been replaced by a movement of knowledge claims and people across physical distance. This is not far from Shapin's view that "SSK has not merely attempted a resuscitation of interest in the 'contexts of discovery' abandoned by philosophers, it has also opened up new curiosity about structures of 'justification' and the translation of knowledge from place to place" ("Here and Everywhere," p. 306).

9. Ludwik Fleck, *Genesis and Development of a Scientific Fact* (Chicago: University of Chicago Press, 1979), pp. 105–44; Harry Collins, *Changing Order: Replication and In-*

tance of distance have been typically limited to how nonspecialists come to attribute high certainty to scientific knowledge.[10] As Collins put it, "Distance lends enchantment: the more distant in social space or time is the locus of creation of knowledge the more certain it is."[11] Distance, therefore, is said to cast its spell on the consumers of knowledge, not on the produc-

---

duction in Scientific Practice (London: Sage, 1985), pp. 144–45; Collins, "Public Experiments and Display of Virtuosity: The Core-Set Revisited," *Social Studies of Science* 18 (1988): 725–48; Donald MacKenzie, *Inventing Accuracy: A Historical Sociology of Nuclear Missile Guidance* (Cambridge: MIT Press, 1990), pp. 370–71.

10. Of all the available literature in science studies and the history of science, the distance-based processes I discuss here resonate best with Ludwik Fleck's analysis of the construction of scientific authority through increasingly schematizing and popularizing levels of literature, and the related difference between what he calls exoteric ("popular") and esoteric ("expert") circles (Fleck, *Genesis and Development of a Scientific Fact,* pp. 105–25). According to Fleck, trust in science cannot be based on the state-of-the-art publications produced by a small group of practitioners belonging to a given discipline simply because no other scientists (either within or without that discipline) would have either the professional competence to understand those texts or the familiarity with the specific (and still unstable) laboratory practices necessary to produce that knowledge. Trust is a function of distance and results from the reading of literature that is one or two steps removed from the scene of knowledge: "The greater the distance in time or space from the esoteric circle, the longer a thought has been conveyed within the same thought collective, the more certain it appears" (ibid., p. 106). The wider and less specialistic reading community—the exoteric circles—is a producer (not just a consumer) of scientific authority because it is at this level of literature that science is presented and communicated as certain and simple. This literature, according to Fleck, is what provides the firm authoritative grounds for the much less stable knowledge claims produced by the esoteric circles: "Certainty, simplicity, vividness originate in popular knowledge. That is where the expert obtains his faith in this triad as the ideal of knowledge. Therein lies the general epistemological significance of popular science" (ibid., p. 115).

Fleck, therefore, casts distance as productive of authority not only for the consumers of science, but for the producers too. In this, his position is closer to mine and distinct from that of SSK discussed below. However, he thinks of distance as the social space between experts and nonexperts as well as the temporal distance between the origin and use of knowledge, whereas I consider measurable geographical distance. To Fleck, the work of distance is that of simplifying and schematizing concepts, not that of framing decisions about epistemological investments. Also, while I think of distance-based negotiations that happen within weeks or months, Fleck tends to look at distance within much longer segments of a "thought collective" life—periods that can span over decades.

11. Collins, *Changing Order,* p. 145. Also: "The irony is that knowledge at a distance feels more certain than knowledge that has just been generated. The degree of certainty which is ascribed to knowledge increases catastrophically as it crosses the core set boundary in both space and time" (ibid., p. 144).

ers.[12] And it does so only after knowledge has been produced at a specific point in time and at a "locus of creation."[13]

Network models give distance a slightly greater role in the production of knowledge. According to Latour and Callon, knowledge is not completely made in one place but is drawn together or calculated in key nodes of the network from inscriptions received from elsewhere.[14] While such a

12. This is confirmed by the kind of metrology Collins applies to this "enchanting distance." At one level, Collins sees distance as the time interval between the moment of creation of knowledge and that of reception or consumption by nonexperts. Then, he talks about "social space," not the geographical distance between two practitioners involved in the production of a knowledge claim. What Collins means by distance in social space is a difference in the level of expertise or in the degree of involvement in the controversy from which that knowledge claim has emerged. In Collins' case, distance could mean the thickness of a door—the door of a laboratory that cannot be crossed by a layperson (the same layperson who is going to be "enchanted" by the knowledge produced inside—a knowledge from whose production s/he has been excluded).

13. According to Collins, the closure of a scientific controversy and the certification of results depends, ultimately, on the negotiations among key participants—what Collins calls a "core set" (ibid., pp. 142–47, 150–52, 154–55). Far from giving distance a productive role in the making of these new knowledge claims, Collins construes the "core set" precisely as an entity that makes the nonlocal local. The members of the core set bring to the site of knowledge production macroscopic (and therefore geographically dispersed) social interests. In a sense, the core set "embodies" the nonlocal context.

14. Bruno Latour, *Science in Action* (Cambridge: Harvard University Press, 1987), esp. pp. 215–37. The actor-network model was introduced in science studies by Michel Callon, "Struggles and Negotiations to Define What Is Problematic and What Is Not: The Sociology of Translation," in Karin Knorr-Cetina, Roger Krohn, and Richard Whitley (eds.), *The Social Process of Scientific Investigation* (Dordrecht: Reidel, 1981), pp. 197–220, and further articulated in Michel Callon and Bruno Latour, "Unscrewing the Big Leviathan," in Karin Knorr-Cetina and Alain Cicourel (eds.), *Advances in Social Theory: Toward an Integration of Micro- and Macro-Sociologies* (London: Routledge, 1981), pp. 277–303. Although the actor-network model involves a kind of movement through space that is distinctly different from SSK-based views—knowledge does not travel from A to B, but is produced by having inscriptions travel from the periphery to a node—it still casts distance as an obstacle in that inscriptions could lose their integrity in transit. This is why the network model deploys categories like "immutable mobiles," "train tracks," etc. Applications of the network model to early modern materials include Steven J. Harris, "Long-Distance Corporations, Big Sciences, and the Geography of Knowledge," *Configurations* 6 (1998): 269–304; Harris, "Expanding the Scales of Scientific Practice through Networks of Travel, Correspondence, and Exchange," in Lorraine Daston and Katharine Park (eds.), *Cambridge History of Science,* vol. III (Cambridge: Cambridge University Press, forthcoming); Harold Cook and David Lux, "Closed Circles or Open Networks?: Communicating at a Distance during the Scientific Revolution," *History of Science* 36

model hinges on the production of inscriptions at geographically dispersed sites, it treats distance simply as that which has to be crossed for inscriptions to reach centers of calculation. The inscriptions and resources that move across distance should not be affected by such a travel: Latour's "mobiles" are supposed to remain "immutable."[15]

Rather than thinking of knowledge as something that either conquers distance or that is constituted by having inscriptions travel unchanged through space from the periphery to centers of calculation and out again as publications, I propose to view knowledge as constituted through a range of distance-based partial perceptions. I do not look at how a knowledge claim travels from A to B, but at how the transactions made possible by the fact that A and B are distant from each other allow for the production of such a knowledge claim.

I illustrate these processes through two historical examples representative of two different configurations of the social infrastructure of early modern natural philosophy. The first example looks at the construction of Galileo's authority around the time of his telescopic discoveries of 1609–10. Here I analyze the specific steps by which a practitioner may be credited with an authority s/he does not yet have, and how geographical distance contributes to the authoritative aura of a scientist by producing perceptions of his/her reputation and status that are necessarily out of synch with the practitioner's predicament in the place where s/he operates. The second example focuses on the Royal Society of London right after its establishment to show that distance continued to frame key aspects of the more corporate world of scientific academies—a world that bears some family resemblance to that of today's science.

(1998): 179–211; and John Law, "On the Methods of Long-Distance Control: Vessels, Navigation, and the Portuguese Route to India," *Sociological Review Monographs* 32 (1986): 234–63.

15. On "immutable mobiles," see Latour, *Science in Action,* pp. 226, 236–37, and "Visualization and Cognition: Thinking with Eyes and Hands," *Knowledge and Society* 6 (1986): 1–40. Distance is to Latour and Callon's model what the void was to the ancient atomists' view of change through recombination of unchanging, elementary particles of matter. As the medium in which atoms can move and recombine, void was a crucial component of that model. Yet it had to be treated as something sterile that could have no effect whatsoever on the atoms themselves.

DISTANT STARS, DISTANT CITIES

On January 30, 1610, a few months after developing his telescope, Galileo wrote the Medici secretary, Belisario Vinta, that he had made a number of important astronomical discoveries (which included four new planets) and that he was in the process of publishing a short report about these findings.[16] The secretary promptly responded that the grand duke had expressed an "extraordinary desire to see those observations as soon as possible."[17] Seizing on the Medici's enthusiastic response, Galileo replied immediately to the secretary that he wished to dedicate his discoveries to the Medici (something he had not mentioned in the previous letter) and that he was holding up the publication of the *Sidereus nuncius* waiting to hear from them about the specific name he should attach to the discoveries.[18]

Galileo was in Padua, the grand duke in Florence—a few days (and a few state boundaries) away from each other. But although Galileo was a professor at the University of Padua, his wit and mathematical skills were known in Florence. Over the years, he had built a good reputation by teaching mathematics publicly at the University of Pisa and privately to young prince Cosimo, by dedicating to him the *Operations of the Geometrical and Military Compass*, and by entertaining the Florentine nobility and literati with his disquisitions about the geometry of Dante's Inferno.[19] Then, in the fall of 1609, he might have traveled to Florence to show the grand duke some of the telescopes he was developing.[20] Based on this preexisting credibility and on the potential benefit the Medici saw in having their name associated with such exceptional findings, Cosimo II went along and, without being able to verify Galileo's claims, allowed Galileo to attach his family name to the discoveries.

There were a number of other things the Medici did not know and, being in Florence, could not know. First, although Galileo had told them on January 30 that "I am now in Venice to have certain observations printed," the observations pertaining to the satellites of Jupiter (soon to become the

16. *GO*, vol. X, pp. 280–81.

17. *GO*, vol. X, p. 281.

18. *GO*, vol. X, p. 283. Galileo asked Vinta, the Medici secretary, whether he should name the new satellites "Cosmici" or "Medicea Sydera." Vinta's answer to that question could have been that the Medici did not wish to be associated with Galileo's discoveries.

19. *GC*, pp. 112–27.

20. *GO*, vol. X, pp. 262, 265, 268, 280.

Medicean Stars) were not even half completed. (The observations included in the *Sidereus nuncius* run from January 7 to March 2). If Galileo thought he needed an observational log of that length to convince his readers of the existence of the satellites of Jupiter, then the Medici accepted the dedication of a product that was still being completed.[21] And the Medici did not even exactly know what that product was. Fearing both competition from other astronomers and the likely failure that would have followed the grand duke's attempt at observing the Stars by himself, the only information Galileo shared with his patrons was that the Medicean Stars were four new planets that orbited a large star.[22] That Jupiter was such a star was mentioned only later, in a letter he sent Vinta *after* the *Sidereus nuncius* came off the press.[23]

Since the very beginning of the negotiations, the Medici had expressed an "extraordinary desire" to witness the discoveries and continued to press Galileo for a demonstration.[24] Galileo reassured them that they would be able to see their Stars within a "short time," but warned them about how difficult these observations could be, and that they might have to wait for Galileo's return to Florence.[25] Although he promised to send them a telescope, there is no evidence that any instrument (had Galileo actually sent it) ever reached the grand duke.[26] All the Medici received were Galileo's of-

21. This is supported by the relentlessness with which Galileo kept observing the satellites even as the book was being printed, obviously in an attempt to lengthen the observational record as much as possible.

22. Owen Gingerich and Albert van Helden, "From Occhiale to Printed Page: The Making of Galileo's *Sidereus nuncius*," *Journal for the History of Astronomy* 34 (2003): 252–56. Not to worry the Medici too much about the Copernican implications of these discoveries—worries that may have cut down on his patronage chances—Galileo did not state that all planets went around the sun, but only that they might do so—"et per avventura li altri pianeti conosciuti" (*GO*, vol. X, p. 280).

23. *GO*, vol. X, p. 289.

24. *GO*, vol. X, pp. 304–5.

25. *GO*, vol. X, p. 284.

26. Galileo repeated his worries about the difficulties the grand duke might have experienced by trying to observe on his own on March 13 (*GO*, vol. X, p. 289) and March 19 (*GO*, vol. X, pp. 299–300). He promised to send a telescope on January 30 (*GO*, vol. X, p. 281), February 13 (*GO*, vol. X, p. 284), March 13 (*GO*, vol. X, p. 289), and March 19 (*GO*, vol. X, p. 299). None of the telescopes promised on January 30, February 13, and March 13 seems to have been sent. The telescope promised on March 19 (which might have been actually sent) was not received by March 30 (*GO*, vol. X, p. 307). There is a reference to a telescope shipped from Florence to the court in Pisa on April 20, but it is

fers to visit the court to show the duke the satellites—a visit that, due to his teaching duties in Padua, would have to wait until the summer or, at the earliest, the Easter recess.[27]

Galileo did not apologize for asking so much while providing so little in return, but instead proceeded to ask the grand duke to help him advertise the Stars he had not yet seen. On March 19, as he was sending a copy of the *Sidereus nuncius* to Cosimo II, he wrote to his secretary that

> it would be necessary to send to many princes, not only the book, but also the instrument, so that they will be able to verify the truth. And, regarding this, I still have ten spyglasses that alone among one hundred and more that I have built with great toil and expense are good enough to detect the new planets and the new fixed stars. I thought to send these to relatives and friends of the Most Serene grand duke, and I have already received requests from the Most Serene Duke of Bavaria, the Most Serene Elector of Cologne, and the Illustrious Cardinal del Monte. . . . I would like to send the other five to Spain, France, Poland, Austria, and Urbino, when, with the permission of the grand duke, I would receive some introduction to these princes so that I could hope that my devotion would be appreciated and well received.[28]

The secretary agreed. He replied: "Our Most Serene Lord agrees that the news [of the discovery] should spread and that telescopes should be sent to princes. He will make sure that they will be delivered and received with the appropriate dignity and magnificence."[29] Only a week later were the grand duke and his family shown their Stars.

Why did the Medici extend so much credit to Galileo? Why did they allow him to attach their name to his discoveries without being able to check them, or without even being told what these discoveries were about? And why did they agree to help publicize a discovery they had not seen? The answer, I believe, lies in the incremental character of Galileo's requests. Between his first mention of the discoveries on January 30, the Medici acceptance of the dedication in February, the publication of the *Nuncius* on March 12, and his request for help with the distribution of telescopes on March 19, the Medici name had become increasingly entwined with Gali-

---

not clear whether this was a telescope by Galileo (*GO*, vol. X, p. 341). Galileo arrived in Tuscany with his own telescope on April 3.

27. *GO*, vol. X, pp. 284, 289.

28. *GO*, vol. X, p. 298.

29. *GO*, vol. X, p. 308.

leo's. By March 19, the *Nuncius* had already sold out and the Medicean Stars were quickly gaining European visibility, at least in printed form. It would have been quite costly for the Medici to pull out at that point.

Still, a question remains as to why the Medici invested in Galileo's claims to begin with; that is, why did they not demand that he come to Florence to show them the Stars before the publication of the *Nuncius?* I argue that Galileo used the distance between Padua and Florence to justify postponing the demonstration requested by the grand duke. Without such a distance, he could have been confronted with a range of difficult problems.

The first concerned replication. Had Galileo been in Florence, most likely the Medici would have wanted to see the Stars. This could have opened up dangerous cans of worms. Judging from the importance Galileo attributed to a two-month sequence of observations of the satellites of Jupiter to demonstrate their physical existence by virtue of their periodic motions, it is not likely that he could have convinced the Medici just with a one-night demonstration. As we will see in the next chapter, he did not believe that one or a few scattered observations would do, and we know that he failed to convince his audience with two nights of observations in Bologna on April 24 and 25. It is not at all clear whether the grand duke and his entourage would have been willing to suspend their judgment and patiently observe with Galileo for several (and hopefully cloudless) nights.[30]

It was also difficult for Galileo to guess the Medici's skill or ineptitude as observers.[31] Had the grand duke or any other influential family member been unable to see the Stars, or had they seen them only unclearly, the court, given its typical appetite for controversy, would have likely amplified their doubts. The acceptance of Galileo's dedication could have hung in the balance. It would have been much easier, instead, to have the Medici see

30. It would have been also quite disruptive for Galileo to have the Medici and their courtiers participate with him in real-time research. The examples of Tycho going to Hven, Descartes to the Netherlands, and Boyle to the countryside to avoid time-consuming and disrupting visits can give us a sense of what kind of inconveniences Galileo would have encountered had he been in Florence, next door to his would-be patrons.

31. Galileo suggested that the grand duke might have conducted early observations of the lunar surface with Galileo himself, probably in the fall of 1609. He claimed, however, that he was able to show the moon to the grand duke only "partially" and "imperfectly," due to the poor quality of the instrument he was using at that time. The fact that Galileo was certain of his observations but that the grand duke was not suggests either that the grand duke was more skeptical than his tutor or that he did not have his same observational skills (*GO*, vol. X, p. 280).

their Stars after having them primed with the long narrative of the satellites' motions in the *Sidereus nuncius* and then shown a direct observation only for "reality effect." Furthermore, even if the Medici were to succeed at seeing the satellites, they could have found them underwhelming: four fuzzy little dots of light around a bigger shiny dot, the whole thing made even less breathtaking by the telescope's various optical aberrations. The Medicean Stars sounded much more impressive when read about in the *Sidereus nuncius* than when observed directly. Galileo's extravagant dedication comparing them to the great monuments of antiquity (but surpassing them in durability) gave them a significance that would have escaped less eloquent observers, and his discovery narrative created an effect of excitement that could be hardly reexperienced by subsequent viewers. The aura of the Medicean Stars was best entrusted to a text.

The distance between Florence and Padua also helped Galileo defend his priority. The hurry with which he composed and printed the *Nuncius* testifies to his fear of being scooped.[32] A distant patron was one whose symbolic capital could help Galileo legitimize his discoveries, but also one whose absence would grant him more time to produce those observations within the narrow window needed to protect his priority claims. That kind of patron was also more likely to be taken in by the sense of urgency communicated by Galileo's January 30 letter, where the Medici secretary was told that Galileo was already in the process of printing a report on his discoveries when, in fact, he was still observing and writing.[33] This sense of great urgency—the danger of losing Galileo's (and the Medici's) claim to astronomical fame—was, I believe, influential in restraining the Medici's requests for a demonstration prior to the publication of the *Nuncius*.

Finally, had Galileo been obliged to go to court to demonstrate his findings to the grand duke, other people would have learned of the discoveries, spread the news, or even rushed to print to claim priority. Courts were not known as oases of confidentiality. Galileo worried so much about priority

32. See Gingerich and van Helden, "From Occhiale to Printed Page," pp. 254–56, and chapter 2 in this book.

33. It is not clear how much material Galileo had given the printer at the beginning of February, but we know that he received the imprimatur only on March 1 (*GO*, vol. XIX, pp. 227–28). Galileo did tell the Medici that he had "written most of [the *Nuncius*] as the first sections were being printed," but this was after the publication of the book (*GO*, vol. X, p. 300). Whether accidentally or intentionally, Galileo did not mention that he was not just writing the text as the printer was at work, but that he was still completing the observations of the Medicean Stars as well.

as to limit the information he gave his patrons, and even to beg their secretary to maintain the utmost secrecy about the little he knew about Galileo's findings.[34]

PARCELING OUT INVESTMENTS

I do not think that the Medici decided to take a chance on Galileo's discoveries simply because of his preexisting reputation and the grand duke's desire for another crown jewel. The initial trust the Medici had in Galileo may not have been sufficient to support him through what turned out to be a bumpy debate over the reality of his discoveries. For sure there were times when the Medici wavered.[35] Some of that hesitation may have resulted from seeing that they were getting increasingly involved in (and seen as supportive of) Galileo's controversial claims. To a large extent, their progressive involvement resulted from Galileo's parceling out his requests for additional installments of credit—a tactic made possible by the physical distance between him and the Medici.

The process through which the Medici's degree of involvement in Galileo's discoveries increased in time suggests that we are indeed talking about investment, not just trust. Trust is commonly seen as something one person attributes to (or withholds from) another in a voluntary fashion.[36] Trust can be earned but not extorted. While the relationship between the Medici and Galileo did not involve extortion—the Medici could withhold trust from Galileo anytime they wished to do so—they also found themselves in a situation where such a withholding could have been costly. That cost was the result of Galileo's previous actions. Probably the Medici did not want to back up Galileo more than they had already, but, having already accepted the dedication, they were pressed into giving him even more credit. This was the result of a double negative: the Medici gave him credit because they could not *not* give him credit or, more precisely, they gave him more credit because it would have been even riskier to give him less.

34. In addition to not telling the Medici that the four planets circled Jupiter (Gingerich and van Helden, "From Occhiale to Printed Page," pp. 254–56), Galileo told Vinta on February 13 that "Due cose desidero circa questo fatto [the discoveries], et di quelle ne supplico V.S.Ill.ma: l'una è quella segretezza che assiste sempre a gl'altri suoi negozi più gravi" (*GO*, vol. X, p. 283).

35. *GC*, pp. 135–36.

36. Steven Shapin, *A Social History of Truth: Civility and Science in Seventeenth-Century England* (Chicago: University of Chicago Press, 1995), esp. pp. 3–41.

Had the Medici dropped Galileo in the middle of the debate over the existence of the Medicean Stars, they would have lost face no matter what the outcome of that debate might have been. The Medici, I believe, were not facing a binary question like "do we or do we not trust Galileo?" but were rather weighing the choice between taking a loss outright or investing more and hoping for the best. The Medici were not moved to lend their credit to Galileo as the result of a voluntary, positive decision—"I trust you because I believe you deserve my trust"—but were acting under some duress—a duress that resulted from their previous smaller investments in Galileo.

PARALLEL INVESTMENTS

Very quickly, Galileo tried to reinvest the credit he gained from people who read the *Nuncius* and its dedication as a sign that his discoveries had been endorsed by the Medici. Asking the Medici to help distribute books and telescopes to princes through their diplomatic networks was part of that tactic. Because they were not in Florence, these princes were likely to overestimate the extent of the credit the Medici had actually granted Galileo.

This was not the result of false advertisement. Neither Galileo nor the ambassadors claimed that the Medici had endorsed the discoveries. Technically speaking, by the time the grand duke had agreed to the distribution of books and instruments, he had not even formalized his acceptance of Galileo's dedication. All Galileo had received was a letter in which the grand duke's secretary expressed his personal opinion that it would have been quite appropriate for Galileo to dedicate his discoveries to the Medici by calling them Medicea Sydera.[37] Of course the Medici were behind that letter, but all this confirms is that geographical distance and the partial perceptions it produced were enough to create an authority-effect without any conclusive evidence of the Medici's endorsement of Galileo.

Interestingly, Galileo tried out these tactics even before the Medici agreed to distribute telescopes and books on his behalf. He did so by mobilizing a friendly ambassador placed at a key court. About two weeks after the publication of the *Nuncius,* Galileo wrote Giuliano de' Medici (a member of a minor branch of the Florentine family), who was then the Medici ambassador in Prague at the court of Emperor Rudolph II. Together with the letter, he sent Giuliano a copy of the *Nuncius* asking him to share it with Kepler—the Imperial Mathematician. He also asked Giuliano to re-

37. *GO,* vol. X, pp. 284–85.

quest Kepler's opinion about the book.[38] Galileo probably hoped that Kepler—a Copernican likely to be pleased by Galileo's discoveries—would endorse them. Kepler's renown as an astronomer and his title of Imperial Mathematician would have conferred high credibility to his testimonial, credibility that Galileo badly needed to convince others elsewhere.

Giuliano de' Medici was an acquaintance of Galileo, but he was, first of all, the Medici ambassador in Prague—a very prestigious diplomatic post at the time. As a result, he was likely to display a certain amount of ambassadorial pomp and circumstance even while taking care of a private matter on behalf of Galileo. That this matter involved a publication about the Medicean Stars—something he saw as an important contribution to the "honor of our fatherland"—only added to Giuliano's tendency to act official.[39] When Giuliano had the copy of the *Nuncius* delivered to Kepler's home on April 8, he included an invitation for him to go to the embassy on April 13.[40] Kepler was not informed of the reason for this meeting. When Kepler arrived, Giuliano read to him aloud a section of Galileo's letter in which he asked Kepler to express his opinion about his discoveries. To this, Giuliano added his own exhortation.[41] He then invited Kepler back to the embassy for lunch on April 16, perhaps to check how his response was coming along.[42]

Kepler's description of the meeting betrays his excitement about the treatment he received, and about being put on the international stage. The day after the meeting at the embassy (but without any apparent knowledge of it) Martin Hasdale wrote to Galileo from Prague reporting on Kepler's enthusiastic reception of the *Nuncius*.[43] Not only was Kepler representing Galileo as a long-time friend (although the two had not corresponded for thirteen years), but he was happily inscribing the *Nuncius* into a Coperni-

38. Galileo's original letter is lost, but Giuliano's April 19 reply to Galileo recapitulates some of Galileo's requests (*GO*, vol. X, pp. 318–19).

39. *GO*, vol. X, p. 318.

40. Johannes Kepler, *Dissertatio cum Nuncio Sidereo* (Prague: Sedesan, 1610), reproduced in *KGW*, vol. IV, p. 285. The five-day delay between the delivery of the book and the appointment may have been designed to gauge Kepler's overall attitude toward the book before asking him for a written response.

41. *KGW*, vol. IV, p. 285.

42. *GO*, vol. X, p. 315. Perhaps the ambassador was trying to monitor Kepler's view of the *Nuncius* by having a conversation with him as he was writing the response.

43. *GO*, vol. X, pp. 314–15.

can genealogical line that included the works of Copernicus, Bruno, Galileo, and, obviously, Kepler himself.[44] Having just published a key Copernican text—the *Astronomia nova*—Kepler was probably eager to use the debate around Galileo's book to call attention to his own work and to Copernicanism in general.[45]

Given the Copernican potential of Galileo's discoveries and Kepler's interest to give visibility to himself and his recent work, as well as his need to respond to the emperor, who wanted to hear about the *Nuncius*, the Medici ambassador's request may have given Kepler a perfect excuse to write a Copernican apologia—one in which he would praise Galileo while praising himself.[46] What we know is that this response—promptly published as the *Dissertatio cum nuncio sidereo*—was ripe with references to Kepler's own books, openly cast Kepler and Galileo as Copernicans, and did its best to argue that Kepler's 1604 book on optics provided the key to understanding (and improving) Galileo's telescope.[47] And while it heaped praises on Galileo as a great observer and instrument maker easily comparable to Tycho, the *Dissertatio* denied him credit as a philosopher and cosmologist—a role that it reserved for Kepler himself. Galileo was the owner of skilled hands and eyes but Kepler was graced with a fine philosophical mind, or so he intimated. Had Kepler not been "obliged" to write to meet an "official" request from the Medici ambassador, his text could have been perceived as unabashedly self-serving.[48] For sure, the short and hastily

44. According to Hasdale's report, Kepler's only complaint was that Galileo had not given sufficient credit to those Copernicans who had open the way to his discoveries (*GO*, vol. X, p. 315).

45. Kepler's *Astronomia nova* was published in 1609, after a long intellectual and legal journey that involved negotiations with the emperor as well as with Tycho Brahe's heirs—the legal owners of Tycho's data used by Kepler (James Voelkel, "Publish or Perish: Legal Contingencies and the Publication of Kepler's *Astronomia Nova*." *Science in Context* 12 [1999]: 33–59). Given his uneasy patronage relationship with the emperor, Kepler could have benefited from the esteem he was receiving from abroad and from showing that he had been attributed the judge's role concerning an unprecedented set of new discoveries.

46. Rudolph II's request is mentioned in *KGW*, vol. IV, p. 289. Other Prague diplomats might have asked Kepler's opinion too (*GO*, vol. X, p. 314).

47. Starting just a few lines into the book, Kepler mentions several times his *Astronomia nova* (1609), *Ad Vitellionem paralipomena* (1604), *Mysterium cosmographicum* (1596), *De stella nova* (1604), *Phaenomenon singulāre seu Mercurius in Sole* (1609), and even a text he never published, the *Hipparchus*.

48. I believe that Kepler's extraordinary praises of Galileo were probably necessitated by Kepler's desire to be equally celebratory of his own work and of how much Galileo owed

composed *Dissertatio* proved an effective advertisement for Kepler in Italy, where his landmark texts on optics and planetary theory had received little attention and virtually no diffusion before then.[49]

It is not clear whether the formality of the event led Kepler to take Giuliano's request as an official act—something that came from Galileo but that was ratified and supported by the Medici themselves—or whether he simply chose to read it that way to confer a certain aura of disinterestedness to his own response.[50] Kepler knew that Galileo taught mathematics at Padua, but probably did not know, until he read the *Nuncius,* that he was also a Florentine and a Medici subject.[51] Galileo's mention, in the dedication, of the four summers spent teaching mathematics to the grand duke when he was still a young prince lent support to his claims of a direct client-patron relationship with the Medici.[52] Kepler was not exactly naïve about patronage matters and probably read all this as a statement of Galileo's intents rather than a description of his employment.[53] But the choreography with which the text and Galileo's request for a comment were delivered to him, and his own personal and intellectual interests, seem to tilt his judgment toward believing that, if Galileo was not on the Medici payroll, he was at least a close client of theirs (and in fact referred to him as such).[54] Be that as it may, Kepler's long response was completed in six days so that

---

to him and other Copernicans. In order to make Galileo's text work for him without sounding envious or grabby, Kepler probably felt he had to be overly positive about it.

49. Massimo Bucciantini, *Contro Galileo: Alle origini dell'affaire* (Florence: Olschki, 1995), p. 104.

50. The fact that Kepler's representation of these events comes from the preface to the printed version of the response suggests that what we are reading is a well-thought-out presentation that may reflect Kepler's strategic framing of the book more than his actual impression of the audience at the Medici embassy.

51. While the book did not state that its author was on the Medici payroll, it did not dispel that possibility either. One line on the title page identified Galileo as a mathematician at the University of Padua, but a longer line (in a font twice the size) above it presented him as a "Florentine patrician" (*SN,* p. 27). Then, the gist of the dedication was that Galileo's discoveries stemmed from him having always been a faithful Medici subject—"I am not only by desire but also by origin and nature under Your dominion" (*SN,* p. 32).

52. *SN,* p. 32.

53. In the printed version of his response to the *Nuncius,* Kepler presents Galileo as both a "Florentine patrician" and a "mathematics professor at the University of Padua" (*KGW,* vol. IV, p. 288).

54. "Galilaeus Mediceorum cliens esset" (*KGW,* vol. IV, p. 285).

it could be sent to Italy by the first available courier. It was printed less than three weeks later.[55]

It is clear, however, that Giuliano was not following instructions from Florence and that he perhaps knew nothing about the negotiations between Galileo and the Medici. On April 8 (the day Kepler received the copy of the *Nuncius* from Giuliano) Galileo was demonstrating or had just demonstrated the Medicean Stars to the grand duke, and no action had yet been taken on the Medici's decision to help distribute Galileo's telescopes to princes—a decision the grand duke had made about a week before.[56] Also, Kepler's response was not sent to the Medici in Florence but to Galileo in Padua.[57] The Medici's extraneity was confirmed when, ten days after Kepler's delivery of his response, the Medici resident in charge of the diplomatic mail between Florence, Venice, and Prague grew worried about the amount of correspondence between Galileo and Giuliano and asked Galileo to clear his further use of diplomatic channels with Florence.[58]

Had he been in Florence or Venice, Kepler would have understood that at that time Galileo had not received any official endorsement by the Medici and that he was not a direct client of theirs. He would have also understood that the ambassadorial pomp displayed by Giuliano while delivering Galileo's letter did not signify the Medici's official commitment to Galileo's work. Instead, a few days after being summoned at the Medici embassy, Kepler (who did not have access to a suitable telescope at the time) *confirmed Galileo's discoveries without having being able to see them himself.* Kepler's response remained the only strong public endorsement Galileo's discoveries were to receive for several months.

Geographical distance and good timing worked again in Galileo's favor. Like the Medici before him, Kepler was not compelled to believe Galileo (though, like them, he had developed some trust in him prior to this exchange and, like them, might have gained from investing in Galileo's dis-

55. The letter version to Galileo is dated April 19 (*GO,* vol. X, p. 319), and the dedication of the *Dissertatio* to Giuliano is dated May 3 (*KGW,* vol. IV, p. 285).

56. Vinta wrote Galileo about the Medici's willingness to distribute his telescopes on March 30 (*GO,* vol. X, p. 308). The April 8 date is mentioned in the dedication of the *Dissertatio* (*KGW,* vol. IV, p. 285).

57. Giuliano did not even read the letter, or make a copy of it, before sending it to Italy (*KGW,* vol. IV, p. 285). It is most unlikely he would have done so had this been a diplomatic correspondence.

58. *GO,* vol. X, pp. 348–50.

coveries).[59] Perceiving (or choosing to perceive) Galileo's request as an official one, he could not simply ignore it. Of course, he could have been more cautious in his endorsement, so much so that he felt obliged to add a second preface to the *Dissertatio* to justify what some friends had judged as an overenthusiastic endorsement of Galileo.[60] He was equally defensive in the 1611 *Narratio*—a short report on his actual observations of the satellites—which he framed as a response to the skeptics who had criticized his premature endorsement of Galileo's claims in the *Dissertatio*.[61]

Of course Kepler could have limited himself to saying that he could not assess Galileo's claims because he did not have a suitable telescope. But the combination of distance, time pressure, Giuliano's choreographing of the request (conscious or accidental as it may have been), the Copernican value of the discoveries, and Kepler's desire for visibility all joined up to make him decide that this was an offer he could not refuse.[62] Like the Medici before him, Kepler was put in a position in which he could either agree to invest (and invest quickly) in Galileo's discoveries or drop the offer (and perhaps contribute to the dismissal of pro-Copernican discoveries, relinquish the international visibility this episode would have given him, etc.). He decided to bet on Galileo, Giuliano, and Copernicus and (not without some anxiety) wrote a ringing endorsement of something he had not seen. Galileo got more credit on an installment plan.

Kepler's situated perception was inscribed in his text.[63] Although a re-

59. Kepler, however, was not a close correspondent of Galileo's, and had no particular reasons to feel warmly toward him. Prior to Galileo's April 1610 letter, the two had not communicated for more than a decade. And that, I believe, was the result of Galileo's failing to answer Kepler's last letter, in 1597.

60. Kepler, "Ad lectorem admonitio," *KGW*, vol. IV, pp. 286–87.

61. Johannes Kepler, *Narratio de observatis a se quatuor Iovis satellitibus erronibus* (Frankfurt: Palthenius, 1611), in *KGW*, vol. IV, pp. 315–25, esp. 317–18.

62. The topos of the "offer one cannot refuse" foregrounds quite nicely the tension within the notion of offer—the same tensions one can find within the notion of trust. The spectrum that goes from "genuine offer" to "threat" is a long and varied one. At one end—the "genuine offer" end—is a situation represented as one in which the person who does not accept the offer relinquishes only the benefits associated with the thing being offered. At the other end, by contrast, is a situation in which the person who refuses the offer risks consequences that far exceeds the loss of the thing being offered.

63. The terms "situated" and "partial" that I use throughout this chapter are a direct reference to Donna Haraway, "Situated Knowledges: The Science Question in Feminism and the Privilege of Partial Perspective," in Mario Biagioli (ed.), *The Science Studies Reader* (New York: Routledge, 1999), pp. 172–188.

sponse to Galileo, Kepler dedicated the *Dissertatio* to Giuliano de' Medici because

> I cannot think of anyone to dedicate this letter but Your Most Illustrious Lordship. In fact, you motivated me to write it, not only by delivering me a copy of the *Sidereus nuncius* on April 8 through Thomas Segett, but also by summoning me in person on April 13. Then, as soon as I appeared in front of you, you read me the explicit request contained in Galileo's letter, and added your exhortation to his. And I, taking notice of this, promised to write something by the date on which couriers usually depart. I have kept my promise.[64]

Kepler's belief that he was responding simultaneously to Galileo and to the Medici emerges in his decision to print what was supposed to be a private letter—a move that he did not see as impolite because, while the *Dissertatio* was a direct response to Galileo, its genealogy and subject matter was more broadly related to the Medici:

> I have first conceived and then printed [this letter] so eagerly because Galileo, to whom it is directed, is a client of the Medici, and even more because it was requested of me by the ambassador of the Medicean Prince who rules the Grand Duchy of Tuscany (and the ambassador himself belongs to the house of Medici), and finally because the subject matter is such that (if I have been told the truth) it pertains, by deliberate intention of the author, to the high honor of the Medici name.
>
> Therefore, receive Most Illustrious Lord this letter that was a private communication to Galileo and is now yours by public dedication. Recognize in this dedication my desire to celebrate, following Galileo's lead, the truth and fame of the Medici rule, fame that is solely based on truth.[65]

DELIVERING THE GOODS

By the time Kepler's long letter reached Galileo in Padua, the Medici had been favorably impressed by Galileo's demonstration at Pisa (where the court resided in April) and had let him know through their secretary that they were considering a court position for him in Florence. Soon after, however, the Medici began to receive word of dismissive critiques of the *Nuncius*. Some of these fast-spreading rumors were associated with presti-

64. *KGW*, vol. IV, p. 285.

65. *KGW*, vol. IV, p. 285.

gious figures like Giovanni Magini, the professor of mathematics at Bologna, and other philosophers from that institution.[66] The grand duke became quite cautious and, without conveying any sign of explicit distrust to Galileo, he slowed down the confirmation of his informal offer of a court position.[67]

Anxious to finalize the negotiations, Galileo wrote the Medici secretary on May 7 (immediately after receiving Kepler's response in the diplomatic pouch from Prague) to discredit the attacks and to let him know of the many endorsements he had received. Most of these endorsements, however, were left nameless. The only person he identified (not by name but as "Imperial Mathematician") was Kepler.[68] Galileo also forwarded Kepler's letter to the Medici:

> Your Most Illustrious Lordship shall see (and Their Most Serene Highnesses through you) that I have received a letter, actually an entire treatise of eight sheets, from the Mathematician of the Emperor. He has written it to approve of all parts of my book, without doubting or contradicting even the smallest little thing.[69]

He then suggested that Kepler's testimonial was all the more relevant because it came from far away. Kepler had no interest in dismissing the Medicean Stars—as may have been the case with other Italian literati who envied Galileo as much as the neighboring princes envied the power of the Medici.[70]

Distance was crucial again, though not for the reasons given by Galileo. Not knowing of the backstage maneuvers, the Medici may have thought that Kepler had freely offered this endorsement.[71] Conveniently, Kepler's

66. *GO*, vol. X, pp. 345, 365. See also chapter 2, this volume, pp. 114–15.

67. *GC*, pp. 133–39.

68. The choice of this designation suggests that Galileo thought that the Medici were more likely to be impressed by the rank of Kepler's patron than by his reputation among mathematicians (a reputation with which the Medici were not likely to be familiar).

69. *GO*, vol. X, p. 349.

70. "Et creda pur V.S. Ill.ma che l'istesso haveriano anco parimente detto da principio i literati d'Italia, s'io fussi stato in Alemagna o più lontano; in quella guisa a punto che possiamo credere che gl'altri principi circumvicini d'Italia con occhio un poco più torbido rimirino l'eminenza et potere del nostro Ser. mo Signore, che gl'immensi tesori e forze del Mosco o del Chinese, per tanto intervallo remoti" (*GO*, vol. X, p. 349).

71. All the letter said about the circumstances of the writing is "While I was thinking the matter [the claims reported in the *Nuncius*] over, your letter to the ambassador of the

early manuscript response to Galileo differed from the later printed version in that it included neither the dedication to Giuliano (in which Kepler spelled out most of the circumstances of the composition of the response) nor the somewhat defensive "preface to the reader," in which he justified what others had perceived as the unreasonably enthusiastic tone of the endorsement.[72] It also did not include a number of appreciative references to Bruno—references that could have rattled the pious Medici.[73]

On May 22, the Medici secretary wrote Galileo that

> I have received all the letters of Your Lordship, and having read all of them to the Most Serene Patrons, they have received infinite pleasure, especially from the last one [Kepler's endorsement], because all the literati and experts (and even those who had been previously skeptical about your opinions) have been persuaded and convinced by the well-founded deductions, reasons, and observations of Your Lordship. And concerning the desire of the Most Serene Patrons to have you here, and to give you the honorable provisions that I mentioned, and the virtuous leisure to finish your studies and perfect all those works that you will offer to the light of the world for the public good and under the patronage and name of this great and Most Serene Prince, Their Highnesses are committed to it and they have given me their word, and they will think of a most honorable title for you.[74]

Notice that Vinta's narrative links the Medici's decision to go ahead with a position for Galileo to the letters they had received, including Kepler's. That the *Dissertatio* is not mentioned in this letter indicates that the Me-

---

Most Illustrious Grand Duke of Tuscany arrived, full of affection for me. You did me the honor of thinking that so great a man in particular should encourage me to write, and you sent along a copy of the book and added your own admonition." *KGW,* vol. IV, p. 290, as translated in Johannes Kepler, *Kepler's Conversation with the Sidereal Messenger,* trans. Edward Rosen (New York: Johnson, 1965), p. 12.

72. I base these considerations on Kepler's original letter to Galileo as published in *GO,* vol. X, pp. 319–40. Galileo himself did not know of Kepler's more cautious addenda ("Ad Lectorem Admonitio," *KGW,* vol. IV, pp. 286–87). In a sense, distance reinforced Galileo's own perception of the strength of Kepler's response. The version of Kepler's letter that Galileo sent to Florence mentioned only that Kepler had received Galileo's letter from the Medici ambassador.

73. The discrepancies between Kepler's letter to Galileo and the *Dissertatio* are discussed in Rosen's notes in Kepler, *Kepler's Conversation with the Sidereal Messenger.*

74. *GO,* vol. X, p. 355.

dici had not seen the book by May 22, when Vinta delivered Galileo the good news.[75]

As I discuss in the next chapter, the position Galileo received from the Medici allowed him to assume an authoritative, if not arrogant, stance toward those who failed to replicate his discoveries. He began to act as if the difficulties some had encountered in seeing the satellites of Jupiter did not discredit his discoveries but only confirmed that his telescope was the best.[76] By coming through with their promises, the Medici gave him additional credit that he could then use to gain more credit elsewhere.

THE POWERS OF PARTIALITY

Galileo started out with some credit in the eyes of the Medici—the reputation that predated his telescopic discoveries. Then, step by step, he increased it through a creative marketing of his discoveries from a distance, first with the Medici themselves and then with Kepler. He finally cashed all this credit back with the Medici, and obtained a position from which he could control his monopoly of telescopic astronomy.[77] The quick pace of these transactions did not simply enhance Galileo's authority but rather constructed it. His success at securing a position at the Florentine court did not simply result from delivering the Medici the goods he claimed to have had ready for them since the very beginning. Rather, he used their credit to develop the very goods they had bought from him. The Medici thought they had purchased something that was ready for delivery, but were simply investing in Galileo and his unfinished observations.

This process did not involve misrepresentations or misunderstandings in

75. Kepler's dedication of the *Dissertatio* to Giuliano is dated May 3. Galileo wrote on May 28 that the *Dissertatio* was being reprinted in Venice (*GO*, vol. X, p. 358). It never was. Rosen has suggested that this was because of the appearance of a pirated version in Florence (Kepler, *Kepler's Conversation with the Sidereal Messenger,* p. 71 n. 84). I believe instead that Galileo could have dropped the publication project after receiving the May 22 letter from Vinta.

76. See chapter 2, this volume, pp. 81–83, 132–33.

77. The production of credit through successive investments at a distance shares the cyclical developmental nature of the model proposed in Bruno Latour and Steve Woolgar, *Laboratory Life: The Construction of Scientific Facts* (Princeton: Princeton University Press, 1986), pp. 187–233. In both cases, one can find a conversion between what Latour and Woolgar call "credit as credibility" and "credit as reward" (ibid., p. 198) and an overall increase of the practitioner's professional capital. However, here I tie "credit as credibility" to the work of distance and partial perception.

the strict sense of the terms. At the same time, the decisions made by the various actors did not rest on so-called rational choices based on calculable risks and benefits attached to each possible move.[78] What we see at play are judgments predicated on distance and guided by the investors' desire, interest, and willingness to invest, that is, the situated and partial perception of the potential benefits they might have obtained from those investments.[79] While I believe that Galileo knew more than the others and used his additional knowledge to plan and time his various moves, he did not trick his investors either.[80] He walked a thin line between partial representations and misrepresentations without technically falling into the latter.

I do not believe that the availability of more background information would have made these exchanges more "rational," but it could have eroded their condition of possibility.[81] While these transactions were rooted in a series of exchanges that appeared reasonable to both sides, those who engaged in these exchanges did not do so because they attributed the same value to what they traded. The possibility of commercial exchange rests on the fact that the two parties involved in the exchange have different interests in (and views of) the objects being exchanged. Exchange signals the local intersection of different (and perhaps even incommensurable) interests.

Of course, Galileo's investment tactics were by no means bound to be successful. Had the initial modicum of credit that Galileo had from both the Medici and Kepler—his start-up symbolic capital—been less than it was, it is not clear whether the cycle of investment could have been set in motion to begin with. The same could have happened if the distance between Galileo, Kepler, and the Medici had been much greater (or smaller) than it happened to be. Also, had the Medici perceived Galileo's discovery as too controversial, they might have declined the dedication. Alternatively, stronger attacks on the Medicean Stars could have led the Medici to drop the investment and take the loss before Kepler's comforting letter could

78. In this sense, Galileo, the grand duke, the Medici ambassador, and Kepler were not even allies in the Latourian sense of the term because, with the exception of Galileo, no one had a comprehensive map of the network, its members, and its goals.

79. Haraway, "Situated Knowledges," pp. 172–88.

80. This was no Ponzi scheme. Galileo was simply looking for investments to fund the development of his product (which he did develop). Although my narrative has presented Galileo as the mastermind of this investment cycle, I believe that the same chain of events and exchanges could have taken place even without attributing to Galileo all the agency and knowledge I have attributed to him.

81. See, for instance, the no-trade theorem in Paul Milgrom and Nancy Stokey, "Information, Trade, and Common Knowledge," *Journal of Economic Theory* 26 (1982): 17–27.

reach them in Florence. One can also think of a scenario in which Galileo would not have had sufficient time to "develop" the value—the epistemological robustness—of the Medicean Stars.

In the absence of a corroboration like the one he received from Kepler, the Stars could have floundered and become a bubble, only to be picked up again by someone else, somewhere else, at a later time. Given the quick pace of development of telescopes and telescopic astronomy in 1610, it would be hard to believe that the satellites of Jupiter would have never made it out of the dustbin of discredited discoveries. But it would be equally difficult to treat all bubbles as inflated investments that deserved to burst. More simply, they could just be claims that were not or could not be successfully developed at that time, within that window, and according to what "acceptable development" meant there at that time.[82]

In any case, as I show in the next chapter, Galileo did not seem to worry about lack of independent replications, but about the fact that those who could reproduce his findings might also make further discoveries—discoveries he wanted to make himself. I believe that Galileo sought Kepler's testimonial so promptly and aggressively because he wanted to close the deal with the Medici and obtain the court position that, in turn, would have allowed him to ignore the remaining critics, and dedicate himself to producing additional discoveries (like the phases of Venus and the appearance of Saturn).

By securing the court position as promptly as possible, Galileo was trying to maximize his chances at further increasing his credit and authority. His tactics were not merely about constructing authority, but about constructing it fast because fast meant more. He cornered the market of telescopic astronomy by getting into it very quickly, thus making it very difficult for others to break his monopoly.[83] The effect of that monopoly was then propagated through the historiographical mythology of Galileo—his aura—that is still with us today.

CORPORATE DISTANCES

Such a productive role of distance was not an accident of the highly personalized network of connections typical of the patronage system in which Galileo operated, or of the exceptionality of his discoveries. While I cannot

82. As shown by how many potentially interesting inventions never make it to the market or fail shortly after having gotten there.

83. See chapter 2, this volume, pp. 79–85.

offer a comprehensive analysis of the role of distance in science across different historical contexts, I can briefly show how it continued to function, in mutated but pervasive forms, in the more institutionalized and bureaucratized practices of an early scientific academy like the Royal Society of London. The permanence of the productive role of distance in the transition from the patronage system to the kind of corporate infrastructures typical of scientific academies is, I believe, a telling piece of evidence. The Royal Society cannot be taken to exemplify modern science in all its complex configurations and forms of organization, and yet it does bear more than a passing family resemblance to it. Key elements of today's social system of science like the registration of claims and discoveries, the publication of dedicated journals, and the introduction of peer-based protocols for the evaluation of knowledge claims and the attribution of credit can be recognized in the practices of the Royal Society. The role of distance in the workings of the Society may give us pointers for looking for signs of its function in today's science, while a comparison of the role of distance in Galileo and the Royal Society may add new dimensions to the complex transition from patronage-based to institution-based frameworks of early modern natural philosophy.

Although it quickly managed to establish itself as a crucial node in the emerging philosophical republic of letters, the Royal Society of London had little authority at the time of its foundation, and received very limited financial support or direct legitimation from the English king. The main resource it received from Charles II in 1662 was its charter.[84] If the legitimation of its philosophical program required some complex sociopolitical bricolages, the managing of the Society's bottom line and its weekly meetings was no easier task.[85]

84. This section expands part of an argument presented in my "Etiquette, Interdependence, and Sociability in 17th-Century Science," *Critical Inquiry* 22 (1996): 193–238, esp. 225–30. I have also benefited from the discussion of the Oldenburg-Hevelius correspondence in Christopher Coulston, "The Bank of the Republic of Letters: Johannes Hevelius and the Royal Society" (unpublished manuscript, Department of History of Science, Harvard University, 2001), for which I thank the author.

85. On the relations between the Society's experimental "form of life" and the culture of Restoration England, see Steven Shapin and Simon Schaffer, *Leviathan and the Air Pump* (Princeton: Princeton University Press, 1985), esp. pp. 283–344. Some aspects of this picture were articulated in the first official history of the Society: Thomas Sprat, *A History of the Royal Society* (London: Martyn, 1667). See also Michael Hunter, *Establishing the New Science: The Experience of the Early Royal Society* (Woodbridge: Boydell, 1989), pp. 45–71.

Despite its name, the Royal Society was a private, voluntary organization with very limited and poorly paid staff.[86] It was not blessed with a particularly productive or competent membership, but a more stringent admission policy might have been unwise given that membership fees were the Society's main source of income. The academy struggled to secure a significant endowment to support its activities, hire staff, find a building to call its corporate home, keep its members interested, maintain a good level of activity at its weekly meetings, and make sure that everyone paid his dues.[87] Despite the enthusiasm that permeated the first few years of the academy's life, crisis was just below the surface, and the corporate survival of the Society never certain.[88] Already in November 1664, the secretary of the Society—Henry Oldenburg—was writing Boyle that

> this Society would prove a mighty and important Body, if they had but any competent stock to carry on their desseins and if all ye members thereof could but be induced to contribute every one their part and talent for ye growth, and health and wellfare of their owne body.[89]

The same themes had emerged in an earlier letter and were to come up again in February 1666:

86. The endowment given by the king to the Society amounted to Chelsea College, which the Society returned to the king a few years later, after having had difficulties securing satisfactory possession of it. Marie Boas Hall, *Promoting Experimental Learning: Experiment and the Royal Society, 1660–1727* (Cambridge: Cambridge University Press, 1991), p. 14. Members paid forty shillings when admitted, and one shilling per week afterwards.

87. No significant endowment was raised after the king's initial grant of Chelsea College. While the membership peaked over at two hundred in the 1670s, the number of active members from 1663 to 1685 hovered around twenty (John Heilbron, *Physics at the Royal Society during Newton's Presidency* [Los Angeles: Clark Library, 1983], pp. 4–6). The collection of the membership fees was, to put it mildly, lax. Around 1670, the society collected about half of what it was owed by the members (ibid., p. 6). The Society moved to a house of its own in Crane Court in 1710, during Newton's presidency (ibid., p. 17).

88. The Society's crisis during its first two decades is discussed in Hunter, *Establishing the New Science,* pp. 185–239, esp. 189–91.

89. *OL,* vol. II, p. 320. In the same letter, talking about the manuscript of Sprat's *History,* Oldenburg writes: "I must confesse, ye style is excellent, even, full, unaffected; but I know not whether there be enough said of particulars, or, to speake more truly, whether there are performances enough, for a R. Society, yt hath been at work so considerable a time" (*OL,* vol. II, p. 321). I do not know whether Oldenburg is referring to Sprat's paucity of examples or to the fact that the Society had not produced sufficient work to provide Sprat with such examples.

> Such persons, as you, Sir, we highly need to assert and promote ye dessein and interest of ye Society, and to suggest ye proper wayes of carrying on their work. There are so few of such, yt, unlesse either they redouble their zeale, or their number encrease; yt Noble Institution will come far short its End. We are now undertaking severall good things [. . .] but ye paucity of ye Undertakers is such, yt it must needs stick, unlesse more come in, and putt their shoulders to the work.[90]

Then in September 1667:

> The R. Society hath not met these 2. months; but I hope, they will shortly meet again. I know not, what deadnes there is upon the members of it.[91]

In the same letter, he expressed his hope that "so noble and usefull an institution may not fall to the ground." A few months later a more cheerful Oldenburg told Boyle that "the fame of the Society riseth very high abroad," but this time it was Boyle's turn to sound skeptical:

> I am not sorry that the reputation of our Society increases abroad; but we shall have cause to be sorry if nothing be done at home (by those whom it most concerns) to enable us to make it good.[92]

It is intriguing that at a time when crucial insiders worried about the Society's ability to survive its corporate childhood, people abroad could have quite a positive view of the Society.

As the crisis continued and domestic productivity further declined (to the point of having a twenty-shilling "prize" awarded to fellows who could produce an acceptable experiment) the amount of out-of-London correspondents increased and the discussion of incoming letters took up much of the time previously dedicated to making and discussing homegrown ex-

90. *OL*, vol. III, p. 45. In September 1664, Oldenburg had already complained to Boyle that while a French visitor he had just entertained "will extoll our Institution and proceedings to ye sky, wheresoever he comes; though I must needs say, we grow more remisse and carelese, yn I am willing to exspatiate upon. Yet this I must say, [. . .] yt nothing is done wth ye king for us; yt our meetings are very thin; and yt our committees fall to ye ground, because tis not possible to bring people together; tho I sollicite, to ye making myselfe troublesome to others, not to say much of ye trouble, wch I create to myself, good store" (*OL*, vol. II, p. 235).

91. *OL*, vol. III, p. 476. In June 1666 Oldenburg wrote Boyle that: "I hope, our Society will in time ferment all Europe, at least; I wish only, we had more zeale, and a great deal more assistance, to doe our work thoroughly" (*OL*, vol. III, p. 155).

92. *OL*, vol. IV, p. 299.

periments.[93] The pattern of high visibility abroad and poor performance at home deepened in the 1670s.[94]

TROUBLES AT HOME, FAME ABROAD

According to Michael Hunter, "It soon became apparent that, so long as there was something to report in the way of research and publication, it was paradoxically irrelevant to the vitality of the correspondence how healthy the Royal Society as an institution actually was."[95] I think that the "paradox" clearly spotted by Hunter was, in fact, a structural feature of the process that helped the Society survive its early relative unproductivity and lack of resources.

93. The minutes of the April 13, 1668, council meeting indicate that a reward was established for those who supplied experiments: "That the President be desired signify to the society, that considering the want of experiments at their public meetings, the council had thought proper to appoint a present of a medal of at least the value of twenty shillings to be made to every fellow, not curator by office, for every experiment, which the President or Vice-president shall approve of." Thomas Birch, *The History of the Royal Society of London* (London: Millar, 1756), vol. II, p. 265. On the relation between experiments and correspondence Hunter argues that: "The rise in bulk of the Society's correspondence and the decline of corporate experiment were connected, because discussion of correspondence occupied an increasingly large share of the Society's time" (Michael Hunter, *Science and Society in Restoration England* [Cambridge: Cambridge University Press, 1992], p. 50). However, the founders of the Society entertained the idea of corresponding with foreign virtuosi from very early on, as reflected in the stipulations about correspondence in the Society's charters (*The Record of the Royal Society* [London: Royal Society, 1940], p. 235 [first charter] and p. 262 [second charter]).

94. For instance, on September 2, 1674, Oldenburg wrote Williamson that "M. Slusius, who is one of our best Correspondents [. . .] hath exprest a great esteem of his Majesties Institution of the R.Society, in ye doing of wch he concurs wth ye most Eminent men in most parts of ye World. Wch, as it adds not a litle to ye renown of England, now admired abroad above all nations for advancing Experimental knowledge as well as Academic Learning, so I hope it will at length induce at home [. . .] to contribute to ye support and encouragement of so excellent a Foundation" (*OL*, vol. X, pp. 175–76). But within a few weeks of Oldenburg's letter, the council tried to revive domestic productivity by mandating "that such of the Fellows, as regard the welfare of the Society, should be desired to oblige themselves to entertain the Society, either per se or per alias, once a year at least, with a philosophical discourse grounded upon experiments made or to be made" (Boas Hall, *Promoting Experimental Learning,* p. 66). On the foreigners' perception of the Society, see also Rob Iliffe, "Foreign Bodies: Travel, Empire and the Early Royal Society of London (pt. 2)," *Canadian Journal of History* 34 (1999): 32.

95. Hunter, *Establishing the New Science,* p. 254.

The wide networks of correspondence developed by Oldenburg were much more than a means to communicate the Society's work abroad or to promote the empirical and collaborative style of natural philosophy it espoused. What Oldenburg variously called "philosophical commerce," "learned trade," or "philosophical trade" was a specific form of exchange.[96] In exchange for the sense of partaking in a prestigious enterprise—a belief that could be sustained through partial perspectives produced by distance—the correspondents sent their reports and observations to London, effectively providing the Society with a blood infusion. This became evident when the inflow of correspondence slowed down to a trickle after Oldenburg's death in 1677, making Flamsteed complain that "Our Meetings at the Royal Society want Mr Oldenburgs correspondencys and on that account are not so well furnished nor frequented as formerly."[97] The impact of the Society's decline, however, went beyond the Society itself. Correspondents needed the Society as much as the Society needed correspondents. Foreigners may have not known that their letters were helping the academy to stay alive, but it was the Society's survival (and the projection of that survival as success) that made their association with it a prestigious-sounding one.[98]

96. "If this way of printing Journals spread over all, we may have a good generall Intelligence of all ye Learned Trade, and its progress" (*OL,* vol. IV, p. 275); "That such expressions in so publick a place and in so mixt an assembly, would certainly prove very destructive to all philosophicall commerce" (*OL,* vol. IV, p. 27); "How large and usefull a Philosophicall trade could I drive, had I but any competent assistance" (*OL,* vol. III, p. 613). Anne Goldgar has argued that in the republic of letters, correspondence was experienced as a contract between two parties who agreed to correspond, answer to each other's letters, send relevant news, and occasionally help each other with errands, books, information, contacts, etc. (Goldgar, *Impolite Learning: Conduct and Community in the Republic of Letters, 1680–1750* [New Haven: Yale University Press, 1995], pp. 12–26]. The presence of these mutual obligations indicates that terms like "commerce" and "trade" were more of a description than a metaphor for correspondence. For instance, Oldenburg felt entitled to use somewhat sharp language with Hevelius to remind him the promise to connect the Society with eastern European and Russian correspondents: "I make only this one request, that as you had promised the philosophical correspondence of experienced and ingenious men you will not shrink from translating your promise into actuality" (*OL,* vol. III, p. 521).

97. John Flamsteed to Richard Towneley, February 1680, cited in Hunter, *Establishing the New Science,* p. 255.

98. Distance, however, cut both ways. As the Society's survival was based on "foreign fame" produced not through direct evidence but distant perceptions, it could be ruined

The use of foreign credit to sustain the Society became more sophisticated with the introduction of the *Philosophical Transactions.* Relying mostly on material he received through the Society's correspondence, in 1665 Oldenburg began to publish one of the very first journals of natural philosophy.[99] The journal format gave Oldenburg a more efficient way to manage his vast correspondence.[100] Even more importantly, it added a crucial element to the Society's "philosophical trade" that was not to be found in its epistolary system of communication. With a print run of about a thousand the *Transactions* gave much greater visibility to the correspondence Oldenburg deemed worth publishing—letters that, until then, he would have shared with the twenty or so members who typically showed up at the Society's meetings, or with a handful of correspondents.[101] This added visibility greatly increased the foreign correspondents' incentive to send their communications to the Royal Society. Symmetrically, the success of the *Transactions* increased the Society dependence on the journal.[102] When, after Oldenburg's death, Robert Hooke took over the journal, with mediocre success, Beale remarked that "the Royal Society does apparantly goe backwards till you got an Industrious & ingenious Person to go on constantly with Phil: Trans:."[103]

---

by distance-bred rumors of its demise—rumors that were as empirically accurate or inaccurate as those on which the Society's foreign fame was based. For instance, in September 1668, Justel wrote Oldenburg from Paris that "[h]ere it is said that your Society no longer works seriously, that the King treats it in a discourteous manner, and that he has no good opinion of it, that most of the members attend no longer, and that soon it will be quite dispersed. This is said with such confidence here that I should be very glad to know the truth in order to be able to reply to those who talk about it" (*OL,* vol. V, p. 39). We do not have Oldenburg's response, but it must have been a pretty anxious one.

99. The *Journal des Sçavans,* published in Paris, preceded the *Philosophical Transactions* only by a few months. On the early history of the *Transactions,* see Adrian Johns, *The Nature of the Book: Print and Knowledge in the Making* (Chicago: University of Chicago Press, 1998), pp. 497–521.

100. Marie Boas Hall argues that since 1663 Oldenburg had thought about making some badly needed profit from his vast correspondence networks by offering a subscription-based international "news sheet" of philosophy, politics, and gossip. These plans eventually evolved into the more specialized *Transactions.* Marie Boas Hall, *Henry Oldenburg: Shaping the Royal Society* (Oxford: Oxford University Press, 2002), pp. 79–86.

101. Boas Hall, *Henry Oldenburg,* p. 86.

102. Hunter, *Science and Society in Restoration England,* p. 51: "Its [the journal's] influence surpassed that of the Society."

103. Ibid., p. 52. Others expressed similar feelings later on in relationship to both correspondence and journal (Michael Hunter, *The Royal Society and Its Fellows, 1660–1700:*

DISTANCE, LETTERS, AND PARTIAL PERCEPTIONS

Like Galileo before his discoveries of 1610, the Society did already have some cultural capital in 1662—a royal charter, a few productive and internationally known members (Boyle being the most visible among them), and several high-ranking courtiers, clergy, and aristocrats on its membership list. It also had a polyglot secretary who was as good at managing and charming a large number of correspondents as Galileo was at playing the patronage game. More importantly, there was an uncanny match between the Society's actual resources and the kind of things that could (or could not) be commonly represented in a long-distance correspondence. While visitors did go to London and were allowed into meetings (usually carefully choreographed ones), correspondence remained the principal medium for information about the Society.[104]

It did not take much to enable a distant correspondent to assume that the "Royal" in the Society's name meant more than it actually did, or to make him believe that the Society was buzzing with action just by sending him a list of recent publications by Boyle and a handful of other productive members. A letter could also include membership lists that, "profusely decorated with the names of (mainly inactive) bishops, statesmen and aristocrats, enjoyed wide circulation, disseminating esteem for the Society among nonmembers at home and abroad."[105]

The Morphology of an Early Scientific Institution [Oxford: British Society for the History of Science, 1994], pp. 44–45).

104. Virtuosi visited London from the Continent, but rather infrequently during the first decade of the academy's existence. Marie Boas Hall remarks that England became a common destination of the grand tour only toward the end of the seventeenth and the beginning of the eighteenth centuries, but in the previous period it flowed mostly the other way (Boas Hall, *Promoting Experimental Learning,* p. 142). Those who visited the Society and participated to its meetings probably came away with an artificially rosy picture of the Society. Oldenburg's correspondence indicates that these visitors were given a special treatment, with more experiments, reruns of older but catchy trials, etc. The best treatment of the philosophical grand tour in the late seventeenth century and the views expressed by English travelers about other experimental cultures, institutions, and key practitioners is Rob Iliffe, "Foreign Bodies: Travel, Empire and the Early Royal Society of London, pt. 1: Englishmen on Tour," *Canadian Journal of History* 33 (1998): 358–85; and "Foreign Bodies: Travel, Empire and the Early Royal Society of London, pt. 2: The Land of Experimental Knowledge," *Canadian Journal of History* 34 (1999): 24–50.

105. Hunter, *Science and Society in Restoration England,* p. 48. Some publications promoted by the Royal Society, like Thomas Sprat's 1667 *History of the Royal Society,* Rob-

Symmetrically, the genre of the letter made it easy for Oldenburg not to dwell on less flattering features of the Royal Society, such as that "Royal" was an almost empty signifier, that active members were a minority, that attendance was poor, or that some of the Society's more productive members were only nominally attached to the institution. The time and space constraints familiar to any correspondent provided Oldenburg with a good excuse for evading more specific questions about the Society's daily activities, what facilities it had, and what kind of endowments the king had "blessed" it with and simply refer the correspondent to a forthcoming book-length account of the Society, its origins, its organization, and its philosophical plans—Sprat's 1667 *History of the Royal Society.*[106] He would neglect to say that the Society had commissioned such a book to begin with and had reviewed the manuscript before publication.[107] However, with very few exceptions, Oldenburg engaged in limited representations, not misrepresentations. Distance did the rest.

He would typically send out variations on a boilerplate letter that opened with a somewhat grandiose description of the Society, its Royal

ert Hooke's 1665 *Micrographia,* and Nehemiah Grew's 1681 *Musaeum Regalis Societatis,* reflected the same concern for publicity one finds behind the membership lists.

106. Oldenburg mentions Sprat's book in *OL,* vol. II, p. 401; vol. III, pp. 193, 416, 621; and vol. IV, pp. 70, 92, 136, 168. On one occasion Oldenburg declined to answer questions about the Society, claiming that such information could not be made public: "You wish to know the rules, statutes, labors, and endowments of our Society. All those things will be made public, I believe, in a short time. Such matters have until now been confined to the Council of the Society and it would be quite improper for me, as Secretary of the Society, to divulge them at this time. As soon as it is allowed I shall be glad to do so. Meanwhile I can, without breach of confidence, let you know that our King bestows remarkable favor upon us and has resolved to endow generously this, his Royal Society. For (as you rightly suppose) if it should lack endowments everything would be hindered. But if they are made rich enough, and if the philosophers themselves remain constant in their independence of mind . . . what can limit their lofty endeavors?" *OL,* vol. II, p. 14.

107. Oldenburg to Boyle, November 24, 1664: "Mr Sprat intends to begin next week to print ye History of our Institution, wch hath been perused by Ld Brounker, Sr R. Moray, Dr Wilkins, Mr Evelyn and others; but we are troubled, yt you cannot have a sight of it." *OL,* vol. II, pp. 320–21. What Oldenburg was presenting as a description of the Royal Society was, in fact, closer to an advertisement. That Sprat's book was published in 1667 (five years after the Society's first charter) but that it was already being reviewed and readied for printing in November 1664 suggests that the Society wanted it published as quickly as possible. On the publication history of Sprat's book see Hunter, *Establishing the New Science.*

genealogy, its goals, its noble membership.[108] For instance, in his first letter to the Polish astronomer Johannes Hevelius in 1663, Oldenburg flattered him by letting him know that "your merits in the republic of letters [were] praised to the sky in that assembly of the Muses [the Royal Society]" and then continued:

> It is now our business, having already established under royal favor this form of assembly of philosophers who cultivate the world of arts and sciences by means of observation and experiment and who can advance them in order to safeguard human life and make it more pleasant, to attract to the same purposes men from all parts of the world who are famous for their learning, and to exhort those already engaged upon them to unwearied efforts. [. . .] This our Fellows are striving for with all their might and for that reason they are developing a wider correspondence with those who philosophize truly.[109]

Oldenburg then told Hevelius that participation in the Society's networks of correspondence would give even greater visibility to his work and allow him to "claim greater glory and rise to the fame of the very greatest—Copernicus, Brahe, Galileo, etc."[110]

Although Oldenburg was asking Hevelius to provide the Society with several fairly tedious observations of eastern European natural history (some of which he would have had to obtain by enrolling some of his own more distant correspondents), Hevelius seemed flattered by Oldenburg's letter. He replied:

> And so I heartly rejoice, and think it splendid for literature, that nowadays kings and princes take an interest in literary matters [. . .] The learned world especially owes humble and eternal thanks to the King of England, a prince

108. See, for instance, *OL*, vol. II, 14; and vol. III, pp. 120–21, 338–39, 384–85, 621. As Hunter put it, "Oldenburg gave a grandiloquent view of the foundation of the Royal Society, armed with Royal patronage and supported by the eminent and the wise, as the embodiment of the enterprise of reforming knowledge. . . . This was not necessarily an accurate representation of the Society as it actually existed, but more a projection of a sort of idealized view of the Society, treading a tightrope between the ideal and what actually underlay it" (*Establishing the New Science*, p. 251).

109. *OL*, vol. II, p. 27. Oldenburg's letter shows that the Society made contact with Hevelius after two of its fellows (Seth Ward and John Wallis) reported having received his 1662 *Mercurius in Sole visus*. It is not clear whether Hevelius knew of the Society at that time.

110. *OL*, vol. II, p. 28.

> worthy of every high praise, because he has founded a unique Assembly of those philosophers who cultivate and advance the arts and sciences by following not tradition, but observation and experiment alone. I am the more conscious that the most renowned Royal Society held me worthy to pursue the same purpose in the very courteous letter you wrote me, in which you did me great honor [. . .] I shall labor with all my might to gather my harvest, lest I should waste any of its time with what I shall have judged to concern the glory and honor of the famous Royal Society.[111]

The Society's request for reports of observations and experiments was presented as part of an international effort to build a repository of observations and experiments on which true philosophy could be built. In exchange for a contribution to the Society's grand project, the correspondent was told that the Society would return the favor and respond to natural philosophical queries it could find an answer for among its members or in its growing natural philosophical "storehouse" or "magazzen."[112] The academy also promised to function as a public register of the practitioners' priority claims, "a perpetual record devoted to the honor of those individuals whose talents and industry deserve it."[113] Additionally, several correspondents were granted the honor of being elected to the Society.

What Oldenburg would not emphasize was that the Society's grand database was still almost empty and that the domestic members were filling it at a slow pace, or that the qualifications for membership were lower than what many foreign correspondents might have imagined. More importantly, he also did not say that while the flow of information from the periphery did indeed fit the Society's grand long-term philosophical plan, these letters were much needed in the short term to keep the Society up and running and to help it turn foreign admiration into domestic respect and, hopefully, material support.[114]

111. *OL,* vol. II, p. 138.

112. *OL,* vol. III, p. 384; vol. X, p. 176. Oldenburg told Peter van Dam that if he were willing "to communicate to me as intermediary whatever seems to you remarkable and worthy of note, I pledge myself that the Society will be most grateful for it and I promise that we will return like things by way of recompense." *OL,* vol. II, p. 15.

113. *OL,* vol. IV, p. 422.

114. Oldenburg did come very close to linking the Society's hopes for institutional longevity to the flow of knowledge from foreign correspondents: "when it seems good for you, freely transmit your contribution to this Royal Society, for which we prophesy an enduring future, God willing, because of the strength and genius of its foundation, and a continuing welcome for the labors of experts of all sorts in any corner of the Earth." *OL,* vol. IV, p. 422.

The glowing responses to Oldenburg's partial representations of the Society suggest that there were people out there who, for various reasons, were willing to believe him (not unlike the way Kepler was willing to buy into the officiality of the ambassador's request). A Roman virtuoso, after reading the *Transactions*, wrote Oldenburg in 1669 that:

> My thought took wing for London, as to a glorious market whence the most rich supply of all philosophical commodities is to be obtained. For, famous Sir, so obvious is the fame of the Royal Society as transmitted to us, that its Fellows need not envy the Stoa of Zeno, the Lyceum of Aristotle, or the Academy of Plato; it is more fortunate than all others in having His August Majesty at its head.[115]

At the less eloquent end of the spectrum, we find an English country schoolmaster who showed his admiration for the "Illustrious Society" by suggesting that a simple acknowledgment of a letter he had sent Oldenburg in 1667 would be a prized gift for him:

> That the observations [I sent] came to yr hand, was a sufficient reward for mee, who had no other designe, nor higher ambition, than to express my service and zeale to ye Illustrious Society [. . .] But for mee to receive a letter of acknowledgement, (I should bee too proud to owne it under the notion of thankes, though you are pleased so to style it) this is a favour far transcending my deserts or expectation.[116]

More aggressively, a German philosopher wrote in 1668 asking whether the Royal Society could finance the establishment of a "small college" in Germany "as a dependency to your larger one" given that "your Society abounds in very rich patrons and the King of England maintains a consul in this town."[117] Another German wrote in 1665 that he had received news of the "Royal Academy" founded

> with princely revenues by the nobility and erudition of the chief men of England [. . .] Surely your illustrious College will excel the rest, as much as the

115. *OL*, vol. IV, p. 259. The writer, Francesco Nazari, meant the flattering things he had to say about the *Transactions*, and in fact helped himself to some of his articles for his own journal, the *Giornale de' Letterati*.

116. *OL*, vol. IV, p. 21.

117. He added: "Unless subsidies can be promised to our people by the Royal Society of England or by other societies . . ." (*OL*, vol. IV, p. 531). Oldenburg commented on this letter to Boyle that: "This is a *fallacia compositionis et divisionis;* There are amongst us many that are rich enough; but where are the rich Patrons?" (*OL*, vol. IV, p. 571).

> slow viburnum does the cypress. For, buttressed by Royal authority and selected from such very able men (in whom England abounds), and doubtless cherished by royal grants, it will be quite furnished with everything necessary for performing experiments and all the funds that are required.[118]

A correspondent from Milan was flattered by Oldenburg's invitation to "co-operate" with the Society—an invitation he saw (or wanted to see) as a sign of his election to the Society. He took that as a "glorious promotion" that made him "endeavor to fulfil the expectation formed of me in order not to seem a goose among swans."[119] A few correspondents took a more blunt approach and asked to be elected, while others exchanged letters with Oldenburg until they were elected, only to drop out as soon as they received the badge of honor they had been corresponding for.[120]

These diploma hunters support Hunter's point that the actual conditions of the Royal Society did not matter much to the correspondents as long as a simulacrum of authority could be projected abroad.[121] One could even say that for these correspondents less information was better information as the value of the diploma they were after might have been spoiled had people known more about the Society's mundane realities. They certainly would not have been flattered to know of the ease with which the academy was granting membership to foreigners.[122] At the same time, the Society's

118. *OL*, vol. II, p. 345.

119. *OL*, vol. III, p. 455. Oldenburg's initial letter does not mention election, only "co-operation" (*OL*, vol. III, pp. 440–41).

120. Birch, *The History of the Royal Society of London*, vol. II, p. 162: "Monsr. Ismael Bullialdus and Monsr. Samuel Petit were upon their desire in a letter proposed candidades by Mr. Oldenburg" (March 28, 1667); and, at p. 401: "Monsr. George Stiernhelm, a Swedish gentleman [. . .] was proposed candidate upon his desire expressed in a letter from Stockholm to the president" (November 18, 1669). Francesco Travagino's self-propelled election is mentioned in Boas Hall, *Promoting Experimental Learning*, p. 143. On fleeting correspondents see Hunter, *The Royal Society and Its Fellows*, p. 120.

121. Adrian Johns has discussed the case of the Athenian Society, a fictitious academy invented (together with its successful journal and "official" printed history) by the London bookseller John Dunton (Johns, *The Nature of the Book*, p. 457). An understanding of the role of distance and partial information in creating authoritative auras makes one appreciate the substantial family resemblances between a "real" academy like the Royal Society and a "fake" one like Dunton's creation.

122. Boas Hall, *Promoting Experimental Learning*, p. 143; Hunter, *The Royal Society and Its Fellows*, pp. 119–20.

willingness to spread membership abroad indicates how eager the academy was to keep correspondents proud and hopefully productive.[123]

A few diploma hunters aside, most correspondents were genuinely interested in natural philosophy, lived in Continental Europe, and tended to have little philosophical community around them.[124] Although some were already connected to epistolary networks, they were eager to be better known and to show what they knew. Mostly, these were people who, so to speak, kept much of their philosophical money under the mattress but were happy to do something with it if given the opportunity.[125] These correspondents too had a specific interest in sending their findings back to the Society without asking probing questions about corporate life in London. Partial perceptions based on distance worked well at both ends: the Society gained from not saying too much about itself, and the correspondents gained as well by not asking too much about the institution they were contributing to. Some questions, however, were asked.

TO ASK OR NOT TO ASK?

Even famous practitioners like Hevelius—a wealthy businessman and senator of Danzig, pensioner of Louis XIV, author of the well-known 1647 *Selenographia,* and owner of probably the best private observatory in Europe—had something to gain from contributing to the Society's discussions and its journal.[126] But unlike more marginalized or lesser known practitioners, he was not so hungry for recognition as to join the Society's networks without asking what kind of institution was at the other end of

123. The inflation of the Society's foreign membership, however, did not devaluate the diploma. That could have happened only if several correspondents compared notes—a most unlikely scenario.

124. Some also lived in England (but outside of London), Ireland, Scotland, and British overseas colonies (like Bermuda, Connecticut, or Massachusetts), or were posted abroad in the foreign service.

125. The two French correspondents (Boulliau and Petit) who asked to be admitted to the Society (see note 120 above) did so, apparently, because they felt unappreciated in their country (*OL*, vol. III, p. 369). Perhaps they wanted to use the membership in the Royal Society as a way to be elected to the then emerging Académie Royale des Sciences.

126. On Hevelius' reputation, see Mary Winkler and Albert van Helden, "Johannes Hevelius and the Visual Language of Astronomy," in Judith Field and Frank James (eds.), *Renaissance and Revolution: Humanists, Scholars, Craftsmen, and Natural Philosophers in Early Modern Europe* (Cambridge: Cambridge University Press, 1993), pp. 97–116.

the epistolary line, or without testing the academy's willingness to return the favors it was asking of him.

Hevelius, who was far from naïve about commerce (either philosophical or mercantile), agreed to "gather the harvest" requested by Oldenburg, but immediately asked for a return gift—a copy of rare astronomical tables "which are to be found (unless I am mistaken) in a certain Persian manuscript at Oxford."[127] And while the tables were prepared, Hevelius asked for more details about the glorious institution to which, in the meantime, he had been elected.[128] The first response sent to him (by John Wallis) was technically correct but not terribly informative:

> The Society has no fixed set or number of members. Besides the Council (which includes the President and other officers, and number in all 21) nominated initially by the King himself in his charter . . . the remaining Fellows—both natives and foreigners who visit our country—are elected as seems fit without regard to number. . . . Thus at present the Society is composed of about 120 Fellows—magnates, nobles, theologians, medical men, lawyers, mathematicians, merchants, and others.[129]

Wallis did not say that of the Society's 120 fellows only a few were competent natural philosophers and even fewer attended its meetings, or that the election of new members was not exactly done "without regard to number," because membership fees were the Society's main source of revenue.[130] While Hevelius wanted to know who all the members were, he was told only that the current president was Viscount Brouncker—probably an attempt to impress Hevelius with the noble membership of the Society.[131] Like other correspondents, he was also told that Sprat's *History of the*

127. *OL*, vol. II, pp. 138–39. In the same letter, Hevelius also asked for some help to recover a credit from the heirs of Samuel Hartlib, a former close associate of the Society's founding group.

128. This first request was in a now lost letter to Wallis. Hevelius had corresponded with Wallis before being contacted by Oldenburg. Hevelius was elected to the Society on March 9, 1664.

129. *OL*, vol. II, pp. 169–70. This letter by John Wallis was never received by Hevelius.

130. Hunter, *The Royal Society and Its Fellows*, pp. 19–20. On average, only about 15 percent of the membership came to meetings, and most of those who came expected only an hour or so of philosophical entertainment. Even among the active fellows, only a few had a noticeable publication record in natural philosophy.

131. *OL*, vol. II, p. 170; Hunter, *Establishing the New Science*, p. 49.

Royal Society would have provided all the information he sought.[132] But when Hevelius eventually received it, he remarked that a book written in English would have been hardly informative to foreigners like himself.[133] Undeterred, he continued to ask Oldenburg for basic facts about the academy he was a member of:

> I particularly beg you to send me, if it's no trouble, the names of all those distinguished men who have been enrolled in our Society until now, so that I might at least know the membership of our Society by name.[134]

While it is not clear whether Hevelius was ever satisfied with the grainy picture he was given of the Royal Society, his frequent correspondence with Oldenburg suggests he appreciated the advantages offered by the Society's networks.

He might have changed his mind when, a few months later, the Society did not support him in his dispute with the French astronomer Adrien Auzout over the comet of 1664, and even went so far as to elect Auzout a member during that dispute.[135] He probably did not understand that, behind his flattery, Oldenburg was eager to continue to receive (and pub-

132. *OL*, vol. II, p. 169. Sprat's book, however, was written primarily with English readers in mind, as it mostly tried to address objections and fears that experimental philosophy might have elicited within the English establishment. Sprat says explicitly, "This I speak, not out of Bravery to *Foreiners* (before whose eyes, I believe this negligible Discourse will never appear) but to the learned Men of this Nation" (p. 70). Although I would not call this a truly xenophobic book, it was no founding text of the republic of letters either. Sprat's unwavering pro-English sentiments, dismissive remarks about the Dutch, Spanish, and French, and his generally cool view of non-English Europeans would not have made him many friends on the Continent. Sprat's treatment of the Society's self-financing structure (pp. 77–78) may have also been detrimental to a perception of the Royal Society as royally endowed. It is therefore surprising to see that Oldenburg worked hard at having French and Latin translations made of it.

133. Hevelius kept asking Oldenburg about corporate formalities (such as if fellows identified themselves as such in print) and about the publication of Sprat's *History* (*OL*, vol. II, pp. 396, 496; vol. III, p. 6). Oldenburg responded briefly (*OL*, vol. II, p. 623). Then, when Hevelius finally received Sprat's volume, he remarked, "I thank you very much for this kindness, but the History would have been far more acceptable to foreigners if written in Latin" (*OL*, vol. IV, pp. 446–47).

134. *OL*, vol. III, pp. 6–7. Oldenburg eventually sent the membership list on March 30, 1666 (*OL*, vol. III, p. 76).

135. The history of the dispute and of the Society's acrobatic defense of its codes of philosophical civility is analyzed in Shapin, *A Social History of Truth*, pp. 260–91.

lish) Avzout's observations as much as those of Hevelius. Taking sides would have threatened the Society's relationship with two of the *Transactions*' good copy providers and their contributions to "philosophical commerce."[136]

The correspondence between the Dutch microscopist Antoni van Leeuwenhoek and Oldenburg indicates that lower-class or lesser known correspondents had a much more deferential attitude toward the Society. Leeuwenhoek initiated the dialogue, was very appreciative to be accepted as a correspondent, and never asked probing questions. His view of the Society was already so extravagantly positive that not even Oldenburg could have enhanced it.

The son of a basket maker, Leeuwenhoek was a draper by profession, with no university training and no languages but Dutch.[137] His correspondence with Oldenburg was initiated in 1673 by an intermediary—Reinier de Graaf—who vouched for Leeuwenhoek's credibility.[138] The first report of Leeuwenhoek's observations, dated April 28, was read at the Society's meeting of May 7. Its swift inclusion in the May 19 issue of the *Transactions* suggests how eager Oldenburg was to get new and newsworthy copy

136. Concerns about the maintenance of academic civility and with securing publishable copy for the Society's journal are related but not identical. For instance, in a 1667 letter to Boyle about "a certain Physitian" who explicitly contested the truthfulness of two reports of medical experiments received from Danzig (via Hevelius) and read to the Society, Oldenburg wrote that "I could not but take him afterwards aside, and represent to him, How he would resent it, if he should communicate upon his owne knowledge an unusual Experiment to ye curious at Danzick, and they in publick brand it wth ye mark of falsehood: That such Expressions in so publick a place, and in so mixt an assembly, would certainly *prove very destructive to all philosophicall commerce, if the Curious abroad should be once informed, how their symbola's were received at ye R. Society*" (*OL*, vol. IV, p. 27, emphasis mine). What Oldenburg sees endangered here is the flow of correspondence—"philosophical commerce"—not the broader conditions for philosophical consensus. As editor (and owner) of the *Transactions,* Oldenburg may have been more concerned with the availability of printable copy than with the dangers posed by impoliteness to the achievement of philosophical consensus.

137. The other jobs he later took—custodian of the local court house and registered surveyor—were equally philosophically undignified (Clifford Dobell, *Antony van Leeuwenhoek and His "Little Animals"* [New York: Dover, 1960], pp. 31–35).

138. "I will communicate to you at this present time what a certain very ingenious person named Leeuwenhoek has achieved by means of microscopes which far excel those we have seen hitherto made by Eustachio Divini and others, of which this enclosed letter [. . .] will give you a specimen" (*OL*, vol. IX, p. 603).

for his journal, even when it came from a lowly draper.[139] At the same time, Leeuwenhoek was so eager to get into the game that he even offered to conduct observations on the Society's behalf.[140] The 1673 letter was the beginning of a large correspondence.[141] Fifty years later, in the same issue of

139. *Philosophical Transactions,* May 19, 1673, no. 94, pp. 6037–38.

140. "Mr. Leewenhoeck [. . .] is ready to receive difficult tasks for more, if the Curious here shall please to send him such: Which they are not like to be wanting in." Ibid., p. 6037.

141. Harold Cook and David Lux argue that despite de Graaf's renown, some members of the Society had lingering doubts about Leeuwenhoek's credibility ("Closed Circles or Open Networks? Communicating at a Distance during the Scientific Revolution," *History of Science* 36 [1998]: 188–89). This claim is also found in L. C. Palm, "Leeuwenhoek and Other Dutch Correspondents of the Royal Society," *Notes and Records of the Royal Society of London* 43 (1989): p. 194. According both to Cook and Lux and to Palm, Constantijn Huygens' August 1673 visit to Leeuwenhoek was a response to these concerns, and that it was only after Huygens' report that Oldenburg invited Leeuwenhoek to correspond. I believe that they overestimate the Society's or Oldenburg's suspicions about Leeuwenhoek and that, instead, this example indicates how willing the Society (or at least Oldenburg) was to take some risk in order to maintain its visibility. For instance, there are no hints of distrust in Oldenburg's May 15, 1673, letter to de Graaf following his delivery of Leeuwenhoek's observations to the Society: "You have done something that was extremely welcome to us in that you decided to impart to us the reflections of your countryman Leeuwenhoek . . . I read over [to the Society] a translation of his observations made into English from the Dutch language and I gathered that our people approved of the man's diligence and outstanding precision." In this letter, Oldenburg stressed how eager the Society was to receive more observations from Leeuwenhoek, listing some that would have been particularly welcome by the London virtuosi (*OL,* vol. IX, p. 654). No cautionary remarks are to be found in Oldenburg's presentation of the observations in the *Transactions.* Leeuwenhoek, therefore, was asked for further observations right after he sent the first ones, not after Huygens had allegedly vouched for his credibility in August. The August 8, 1673, letter by Constantijn Huygens to Hooke cited by Cook and Lux as evidence of the Society's desire to double-check Leeuwenhoek's credibility indicates instead that it was Leeuwenhoek who had contacted Huygens, not the other way around: "Our honest citizen, Mr. Leewenhock . . . having desired me to peruse what he hath set down of his observations about the sting of a bee, at the requisition of Mr. Oldenburg and by order, as I suppose, of your noble Royal Society, I could not forbear by this occasion to give you this character of the man, that he is a person unlearned both in sciences and languages, but of his own nature exceedingly curious and industrious" (J. A. Worp, *De Briefwisseling van Constantijn Huygens* [The Hague: Nijhoff, 1917], p. 330). The observation of the bee sting mentioned by Huygens is precisely one of those included in the May 15 letter from Oldenburg to de Graaf (mentioned above) listing the observations the Society wanted Leeuwenhoek to conduct. Leeuwenhoek, it seems, called upon Huygens to check the observation (or maybe the textual description) he was going to send to the Society, worrying perhaps that his work may not have met the reporting standards of Oldenburg or the So-

the *Transactions* that included Leeuwenhoek's last observations, the Society's vice-president, Martin Folkes, stated that his discoveries

> are so numerous as to make up a considerable Part of the *Philosophical Transactions,* and when collected together, to fill four pretty large Volumes in Quarto [. . .] and of such Consequence, as to have opened entirely new Scenes in some Parts of Natural Philosophy.[142]

But despite having provided material to fill many of the Society's meetings, and much copy for the journal, he was not elected to the Society during Oldenburg's tenure.[143] Leeuwenhoek's candor about his low social status and education in his first letter may not have maximized his chances.[144] While Oldenburg was glad to publish Leeuwenhoek's observations in the *Transactions,* he did not always acknowledge the microscopist's letters and did not answer him with the flattery he used for more socially respectable correspondents.

He probably knew that Leeuwenhoek was something of a captive correspondent who owed his visibility—very high toward the end of his life—to having been published in the *Transactions.*[145] At the other end of the line, Leeuwenhoek understood that if Peter the Great stopped (briefly) in

---

ciety. Probably, he contacted Huygens because de Graaf (his habitual broker) was ill (he died on August 17). The content and tone of Huygens' letter to Hooke indicates that Huygens is not reporting on Leeuwenhoek because of a request he might have received from Oldenburg (or from Hooke). Rather, he is writing to Hooke (and not to Oldenburg) because some of Leeuwenhoek's work is related to what Hooke had published in his 1665 *Micrographia.*

142. *Philosophical Transactions,* November–December 1723, no. 380, p. 449.

143. Leeuwenhoek sent more than 190 letters to London, and 116 of them were published in full or abridged form in the *Transactions.* Palm, "Leeuwenhoek and Other Dutch Correspondents of the Royal Society," p. 193. That most of the later letters were not published had to do with the declining interest in and perceived repetitiousness of Leeuwenhoek's later work, not with distrust for the Dutch microscopist.

144. In the very first letter he wrote directly to Oldenburg on August 15, 1673, Leeuwenhoek told him, "I have several times been pressed by various gentlemen to put on paper what I have seen through my recently invented microscope. I have constantly declined to do so, first because I have no style or pen to express my thoughts properly, secondly because I have not been brought up in language or arts, but in trade, and thirdly because I do not feel inclined to stand blame or refutation from others." *LE,* vol. I, p. 43.

145. Palm argues that the Royal Society was Leeuwenhoek's only true philosophical correspondent, and that it was "one of the continuous factors during his active life in science." Palm, "Leeuwenhoek and Other Dutch Correspondents of the Royal Society," pp. 192, 199.

Delft to see him in 1698, or if virtuosi curious to take a peek at his famous microscopes mobbed his house, it was because of the reception and visibility the Society had given to his work.[146] He was so eager to publish in the *Transactions* that, when he realized that Oldenburg did not have time to translate his letters into English, he offered to have Latin translations done at his own expense.[147]

When Hooke took over the Society's correspondence after Oldenburg's death in 1677, he struck a chord by conveying to Leeuwenhoek his surprise at not seeing his name on the membership. The microscopist replied:

> I also saw that you wonder that my name is not in the list of the Royal Society. Personally I have never thought of expecting this from Mr. Oldenburgh. If, during his lifetime, he had brought the matter up, I would gratefully have accepted the prospect. Seeing that an honourable mind should always consider an increase of honour important and that your offer to make me a fellow of your Society would confer on me the greatest honour in the world, I should be greatly obliged to you in case you could render me the service of procuring for me such a high distinction.[148]

When news of the election came, he gushed:

> I was very surprised to hear that the Fellows of the Royal Society have been pleased to confer on me so great but unmerited an honour and dignity by

146. Peter the Great stopped in Delft in 1698 and invited Leeuwenhoek to his boat (moored in a canal) to show him his microscopes and observations. An almost contemporary description of the visit is reproduced in Dobell, *Antony van Leeuwenhoek,* p. 55. Toward the end of his life he complained about the number of visitors he had to receive, and that, in some cases, he was forced to turn them away (ibid., pp. 78–79). The Royal Society's interest in Leeuwenhoek's observations declined over the years, as they lost part of their original novelty. Several of his later letters were neither acknowledged nor published by those who inherited Oldenburg's secretarial and editorial position. It was at that point that Leeuwenhoek, already well known, published his work in Dutch journals like *Boekzaal van Europe* and in a number of collections (K. van Berkel, "Intellectuals against Leeuwenhoek," in L. C. Palm and H. A. M. Snelders [eds.], *Antoni van Leeuwenhoek, 1632–1723* [Amsterdam: Rodopi, 1982], p. 201; Palm, "Leeuwenhoek and Other Dutch Correspondents of the Royal Society," p. 197.

147. *LE,* vol. II, p. 207. He was so flattered to see himself in print that, when Oldenburg sent him the *Transactions* that included his reports, Leeuwenhoek wanted to pay for them (*LE,* vol. II, p. 5). He also modified his observational narratives according to Oldenburg's instructions so as to tailor them to the Society's expectations, and those of its journal's readership (*LE,* vol. II, p. 171).

148. *LE,* vol. III, pp. 198–99. However, it was not Hooke who nominated Leeuwenhoek.

> admitting me as Fellow of that honourable College; first from a letter written by the Honorary Secretary Thomas Gale and a few days after through the receipt of a sealed diploma. Both were full of expressions far exceeding my merits. However, while protesting, I declare myself extremely obliged to the Fellows of the said Society for the extraordinary favour bestowed on me. It is my fixed purpose and firm promise to exert all my powers and energy, my life long, to be still more worthy of the honour and favour conferred upon me.[149]

He expressed similar feelings to Hooke as well as in all the correspondence with the Society in the following months.[150] Around that time, Constantijn Huygens (Christiaan's older brother) had this unkind description of Leeuwenhoek's state of mind:

> Everybody here is still rushing to visit Leeuwenhoek, as the great man of the century. A few months ago the people of the Royal Society in London received him among their number, which gave him some little pride; and he even seriously inquired of Signor Padre [the Huygenses' father] if, being now invested with this dignity, he would be obliged in future to take a back seat in presence of a doctor of medicine.[151]

CONSTRUCTING THE CENTER FROM THE PERIPHERY, AND VICE VERSA

From Hevelius to Leeuwenhoek, distant correspondents were crucial in constructing the Society's authoritative image—an authority the Society then recycled back to their correspondents by making their names and works visible to a large readership they could not have reached

149. *LE,* vol. III, pp. 222–23.

150. *LE,* vol. III, pp. 227, 271, 341. When Hooke told him that the king had favorably mentioned his name, Leeuwenhoek could not contain himself: "You have written to me . . . that His Royal Majesty has mentioned my name with great respect. In case that it should happen that you come to speak to His Royal Majesty about my person, my humble prayer is that you will offer His Royal Majesty my humblest service, which I also offer to His Royal Highness the Duke of York." *LE,* vol. III, p. 341. Similar words are in a April 25, 1679, letter to Nehemiah Grew: "Your colleague, Mr Robert Hooke [. . .] wrote to tell me that His Royal Majesty, seeing the little animals, contemplated them in astonishment and mentioned my name with great respect." *LE,* vol. III, p. 23.

151. Constantijn Huygens to Christiaan Huygens, August 13, 1680, cited in Dobell, *Antony van Leeuwenhoek,* p. 50.

by themselves. As suggested by Sprat, the Society's official historian, the members

> will be able to settle a constant intelligence, throughout all Civil Nations, and make the Royal Society the general Banck and Free-port of the World.[152]

The resources of a bank, however, do not predate the deposits it receives from its clients and investors. Both in natural philosophy and in money markets, deposits constituted the basis for a bank's lending power—its ability to give out credit.[153] Analogously, the Royal Society started out with few resources and depended on those who deposited their observations and

152. Sprat, *History of the Royal Society,* p. 64. Sprat sees the Society as a "banck" in the sense of a place where knowledge can be deposited (and deposited safely), but also where knowledge can be made available to others. For example, elsewhere in his *History* Sprat talks about the problems posed by secrecy—a trait he attributes to artisans and inventors—and refers to the Society's function with yet another financial term: "publick Treasure" (p. 74). By this he means a form of capital that is co-owned by those who contributed to it, and to be used for the good of all. This he opposes to "Domestic Receipts"—the knowledge that some artisans are unwilling to share (p. 74). In another passage he uses "common stock" for "publick Treasure" (pp. 75, 115). He also mentions "Experimental treasure" (p. 129) and "Treasure and Repository" (p. 130). In slightly different ways, "banck," "Free-port," "common stock," "Experimental treasure," and "publick Treasure" all describe the accumulation of publicly constituted and publicly accessible philosophical capital that Sprat sees at the core of the Society's project. Sprat uses "bank" one more time in his *History,* this time to refer to knowledge stored up but yet unwritten, like some kind of uninvested capital (p. 44). On the use of credit and financial categories outside of monetary economies in late seventeenth-century England, see J. S. Peters, "The Bank, the Press, and the 'Return to Nature,'" in John Brewer and Susan Staves (eds.), *Early Modern Conception of Property* (London: Routledge, 1996), pp. 365–88. I thank Christopher Coulston for having alerted me of Sprat's use of "banck" and other financial terminology.

153. Describing the development of the money market in London around 1640, and the emergence of the goldsmiths as bankers and lenders, Bisschop writes: "The Goldsmiths made no charge for their services in this connection [the safekeeping of their clients' cash], but any deposit made in any other shape than ornaments was looked upon by them as a free loan." While these funds remained "at call" and could be withdrawn by the depositors at any time—the beginning of the current account system—the balance was quickly lent out at varying interest rates (W. R. Bisschop, *The Rise of the London Money Market* [New York: Kelley, 1968], pp. 44–45). On the fast-developing banking practices in seventeenth-century England, the role of some of the Society's members in it (especially Petty and Montagu), and mechanisms of credit and investment, see also Walter Bagehot, *Lombard Street: A Description of the Money Market* (London: Murray, 1919); John Clapman, *The Bank of England: A History* (Cambridge: Cambridge University Press, 1944); Ellis Powell, *The Evolution of the Money Market* (London: Financial News, 1915).

discoveries in its archives to be able to function like an authoritative and credit-giving node in the developing "philosophical trade."

A focus on how the center was constructed through the periphery reframes what Hunter has seen as a "tension between institutional weakness and public success" of the Society, and his perceptive claim that

> the paradox of the early Royal Society is that this [its international visibility] was not separable from its London meetings. The Society could not have consisted merely of a correspondence secretary and the editor of the *Philosophical Transactions,* even if they were the chief agents of its success: for their authority derived from associations with regular meetings of a chartered institution. The two parts of the Society's operations had to coexist, and this explains the tension between achievement and failure that it experienced throughout the century.[154]

True, it would have been very difficult to start the *Philosophical Transactions* without the Royal Society. Oldenburg had remarkable skills, but it was the fact that he spoke as the secretary of the Society that gave him credibility in the eyes of foreign, poorly informed correspondents. But there is no paradox or tension in this arrangement. The difference between the insiders' view of the Society and that of the foreign correspondents was quite productive. This arrangement required a center, but such a center did not need to be a powerful one and did not need to match the flattering perceptions foreigners had of it. All that was required was a center that, thanks to distance, could be projected as authoritative. An effect of authority was sufficient to set in motion a flow of contributions toward London.[155] It was also very important to have a journal (as distinct from a correspondence network) so that the contributions sent to the center could be quickly distributed out again to credit the correspondents and ultimately sustain a flow that produced credit for both the center and the periphery.[156]

154. Hunter, *Science and Society in Restoration England,* pp. 57–58.

155. This did not need to be the case later on when the *Transactions* took off and became an institution in and of itself, like today's noninstitutional scientific journals. The success of the *Transactions* and its relative independence from the Society became clear fairly early. Hooke, for instance, was unhappy about what he saw as the readers taking advantage of the Society as a kind of philosophical intelligence service. They could simply buy the *Transactions* without having to contribute to the Society's activities. To counter this, he thought that only the members of the Society should receive the journal and the benefits it carried (Hunter, *Science and Society in Restoration England,* p. 57).

156. This perception lasted for several decades: "Both in Germany and elsewhere an exalted idea of this Society has been formed, both of it and of the collections they have in

The remote contributors to the *Transactions* (and those among them who became foreign members) were not semiexploited gatherers of matters of fact that were not accessible to the London virtuosi. Neither were they conscious allies who sent knowledge from the periphery to an authoritative center in exchange for scientific credit. Correspondents were not an appendix to a preexisting core. Without knowing it, they constructed the center and its authority to turn those reports into public knowledge.[157] The distance between them and the center allowed for credit transactions based on situated perceptions, not just for a transfer of resources and blackboxes through standardized Latourian "traintracks."[158] What the correspondents did not know or did not ask about the Society (or what the Society did not want to know about them) was as important as what they did know.[159]

LEAVING THE BODY OUT, SELECTIVELY

In the same way Oldenburg's letters to foreign correspondents gave only a partial representation of the corporation to which he was secretary, a correspondent's letter to Oldenburg provided only a partial representation of the body that wrote that letter and produced the knowledge claims contained in it. This, I believe, was good for philosophical commerce.

Although the Society did not hold a unified corporate stance on the relationship between a practitioner's truthfulness and his social background—Sprat taking a more tolerant position and Boyle a stricter one—there is no doubt that gentlemen were seen as more trustworthy than manual workers

their Museum, especially when one looks at the *Transactions* of this Society and the fine description of the Museum by Grew." W. H. Quarrell and Margaret Mare (eds. and trans.), *London in 1710: From the Travels of Zacharias Conrad von Uffenbach* (London: Faber & Faber, 1934), p. 98.

157. This matches the unplanned and haphazard nature of the development of the Society's channels and networks abroad. These were often expanded through the work of foreign middlemen who followed their financial interests rather than the Society's will. The occasional production and circulation of unauthorized translations of the *Transactions* is an example of such "unplanned" developments (Johns, *The Nature of the Book*, pp. 514–21.)

158. Haraway, "Situated Knowledges," pp. 172–88.

159. As in Galileo's case, no literal misrepresentations took place in the correspondence between the Society and its remote correspondents, only partial perceptions and information. The Royal Society was indeed a royal institution (though in the quite limited sense spelled out by its charter).

and that the Society tried to present itself as a company of gentlemen.[160] Much knowledge about nature, however, lay outside of the social world of gentlemen, and had to be imported into it to enable the Society to sustain its activities.[161] But if the reliability of a knowledge claim was inseparable from the social qualifications of the person who produced or reported it, how could the Society use reports and observations from nongentlemen as unproblematically as it did the overwhelming majority of the time?[162]

I believe that the partial representations involved in epistolary commerce allowed a knowledge-hungry Society and a copy-hungry Oldenburg to take the correspondents' reports while checking their bodies at the door—socially connoted bodies whose visibility might have damaged the credibility of the very claims they had produced. It is not clear how many of the Society's foreign correspondents would have been allowed to join the academy

160. In a few passages of his *History* Sprat presents a particularly "democratic" (if unrealistic) view of the Society membership: in the Society, "the Soldier, the Tradesman, the Merchant, the Scholar, the Gentleman, the Courtier, the Divine, the Presbyterian, the Papist, the Independent, and those of Orthodox Judgment, have laid aside their names of distinction, and calmly conspir'd in a mutual agreement of labors and desires" (p. 427). Similar views are expressed with less flair at pp. 71–74. In general, however, the social group that Sprat sees as closest to the ethos of the Society and its program are the merchants (pp. 86–88, 93, 113, 129). For Shapin's argument about the relationship between English gentlemanly culture and the Society see *A Social History of Truth*, esp. pp. 65–125, 193–242.

161. On the problems associated with the acceptance of knowledge claims produced by nongentlemen, see Shapin, *A Social History of Truth*, pp. 243–66, 355–89.

162. "The overwhelming majority of factual testimonies arriving at the Royal Society were not subject to dispute or even to process of deliberative assessment. Skepticism, like formal reflective appraisal, was something that happened on the margins of a well-working, routinely trusting system." Shapin, *A Social History of Truth*, p. 291. The claim is repeated at p. 303. Similarly, in his study of the Society's uses of reports of travelers of unimpressive social backgrounds, Daniel Carey discusses the relative ease with which the Society accepted many of these communications (Carey, "Compiling Nature's History: Travellers and Travel Narratives in the Early Royal Society," *Annals of Science* 54 [1997], pp. 271, 275). Carey sees this as an effect of the specific culture of naturalists that tended to "embrace the singular, the novel, and the curious," thereby lowering the standards of credibility. Shapin, on the other hand, suggests that gentlemanly philosophical culture erred on the side of acceptance rather than skepticism (*A Social History of Truth*, p. 244) perhaps because acting too probing or distrustful might have appeared rude and perhaps insulting—a prelude to giving the lie. At a more mundane level, I believe that Oldenburg's need for fresh copy for his journal and the Society's need for correspondence to discuss and distribute only enhanced a tendency to accept rather than question.

had they lived in London. Hevelius would have been quite presentable, but most likely Leeuwenhoek would have not. Correspondents still needed to be introduced by other credible people known to the Society, but they did not need to be gentlemen. In any case, being foreigners, they certainly could not embody the specific qualities of the English gentleman that Boyle had turned into the exemplar of the natural philosopher. They needed, however, to write with some decorum (and there were books one could buy and people one could hire to do that), know Latin or English (or know someone who knew those languages), and adopt the kind of reporting narratives deemed acceptable by the Society (rich in detail about the time and place of the observation, the instruments that may have been used, the witnesses' names and qualifications, the conditions of the experiment or observation, etc.).[163]

Correspondence did not do away with problems of credibility but reframed them within an epistolary etiquette that was necessarily different from the bodily etiquettes that regulated short-distance, face-to-face interactions among English gentlemen or Continental courtiers. In particular, letter writing allowed for a kind of identity play continuous with other practices such as pseudonymity, anonymity, or mask wearing—early modern versions, perhaps, of today's online personae.[164] It would be hard to

163. On the various types of letter-writing books (especially French ones), see Roger Chartier, "*Secretaires* for the People?" in Roger Chartier, Alain Boureau, and Cécile Dauphin, *Correspondence* (Cambridge: Polity Press, 1997), pp. 59–111. On the literary genre of experimental or observational reports at the Royal Society, see Peter Dear, "Totius in verba: Rhetoric and Authority in the Early Royal Society," *Isis* 76 (1985): 145–161, esp. pp. 152–56.

164. A discussion of the range of acceptable manipulation in matters of philosophical correspondence (especially printed collections of letters) is in Adam Mosley, Nicholas Jardine, and Karin Tybjerg, "Epistolary Culture, Editorial Practices, and the Propriety of Tycho's *Astronomical Letters*," *Journal for the History of Astronomy* 34 (2003): 419–51, esp. pp. 424–27. Lisa Jardine's *Erasmus, Man of Letters: The Construction of Charisma in Print* (Princeton: Princeton University Press, 1993), is also full of interesting discussions of self-fashioning through the careful construction, editing, and publication of letters. My concerns are contiguous with those of these authors, but I focus more specifically on the partial representations that happen in individual manuscript letters than on those in collections that are edited and printed. The literature on this specific issue is surprisingly slim regarding early modern material, but substantial in relation to internet culture. I find Sherry Turkle's *Life on the Screen: Identity in the Age of the Internet* (New York: Simon & Schuster, 1995), esp. pp. 177–254, to provide important insights applicable to much earlier scenarios.

imagine the existence of the so-called republic of letters without the virtual renegotiations of social and gender boundaries allowed by letter writing.[165] Without those negotiations, it would have been a small republic indeed.

The virtual identities allowed by letter writing worked well with the ambiguous identity of the journal in which many of these letters were published. Not only did distance help create an idealized image of the Society in the eyes of its correspondents and a nonthreatening image of the correspondent in the eyes of the Society; it also created a fuzzy perception of the authorial status of the *Transactions*—a fuzziness that increased their authority and allowed it to publish potentially risqué reports.[166]

The journal was not presented as the official organ of the Society (in the same way that neither Galileo nor the Medici ever said that the Medicean Stars had been initially endorsed by the grand duke). In the very first issue, Oldenburg justified dedicating the journal to the Society by casting it as a direct offshoot of the activities of the academy and a forum for the kind of the philosophy it promoted. But he also stated that the *Transactions* were his own private venture, "onely the Gleanings of my private diversions in broken hours." Such disclaimers did not reveal the extent of the Society's actual involvement in the journal and, in any case, were lost on most distant readers who tended to blend the journal and the academy into one, or who may have missed those disclaimers because they did not read the journal regularly.[167] That Oldenburg was the secretary of the Society and the editor of the *Transactions*, and that there was no clear distinction between submissions to the journal and letters to the academy helped conflate the

165. I believe that the decoupling of professional competence from social background that we see at the roots of today's social system of science may have started with this transition from sites where practitioners interacted directly and face-to-face to systems of interaction at a distance such as early correspondence networks—systems that did not involve the practitioner's whole body and its many social connotations.

166. On the peculiar relationship between the *Philosophical Transactions* and the Royal Society, see my "From Book Censorship to Academic Peer Review," *Emergences* 12 (2002): 11–45, esp. 28–31.

167. For instance, the number 12 (May 2, 1666) issue of the *Philosophical Transactions* informed the reader that: "Whereas 'tis taken notice of that several persons persuade themselves that these *Philosophical Transactions* are published by the Royal Society, notwithstanding many circumstances to be met with in the already published ones that import the contrary; the writer thereof hath thought fit expressly here to declare, that the persuasion, if there be any such indeed, is a mere mistake."

journal with the institution, as well as Oldenburg's secretarial role with his editorial one. As a result, the Society could reap the benefits of the credit it received from the *Transactions* while limiting the risk of having its authoritative aura tainted by the disputes that may have emerged (and did emerge) as the result of the articles published in the journal.

The circles of confusion surrounding the relationship between the Society, Oldenburg, and the *Transactions* produced additional options for the management of correspondents. As secretary of the Society and editor of the *Transactions,* Oldenburg could steer incoming correspondence toward either the Society, the journal, or both. For instance, Leeuwenhoek's first 1673 communication (vouched for by de Graaf) was discussed at the Society's meeting and then published in the *Transactions,* but many of his subsequent letters were simply printed by Oldenburg in the journal without any prior public reading or discussion.[168] While we do not know the specific reasons behind Oldenburg's decision to sort incoming letters this way, it is clear that the Society benefited from the visibility the journal received from publishing striking reports from socially liminal correspondents, and from the possibility of presenting those publications as something it did not vouch for—publications that were Oldenburg's sole responsibility. In different ways, the Society, Oldenburg, and the correspondents all engaged in productive partial representations enabled by distance and by technologies of communication.

FROM AD HOC TO ROUTINE

Distance played different but related roles in the cases of Galileo and the Royal Society. Galileo set up a sequence of transactions that were tailored to one set of claims—his early telescopic discoveries. Because his claims were specific and their value time-sensitive, his marketing options were correspondingly narrow. He had a limited range of people he could interest in investing in his discoveries, and little time to negotiate the dedication. Galileo's options remained limited even past this initial phase. He needed to articulate his investment sequence through a set of people, in remote locales, who might have had personal interests in his products: the Medici because of the "Medicean Stars" connection, and Kepler because of his personal interest in Copernicanism and in increasing his own visibility.

168. Palm, "Leeuwenhoek and Other Dutch Correspondents of the Royal Society," p. 195.

Everything in Galileo's one-time investment sequence was ad hoc, specific, and personal. In the Royal Society, by contrast, we find a system of distance-based, repeatable cycles of credit. These cycles were repeatable precisely because they did not center on the specific features of the claims or on the investment that specific people might have in those claims.[169]

The Society wished to become a philosophical bank, not to invest in specific claims (as the Medici, for example, did in Galileo's discoveries). Its deliberate commitment to a broadly defined, nondogmatic, polite natural philosophy that would minimize disputes and disruptions of philosophical trade indicates that the Society was promoting a whole socio-epistemological framework for natural philosophy, not a set of specific claims, models, or theories. The Society's commitment to considering discourses that relied on the "matter of fact" as their form of evidence indicates that it saw the matter of fact as the fundamental "currency" of its credit system. Insofar as different philosophical discourses relied on matters of fact, their factual claims could be negotiated and credited independently of the interpretations and hypotheses they might have built on those matters of fact. One could welcome and publish Leeuwenhoek's microscopic observations without engaging with his philosophical ruminations. The authority of the Society as a philosophical bank was maximized by accepting matters of fact in deposit (and in making them available to other "investors") while minimizing its investment in the specific interpretations of those claims.

Another relevant difference between the systems developed by the Society and by Galileo hinges on ownership. The Royal Society did not own the claims the philosophers deposited in its registers, but held them like a bank held its clients' funds.[170] But as we will see in the next chapter, Galileo,

169. For instance, Dear has argued that the style and literary form of the reports espoused by the fellows of the Royal Society were "more important than the substance of that science" (Dear, "Totius in verba: Rhetoric and Authority in the Early Royal Society," p. 159).

170. While there were discussions about how accessible the Society's registers should be to nonmembers, I have found no evidence that the Society ever considered itself the owner of the discoveries and claims that had been sent to the academy and then entered in its registers. The Society was the beneficiary of patents on inventions developed by its members, but this was a conscious attempt (approved by the treasurer, Abraham Hill) at improving the institution's finances (Hunter, *Establishing the New Science,* p. 89). The dispute between Hooke and Huygens on the invention of the spring watch, and the controversies triggered by Huygens' promise of the English patent on his device to Oldenburg, shows

casting himself as the original discoverer of the satellites of Jupiter, could offer the Medici some kind of symbolic ownership of their Stars. The difference between what we could call holding in trust and symbolic ownership explains why the Society set up a system of credit that was based on the steady flow of generic communications rather than on custom dedications of spectacular but occasional discoveries.[171]

Because the Society did not take ownership of the claims it received, it benefited from the number of transactions that took place through its correspondence and the *Transactions*. Like a bank that earns money from lending funds it receives in deposit from its investors, the more submissions the Society received and recycled as articles, the more credit it generated for itself. Similarly, the correspondents submitted their claims and reports to the Society because of the visibility they could gain by doing so, not because their specific claims would "fit" the Society the way the Medicean Stars could be made to fit the Medici.[172]

Galileo operated in a system that was as distance based as that of the Society, but not quite as cyclical. Galileo gained credit through one cycle, not many. And the cycle ended with something like a payment for his "sale" of one discovery to his patrons.[173] Of course, Galileo could come up with other distance-based loops of credit centered on his patrons, but these would have to involve claims that, in addition to being novel, had to fit, in some way, the image and interest of the patron.[174] To put it slightly differently, once Galileo became an official client of the Medici, his patrons had

how important the firewall between deposited claims and ownership was for the Society, and how dangerous its breach could be (Rob Iliffe, "'In the Warehouse': Privacy, Property, and Priority in the Early Royal Society," *History of Science* 30 [1992]: 29–68).

171. I do not use the term "generic" in a negative sense, but only to contrast the patron-specific strategies of Galileo with those of the Society—strategies that were explicitly aimed at widening the range of philosophically relevant claims while "standardizing" them through the emphasis on the matter of fact as their evidentiary standard.

172. On the relationship between kinds of claims or objects and kinds of patrons, see also Mario Biagioli, "Scientific Revolution, Social Bricolage, and Etiquette," in Roy Porter and Mikulas Teich (eds.), *The Scientific Revolution in National Context* (Cambridge: Cambridge University Press, 1992), pp. 11–54, esp. pp. 18–26.

173. Galileo's various dedications to the Medici throughout his career should be taken seriously as indications that Galileo was operating according to a credit model in which his work was in some way "owned" by the patron.

174. Another obvious difference is that in Galileo's case it is the client who keeps the investment process alive while in England it is the Royal Society that runs it.

an ownership-like relationship with his work (the way they did with their court artists), and it would have been difficult for Galileo to produce work the Medici would have not liked to "own."[175]

Not only did Galileo's credit system hinge on one specific patron, but his patron related to him as a very specific, unique individual, not as just another correspondent.[176] By contrast, the cyclical, cumulative nature of the Society's system of credit rested on the remoteness of the correspondents—a remoteness that narrowed the range of the personal interactions between the members located in London and the foreign fellows. The Society's system hinged not only on a relative disinvestment in the specificity of the claims it received, but also a relative disengagement from the specific person, the specific body that produced those reports. Both moves proved highly effective at expanding the applicability of that system.

Despite the differences (or maybe because of them), both examples I have outlined here indicate that the notion of local knowledge is something of an oxymoron. Once we consider the productive roles of distance, knowledge appears as something that is never completely local, not even at its so-called moment of origin (a temporal beginning that is as problematic as its spatial counterpart). A knowledge claim is not something that has a certain value in its place of origin and becomes nonlocal by having its value recognized elsewhere. Such a value is always constituted by the actions of remote investors who may or may not bet on it based on the inherently partial information to which they have access.

The sociology of scientific knowledge has been remarkably effective at exposing the problems and limitations of mentalistic views of knowledge—of ideas that seem to spring up into people's minds independent of their sociocultural milieu. It is therefore ironic that, in so doing, it has replicated the same metaphysics of presence that underlies that older picture.[177] The

175. One issue that does not differentiate the two systems is the impact of disclosure. Once they realized that the Society's credit system benefited them, the correspondents should not have been disturbed by realizing that perhaps the Society was not as Royal as they thought. Similarly, once the Medicean Stars were celebrated throughout Europe, the Medici may have not cared that Galileo had engaged in some creative financing.

176. *GC*, pp. 84–90.

177. Mario Biagioli, "From Difference to Blackboxing: French Theory versus Science Studies' Metaphysics of Presence," in Sande Cohen and Sylvere Lotringer (eds.), *French Theory in America* (New York: Routledge, 2001), pp. 271–87; Hans-Jörg Rheinberger, "Experimental Systems, Graphematic Spaces," in Timothy Lenoir (ed.), *Inscribing Science* (Stanford: Stanford University Press, 1998), pp. 285–303, esp. pp. 285–87.

idea that knowledge starts local and then becomes nonlocal reflects the assumption that knowledge is an object with origins—origins that have now been displaced from the mind to geographical space.[178]

178. The fact that SSK presents the production of knowledge as a potentially lengthy process does not change the fact that its logic still hinges on a picture of knowledge that has a site of origin.

CHAPTER TWO

Replication or Monopoly?

The Medicean Stars between Invention and Discovery

I PROPOSE an account of the production and reception of Galileo's telescopic observations of 1609–10 that focuses on the relationship between credit and disclosure. The historiography on Galileo's discoveries has traditionally clustered around two very different views of evidence. Some have treated telescopic evidence as unproblematic, dismissing Galileo's critics as stubborn and obscurantist.[1] Others have argued instead that Galileo's discoveries were not self-evident and that their making and acceptance depended on specific perceptual dispositions (possibly connected to his train-

1. Stillman Drake contended that "the arguments that were brought forward against the new discoveries were so silly that it is hard for the modern mind to take them seriously" (Galileo Galilei, *Discoveries and Opinions of Galileo,* trans. Stillman Drake [New York: Doubleday, 1957], p. 73). In his other publications on the subject, he focused on Galileo's process of discovery but did not discuss the difficulties others may have faced in trying to replicate them. See especially Stillman Drake, "Galileo's First Telescopic Observations," *Journal for the History of Astronomy* 7 (1976): 153–68. He only remarked that astronomers had problems corroborating his claims because suitable telescopes were hard to come by in 1610, and that philosophers were so committed to their bookish knowledge that they could not deal with Galileo's observations (Stillman Drake, *Galileo at Work: His Scientific Biography* [Chicago: University of Chicago Press, 1978], pp. 159, 162, 165–66, 168). The telescope's epistemological status is treated as a nonproblem, and perceptual issues are mentioned only in one case, to say that Galileo, because of an eye condition, had learned to peer through his clenched fist or between his fingers to improve his sight and that this may have given him the idea to stop down the objective lens to improve its performance (ibid., p. 148).

ing in the visual arts), commitments to heliocentrism, or unique (and possibly tacit) skills at telescope making.[2]

By questioning the transparency of the process of observation and discovery, the perceptual relativists have produced the more intriguing interpretations of these events. Yet they do not seem able to account for the fact that, despite all the perceptual and cosmological implications they find in

2. Feyerabend looked at how Galileo's telescopic evidence (mostly about the moon) could convince other observers and readers (or rather how they could not convince them without additional ad hoc hypotheses and "propaganda" tactics). However, unlike Drake, he did not analyze Galileo's process of discovery or his own reasons to believe in what he saw. Feyerabend saw Galileo's telescopic evidence as simultaneously problematic and productive. In his view, Galileo's evidence was deeply problematic but it was only by being so that it triggered conceptual change. It could become unproblematic only later, once it was framed within a new set of "natural interpretations" (Paul Feyerabend, *Against Method: Outline of an Anarchistic Theory of Knowledge* [London: Verso, 1978], pp. 99–161). Like Feyerabend, Samuel Edgerton has studied Galileo's visual representations of the moon and concluded that he was able to read the bright and dark patterns on its surface as pointing to physical irregularities (and to represent them in wash drawings that were then translated into engravings) because he had been trained in the artistic technique of chiaroscuro. Because of that training, Galileo saw the moon as a "landscape" and pictured it as such. Astronomers like Harriot (who had observed the moon with a telescope a few months before Galileo), on the other hand, did not have the same artistic training, did not see what Galileo saw, and pictured the moon not as rugged but just as spotted (Samuel Edgerton, "Galileo, Florentine 'Disegno,' and the 'Strange Spottedness' of the Moon," *Art Journal* 44 [1984]: 225–32). In part, Edgerton has relied on the work of Terrie Bloom, who has argued that Harriot was able to "see" the spottedness of the moon as an index of its morphological irregularities only after he read Galileo's *Sidereus nuncius* and viewed its engravings. The *Nuncius* provided Harriot with the "theoretical framework" he needed to see what he couldn't see before (Terrie Bloom, "Borrowed Perceptions: Harriot's Maps of the Moon," *Journal for the History of Astronomy* 9 [1978]: 117–22). Drawing a difference between encountering (or looking) and discovering, Bernard Cohen has argued that Galileo discovered what he did because of a theoretical mind set informed by a mix of anti-Aristoteleanism and incipient Copernicanism (Bernard Cohen, "What Galileo Saw: The Experience of Looking through a Telescope," in P. Mazzoldi [ed.], *From Galileo's "Occhialino" to Optoelectronics* [Padua: Cleup Editrice], pp. 445–72; and Cohen, "The Influence of Theoretical Perspective on the Interpretation of Sense Data: Tycho Brahe and the New Star of 1572, and Galileo and the Mountains on the Moon," *Annali dell'Istituto e Museo di Storia della Scienza di Firenze* 5 [1980]: 3–14). Van Helden, on the other hand, has focused on the practical and perceptual challenges posed by early telescopes to argue that the making and replicating of Galileo's observations was a remarkable achievement, not a problem-free task. The conditions for such an achievement included suitable telescopes, considerable labor, appropriate observational setups, good eyesight, and, ultimately, a tacit "gift" at observing.

Galileo's discoveries and the ambiguous epistemological status of the instrument that produced them, his claims were commonly accepted within nine months of their publication in the *Sidereus nuncius* in March 1610.[3] This is all the more remarkable considering that the satellites of Jupiter were not visible for about two months during that summer, contemporary networks of philosophical communication were neither broad nor fast, and the corroboration of Galileo's claims required learning how to construct and use a brand-new kind of instrument.[4]

A different picture emerges when we focus on Galileo's own observational protocols and how he did (or rather did not) help others replicate his discoveries. He acted as though the corroboration of his observations were easy, not difficult. Galileo's primary worry, I argue, was not that some people might reject his claims, but rather that those able to replicate them could too easily proceed to make further discoveries on their own and deprive him of future credit.[5] He tried to slow down potential replicators to prevent them from becoming competitors. He did so by not providing other practitioners access to high-power telescopes and by withholding detailed information about how to build them.[6]

3. I take the Roman Jesuits' confirmation of Galileo's claims on December 17, 1610, as a conservative date for the closure of the debate. For a summary of the controversial nature of Galileo's discoveries and instrument, see *SN*, pp. 88–90. Elsewhere, van Helden has remarked, "Now much has been made of the conservative opposition to these discoveries, but I should like to suggest that in view of the circumstances, the time it took Galileo to convince all reasonable men was astonishingly short" (Albert van Helden, "The Telescope in the Seventeenth Century," *Isis* 65 [1974]: 51).

4. Galileo's manuscript log shows a gap in his observations of the satellites between May 21 and July 25 during Jupiter's conjunction with the Sun (*GO*, vol. III, pt. 2, pp. 437–39).

5. *SN*, p. 17. Probably Galileo communicated in letters his major discoveries since the *Sidereus nuncius*—the changing appearances of Saturn and the phases of Venus—but did not rush to print because his priority was widely recognized (and publicized by the Jesuits' corroboration of his key claims in 1611). In the case of Saturn, Galileo may have waited to publish a text on it until he could explain the reasons behind such peculiarly changing appearances. In the case of sunspots, he may have delayed making his discovery public, worrying about the cosmological implications of that finding.

6. Galileo's concerns with priority and monopoly have been noticed before. Drake has remarked on Galileo's reluctance to give out information about the telescope as an "unwillingness to give away advantages" (Stillman Drake, *Galileo Studies* [Ann Arbor: University of Michigan Press, 1970], p. 155). Albert van Helden and Mary Winkler have argued that Galileo "was able to monopolize telescopic astronomy for the first several years and make almost all the important discoveries" (Mary Winkler and Albert van Helden, "Represent-

These tactics were continuous with the secrecy about his discoveries that he maintained before their publication in the *Nuncius*. As we have seen, Galileo even kept his patrons in the dark about the location of the Medicean Stars to prevent them from accidentally leaking any information his competitors could have used to scoop him, and asked the Medici secretary to treat his correspondence about the discoveries at the same level of confidentiality as important diplomatic matters.[7] Gingerich and van Helden have plausibly speculated that Galileo might have sworn his printer to secrecy and may have given him the section on the Medicean Stars only at the very last moment.[8] They have also pointed to the fact that, in the opening pages of the book (the section that was first delivered to the printer) the Medicean Stars were still referred to only as "four wandering stars" orbiting "a certain star notable among the number of known ones."[9] Galileo's substantial drawing skills further contributed to his secrecy by allowing him not to hire an artist to produce the illustrations of the book.[10]

But as important as it was for Galileo to keep his fellow astronomers in the dark, such negative tactics alone would not have allowed him to gain credit from his discoveries and move from his post at the University of Padua to a position at the Medici court in Florence as mathematician and philosopher of the grand duke—goals clearly on his mind in 1610. He

---

ing the Heavens: Galileo and Visual Astronomy," *Isis* 83 [1992]: 214–16). Elsewhere van Helden has remarked that "because he won the instrument race, Galileo was able to monopolize the celestial discoveries" (Albert van Helden, "Galileo and the Telescope," in Paolo Galluzzi [ed.], *Novità celesti e crisi del sapere* [Florence: Giunti, 1984], p. 155). However, they have not seen these monopolistic tendencies as central to the story of the making and acceptance of Galileo's discoveries, or to the narrative structure of the *Nuncius*.

7. Gingerich and van Helden, "From Occhiale to Printed Page," pp. 254–56. Galileo told Vinta on February 13, "Due cose desidero circa questo fatto [the discoveries], et di quelle ne supplico V.S.Ill.ma: l'una è quella segretezza che assiste sempre a gl'altri suoi negozi più gravi" (*GO*, vol. X, p. 283).

8. Gingerich and van Helden, "From Occhiale to Printed Page," p. 254.

9. *SN*, p. 36; Gingerich and van Helden, "From Occhiale to Printed Page," p. 254.

10. One could ask whether there was a relationship between Galileo's use of woodcuts for the illustrations of the satellites of Jupiter and his concern with secrecy. Engravings were more expensive, but would have also taken extra time to produce thus increasing the chances of a leak (and postponing the publication date of the book). On Galileo's drawing skills, see Horst Bredekamp, "Gazing Hands and Blind Spots: Galileo as Draftsman," in Jürgen Renn (ed.), *Galileo in Context* (Cambridge: Cambridge University Press, 2001), pp. 153–92.

needed proactive tactics as well. First, he secured the support of the grand duke through the maneuvers analyzed in the previous chapter as well as through a skillful management of patronage dynamics.[11] Second, through the prompt publication of the *Sidereus nuncius* in March of 1610 he tried to establish priority and international visibility—resources he needed to impress his prospective patron, not just the republic of letters.

The *Nuncius* was carefully crafted to maximize the credit Galileo could expect from readers while minimizing the information given out to potential competitors. Although it was researched, written, and printed in less than three months, it offered detailed, painstaking narratives of Galileo's observations and abundant pictorial evidence about his discoveries. It also said precious little about how to build a telescope suitable for replicating his claims.

Galileo gave a synthetic narrative (rich in dates and names but poor in technical details) of how he developed his instrument. He did not tell his readers how he ground lenses—the distinctive skill that gave him an edge over other early telescope makers—nor did he mention the dimension of his telescopes, the type of glass or the size and focal length of the lenses he used, or the diaphragm he had placed on the objective lens to improve its resolution.[12] He provided only a bare diagram of the instrument and mentioned that his optical scheme involved a plano-convex objective and a plano-concave eyepiece (fig. 1). He also told his readers that unless one had at least a good twenty-power telescope, "one will try in vain to see all the things observed by us in the heavens."[13] He then proceeded to tell how to measure the enlarging power of telescopes, allegedly to prevent his readers from wasting their precious time trying to observe what they could not possibly see.[14] While he promised his readers a forthcoming book on the workings of the telescope, he never published it, nor do we have any manuscript evidence of such a project.[15] Like many discoverers after him, Galileo presented his instrument as the standard of reference but—unlike many of

11. *GC*, pp. 103–57.

12. *SN*, p. 37. While other mathematicians besides Galileo were able to figure out the relationship between the focal length of the lenses and the enlarging power of the instrument, he was quickly able to develop remarkable skill at grinding lenses for telescopes—lenses that were outside of the standard repertoire of glassmakers (van Helden, "Galileo and the Telescope," pp. 154–55).

13. *SN*, p. 38.

14. *SN*, p. 38.

15. *SN*, p. 39.

ſpicillis ferantur ſecundum lineas refractas E C H.
E D I. coarctantur enim, & qui prius liberi ad F G.
Obiectum dirigebantur, partem tantummodo H I. cõ-

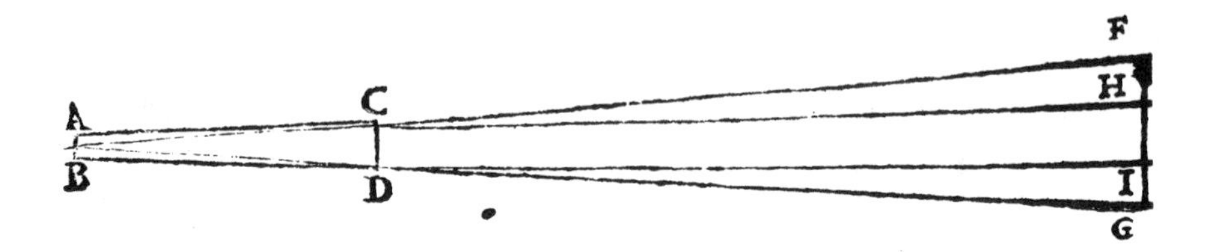

præhendent: accepta deinde ratione diſtantiæ E H. ad
lineam H I. per tabulam ſinuum reperietur quantitas
anguli in oculo ex obiecto H I. conſtituti, quem minuta quædam tantum continere comperiemus. Quod
ſi Specilio C D. bracteas, aliàs maioribus, aliàs verò mi
noribus perforatas foraminibus aptauerimus, modo
hanc modo illam prout opus fuerit ſuperimponentes,
angulos alios, atque alios pluribus, paucioribuſquè
minutis ſubtendentes pro libito conſtituemus, quorũ
ope Stellarum intercapedines per aliquot minuta ad-
inuicem diſſitarum, citra vnius, aut alterius minu-
ti peccatum commodè dimetiri poterimus. Hæc ta-
men ſic leuiter tetigiſſe, & quaſi primoribus libaſſe
labijs in præſentiarum ſit ſatis, per aliam enim occaſio
nem abſolutam huius Organi theoriam in medium pro-
feremus. Nunc obſeruationes à nobis duobus proximè elapſis menſibus habitas recenſeamus, ad magnarũ
profectò contemplationum exordia omnes veræ Philo-
ſophiæ cupidos conuocantes.

De facie autem Lunæ, quæ ad aſpectum noſtrum
vergit

FIGURE 1. Diagram of the telescope in Galileo's *Sidereus nuncius* (1610).

them—he effectively witheld such a reference by not disclosing its specifications or the procedures to be followed to produce such an instrument.[16] And yet, as I will show in a moment, such a lack of disclosure did not necessarily destabilize his claims.[17]

This narrative seems to clash with my previous claim that Galileo distributed several telescopes throughout Europe shortly after the publication of the *Nuncius*. Those instruments, however, were sent to princes and cardinals, not to mathematicians. Princes and cardinals were not Galileo's peers but rather belonged to the social group of his prospective Medici patron. While their endorsements could strengthen Galileo's credibility with the grand duke, their social position prevented them from competing with

16. Much historiography seems to confirm Harry Collins' argument about the experimenter's regress, that is, the presence of a vicious circle linking the determination of the correct outcome of an experiment to the determination of the winner of a dispute (Collins, *Changing Order*, pp. 83–84). As a result, he argues that there is no alternative to the use of the winner's findings to calibrate the instruments of other replicators. Collins' examples show that the scientists who wish to win an experimental dispute and have their claims and instruments "canonized" are not stingy at disclosing details about their instruments and even at helping other replicators build their own instruments. In the early modern period, we see that people like Boyle (in the *New Experiments*), Hooke (in the *Micrographia*), and Hevelius (in the *Selenographia*) fit Collins' narrative in that they provided substantial disclosure of the instruments and procedures they used in making their discoveries. Galileo appears to act like the winner of a Collins-style experimental dispute in that he claims his telescope and discoveries to be the standard against which other observers should tune their eyes and instruments, but he does not match Collins' narrative in the sense that he does not disclose the information necessary to replicate his instrument. More precisely, he tries to help people believe in his findings (or to replicate them with their own instruments) but does not help them reproduce his instruments (which they might then use to make more discoveries). As I discuss later on, the difference between Galileo's actions and the typical actions of a Collins-style experimenter have to do both with the specificity of Galileo's findings (and the fact he thought there were more findings one could make and claim credit for), and with the specificity of the reward system in which he operated—a system where the winner was not determined by the claimant's colleagues.

17. At first, Galileo's tactics seem to resemble those of Newton during the debate on his theory of light and colors. Like Galileo, Newton withheld a great deal of information about the instruments he used in his early experiments (Simon Schaffer, "Glass Works: Newton's Prism and the Uses of Experiment," in David Gooding, Trevor Pinch, and Simon Schaffer [eds.], *The Uses of Experiment* [Cambridge: Cambridge University Press, 1989], pp. 67–104). But if Newton's actions (while not motivated by priority concerns) clashed with the philosophical sociabilities of that time and put him at risk of being "given the lie," I argue that Galileo's tactics were socially acceptable in the field in which he operated and were epistemologically justified by the specific observational practices within which he used the telescope.

him in the hunt for astronomical novelties. Furthermore, most princes and cardinals were already familiar with low-power telescopes because since 1609 glassmakers had been peddling these instruments to them, not to astronomers or philosophers.[18] The first two instruments to come to Italy in 1609 were owned by Count de Fuentes in Milan and Cardinal Borghese in Rome.[19]

By the end of 1609, low-power telescopes went from being wondrous devices to cheap gadgets (by nobles' standards) produced in several Italian cities by traveling foreign artisans and local spectacle makers.[20] Princes sought and used telescopes on terrestrial and, more rarely, celestial objects well before rumors of Galileo's discoveries had begun to circulate, and before most astronomers had developed any serious interest in telescopes.[21]

18. Paolo Sarpi, the theological (and often technical) advisor to the Venetian senate, was Galileo's primary source of information about early telescopes, before one of them actually arrived in Venice in August 1609. Through his diplomatic connections, Sarpi had heard of the Dutch invention of the telescope in November 1608, and wrote about it to a number of correspondents in France. One of them, Jacques Badoer, wrote back from Paris in the spring of 1609 with a more detailed description of the instrument, which by that time was commonly sold by Parisian glassmakers. Sarpi probably showed Galileo this letter in July 1609 (*SN*, p. 37). On Sarpi's correspondence about early telescopes, see Drake, *Galileo Studies*, pp. 142–44.

19. On August 31, 1609 (a few days after Galileo presented his telescope to the Venetians), Lorenzo Pignoria wrote from Padua to Paolo Gualdo in Rome that Galileo's telescope was "similar to the one that was sent to Cardinal Borghese from Flanders." *GO*, vol. X, p. 255. Girolamo Sirtori reported that an instrument was delivered to Count de Fuentes in Milan in May 1609 by a Frenchman (Girolamo Sirtori, *Telescopium, sive ars perficiendi* [Frankfurt: Iacobi, 1618], pp. 24–25). The presence of telescopes in Naples was already mentioned in an August 28, 1609, letter from Giovanni Battista della Porta to Federico Cesi (*GO*, vol. X, pp. 230, 252). Porta did not say he owned a telescope, but that he had seen one, probably an instrument owned by a local noble.

20. *GO*, vol. X, pp. 264, 267, 306. By March, telescopic observations of the moon were being conducted in Siena by Domenico Meschini, a gentleman who claimed to be in contact with other people in Rome who were also observing it with their own instruments (*GO*, vol. X, p. 314). In March 1610, Giovanni Battista Manso, a noble, wrote from Naples saying that, while the *Nuncius* had not yet arrived there, low-power telescopes were available and were being used, with moderate success, to observe the irregularities of the moon (*GO*, vol. X, p. 293).

21. On April 1610, a diplomat from Modena wrote Count Ruggeri that Prince Paolo Giordano Orsini was back from the Netherlands, where he had purchased a number of telescopes, probably to give them as gifts to other Italian princes who did not have them yet (*GO*, vol. X, p. 347).

blication and disclosure need to be reframed within this hybrid econ-
ny—an economy that was different, in scale and structure, from that
nich emerged in late seventeenth-century natural philosophy, especially
ound the use of experiments.

UNAIDED CORROBORATIONS

one of the astronomers or savants who reproduced Galileo's observations
the satellites of Jupiter by the end of 1610 did so with his direct help. The
st independent confirmation came in May from Antonio Santini. A Ve-
tian merchant with no particular background in astronomy, optics, or
strument making, Santini was able to build a high-power telescope and
serve the satellites of Jupiter within two months of the publication of the
incius.[24] He conducted more successful observations in September, when
e satellites became visible again.[25] Although he knew Galileo and was

Santini was originally from Lucca. After his mercantile phase in Venice, he became
monk, and finally a mathematics professor in Rome (*GO*, vol. XX, pp. 531–32). In a
e 1610 letter to Galileo, Santini mentioned the observations of the satellites he had
nducted some time before. The letter itself is a plea on behalf of Giovanni Magini to
nvince Galileo that Magini was not involved in Martinus Horky's printed attack on Ga-
o. The letter ends by saying that Magini had endorsed Santini's corroboration of Gali-
's observations despite the fact that Magini, because of poor eyesight, had been unable
see them (*GO*, vol. X, p. 378). By the end of May, Jupiter was too close to the Sun to
observed, which means that Santini's observations must have been carried out in the
ond half of May at the latest. On September 25, Santini confirmed that he had clearly
n the satellites of Jupiter before conjunction—"Giove vespertino." *GO*, vol. X, p. 435.
at Santini's observations were not made through Galileo's telescope but through an in-
ument of his own production is supported by the fact that by June Santini was already
upplier of good lenses and telescopes to Magini (*GO*, vol. X, pp. 378–79) and that,
he several letters exchanged with Galileo during 1610 he never mentioned having ob-
ved with him. Santini's corroboration was made public in Roffeni's *Epistola apologetica
tra caecum peregrinationem cuiusdam furiosi Martini* (Bologna: Rossi 1611; repro-
ed in *GO*, vol. III, pt. 1, p. 198).

Santini's observations are reported several months after they took place (or perhaps
lier letters mentioning them are lost). In a September 25 letter to Galileo, Santini wrote,
nalmente mi risolsi di rivedere Giove mattutin, se bene, per quello aspetta a me, haveo
ta confermassione dall'averlo visto vespertino, che non dubitavo se li pianeti intorno a
da lei scoperti vi fossero o no (se però non si desse là sopra qualche alterassione). Lo
detti lunedì mattina, alle ore 10, giorno che fu de' 20 stante, e trovai li 4 pianeti tutti
ntali. Alli 23 poi li riveddi del modo che notirò da basso [one to the left and three to
right of Jupiter]" (*GO*, vol. X, p. 435). That Santini's September observations were

Emperor Rudolph II, for instance, observed the Moon
fore the *Nuncius* was published.[22] In Galileo's eyes, pr
constituted a low-risk, high-gain audience. Being mor
telescope than philosophers or astronomers, they wer
preciate the superior quality of his instruments and to
coveries. At the same time, they were not going to com
having little professional and philosophical stake in h
were less motivated to oppose them.

Galileo's differential treatment of his various audien
ful. He did take some short-term risk by relinquishing
have received from other mathematicians and astrono
widespread replications. But by the end of 1610 he ha
nopoly on telescopic astronomy that he then maintaine
available to him as mathematician and philosopher of
Tuscany.[23]

I look at the development of Galileo's monopoly to
relationship between disclosure and credit changed as
ing an instrument maker to becoming a discoverer and
philosopher. My narrative does not follow a chronolog
ganized by a set of interrelated questions: How was
make his observations? What kinds of textual informa
necessary to reproduce them? How could he justify
practices and yet have his findings accepted? What kin
narratives could he develop to minimize disclosure an
What was the relationship between the tactics of Galile
telescope and Galileo the author of the *Nuncius*? By fol
jectory from the development of the telescope in 1609
of a Medici-based monopoly on telescopic astronomy
he drew resources from various economies (of inventio
sual arts) without fitting completely into any one of th

22. *KGW*, vol. IV, p. 290. Since September 1609, the emperor had
scopes from Venice (*GO*, vol. X, p. 259) and perhaps others from

23. Van Helden has remarked on how quickly Galileo reacted to
challenges to his status as the leading telescopic astronomer, and h
havior as an expression of Galileo's concern with maintaining Me
Helden, "Galileo and the Telescope," pp. 156–57). However, the
the other direction as well; that is, Galileo's monopoly had been m
patronage.

perhaps among those who performed observations with him at Padua or Venice, there is no evidence that Santini received either telescopes or instructions as to how to grind lenses from Galileo.[26] Because he resided in a glassmaking center and had probably seen the low-power telescopes that circulated in Venice since the summer of 1609, Santini may have been in a position to replicate Galileo's instrument-making skills by himself.

Kepler was the second to see the satellites in late August and early September 1610 with one of Galileo's instruments.[27] However, that was not according to Galileo's plans because, as I discuss later, the telescope used by Kepler was not intended to go to him.

The third replication came on December 17 from the Jesuit mathematicians at the Collegio Romano.[28] They too had not received telescopes or instructions from Galileo, but used instruments sent them by Santini or produced locally by one of Clavius' students, Paolo Lembo, and perfected

---

cast as a belated rechecking—"I finally decided to see Jupiter again"—not as an urgent matter, confirms Santini's confidence in his earlier corroborations.

26. It is possible that Santini inspected one of the many low-power telescopes available in Venice since the autumn of 1609. However, he didn't necessarily need to have access to that information. It appears that several of the mathematicians and glassmakers who produced early low-power telescopes (Harriot, Marius, Galileo, Lipperhey, Janssen, and Metius) did so after receiving only a verbal description of them (van Helden, "The Telescope in the Seventeenth Century," p. 39 n. 3).

27. Kepler reports observations conducted from August 30 to September 9, 1610. Johannes Kepler, *Narratio de observatis a se quatuor Iovis satellitibus erronibus* . . . (Frankfurt: Palthenius, 1611), reprinted in *KGW*, vol. IV, pp. 315–25.

28. Clavius confirmed the existence of the satellites in a December 17 letter to Galileo (*GO*, vol. X, pp. 484–85). The Jesuits had been recording their sightings of the satellites since November 28 (*GO*, vol. III, pt. 2, p. 863), but had observed them also on November 22, 23, 26, and 27—as reported by Santini to Galileo in a December 4 letter in which he also included diagrams of the Jesuits' observations (*GO*, vol. X, pp. 479–80). The Jesuits seemed particularly cautious. Clavius had written Santini that, even after the November 22–27 observations, "we are not sure whether they are planets or not" (*GO*, vol. X, p. 480). Others seem to have observed the satellites in Rome before the Jesuits. In a November 13 letter to Galileo, Ludovico Cigoli stated that Michelangelo Buonarroti, a friend of Galileo's and Cigoli's, had been an eyewitness (*testimonio oculato*) of the satellites on several occasions, and that his testimonials had been able to convince a few skeptics (*GO*, vol. X, p. 475). Moreover, in a June 7 letter to Galileo, Martin Hasdale wrote from Prague that he had received a letter from Cardinal Capponi in Rome saying that Roman mathematicians approved of Galileo's discoveries, though he did not mention names (*GO*, vol. X, p. 370).

by Christoph Grienberger, a fellow Jesuit.[29] Neither Lembo nor Grienberger had previous experience in manufacturing optical instruments.[30] Two other successful observations of the satellites were achieved in 1610 in France (Peiresc and Gaultier, November 1610) and England (Harriot, October 1610)—both of them without any direct help from or telescopes by Galileo.[31]

Galileo's tendency not to share telescopes or information about their construction was most striking in Kepler's case. In the *Dissertatio* Kepler had publicly endorsed Galileo's discovery of the satellites of Jupiter despite the fact that he was not able to replicate them because he had access only to low-power instruments owned by his patron, Rudolph II.[32] The emperor's instruments were powerful enough to observe the irregularities of the lunar surface, but their magnification and clarity did not allow Kepler to detect the satellites of Jupiter. A few months after delivering the endorsement, Kepler pleaded with Galileo to send him a telescope, saying "You have aroused in me a passionate desire to see your instruments, so that I at last, like you, might enjoy the great spectacle in the sky." [33]

29. *GO*, vol. XI, pp. 33–34.

30. The mathematicians of the Collegio Romano, however, had extensive experience in other kinds of instrument making. On the topic, see Michael John Gorman, "The Scientific Counter-revolution: Mathematics, Natural Philosophy, and Experimentation in Jesuit Science, 1580–1670," Ph.D. diss., European University Institute, 1998, and his "Mathematics and Modesty in the Society of Jesus: The Problems of Christoph Grienberger," in Mordechai Feingold (ed.), *The New Science and Jesuit Science* (Dordrecht: Kluwer, 2003), pp. 1–120.

31. These replications had no historical role in the story I am telling here. On these observations, see John Roche, "Harriot, Galileo, and Jupiter's Satellites," *Archives internationales d'histoire des sciences* 32 (1982): 9–51; Pierre Humbert, "Joseph Gaultier de la Vallette, astronome provençal (1564–1647)," *Revue d'histoire des sciences et de leurs applications* 1 (1948): 316. In 1614, Simon Marius, a German mathematician, claimed to have discovered Jupiter's satellites earlier than Galileo, but his claims have been disputed since (*SN*, p. 105 n. 61). Marius' priority claims are in his *Mundus Jovialis,* translated in A. O. Prickard, "The 'Mundus Jovialis' of Simon Marius," *Observatory* 39 (1916): 367–503.

32. A telescope used by Rudolph II to observe the Moon is mentioned in *KGW*, vol. IV, p. 290. For a discussion of some of Kepler's reasons for endorsing Galileo's claims without being able to replicate them, see chapter 1.

33. *GO*, vol. X, pp. 413–14. Kepler continued, "Of the spyglasses we have here, the best ones are ten-power, others three-power. The only twenty-power one I have has poor resolution and luminosity. The reason for this does not escape me and I see how I could make it clearer, but we don't want to pay the high cost."

Although by this time Galileo had already given instruments to princes and cardinals (and was in the process of sending more), he did not send one to Kepler. He excused himself by suggesting that Kepler deserved only the best of telescopes, which unfortunately Galileo no longer owned because it had been placed "among more precious things" in the grand duke's gallery to memorialize the discovery of the Medicean Stars.[34] Galileo also intimated that he was temporarily unable to produce more instruments because, being in the process of moving from Padua to Florence, he had disassembled the machine he had constructed to grind and polish lenses.[35] The emperor too had requested (with some insistence) an instrument through the Medici ambassador in Prague and had vented his frustration at not being given priority over cardinals whom he knew Galileo had provided with telescopes.[36] However, the imperial pleas, like Kepler's, went unanswered. When Kepler finally observed the satellites of Jupiter in the late summer of 1610 and published his findings in the *Narratio de observatis a se quattuor*

34. *GO*, vol. X, p. 421.

35. *GO*, vol. X, p. 421. On October 1, Galileo reported that his lens-grinding machines (which he said had to be set in place with mortar) were still inoperative (*GO*, vol. X, p. 440).

36. Giuliano de' Medici to Galileo, April 19, 1610 (*GO*, vol. X, p. 319). In July, better telescopes reached Prague from Venice, but none of them was made by Galileo (*GO*, vol. X, pp. 401–2). On July 19, Giuliano de' Medici acknowledged the arrival of additional ephemerides of the satellites (not telescopes) Galileo had sent to Kepler (*GO*, vol. X, p. 403). In the same letter, he urged Galileo to send an instrument to the emperor (*GO*, vol. X, p. 404). On August 9, Galileo was told that the emperor had received a better telescope from Venice, but that Galileo's instrument (that some thought had been received by the Medici ambassador) had not yet been seen (*GO*, vol. X, p. 418). It does not appear it was ever there, as on August 17 Galileo was told of the emperor's aggravation (*GO*, vol. X, p. 420). Interestingly, the imperial court at Prague was not on the first list of potential recipients of telescopes Galileo submitted to the Medici on March 19 (*GO*, vol. X, pp. 298, 301), but was added only in May (*GO*, vol. X, p. 356). While much of the evidence points to the fact that Galileo did not wish Kepler to have a telescope, on May 7 he asked the Medici for permission to send one in the diplomatic pouch from Venice to Prague. He also remarked that he did not have any good telescopes ready (*GO*, vol. X, pp. 349–50). The Medici authorized the shipment on May 22 (*GO*, vol. X, p. 356), but on May 29, the Medici resident in Venice expressed worries that the telescope could get damaged during shipping (*GO*, vol. X, p. 364). That does not seem to have been a problem, as telescopes were often shipped disassembled. It could be that Galileo thought it would be useless to send a telescope to Prague at the end of May, as Jupiter was no longer observable, and that he could have taken the two months before it became visible again to produce a better instrument. However, he never sent such an instrument, probably because, by the time Jupiter was visible again, he had already received a contract from the Medici.

Iovis satellitibus erronibus, he did so with a telescope Galileo had sent to the elector of Cologne—not to him or to the emperor.[37]

Galileo's behavior may seem particularly ungrateful, as Kepler's *Dissertatio* was the first and only strong endorsement he had received from a well-known astronomer before he obtained his position at the Medici court in the summer of 1610. However, in the *Dissertatio*, an enthusiastic Kepler exclaimed: "I wished there were a telescope ready for me, with which I could anticipate you in discovering the satellites of Mars (that, according to the proportion, should be two) and those of Saturn (that should be six or eight)."[38] The use of the verb "anticipate" may have drastically decreased Kepler's chances of receiving the instrument he sought.

Galileo displayed a similarly uncooperative attitude toward other potential allies.[39] On April 17, Ilario Altobelli asked him for lenses or a telescope so that, he claimed, he could help Galileo with testimonials against his critics. But he seemed a bit too eager to determine the periods of the Stars and received nothing in the end.[40] Magini, who had been one of Galileo's early detractors but had slowly changed his mind, asked him for an eyepiece on October 15, but received none.[41]

Galileo did not help the Jesuit mathematicians of the Collegio Romano either. Right after moving back to Florence from Padua in September 1610, he wrote Clavius that he had heard the Jesuits were having problems seeing the satellites of Jupiter. That did not surprise him, Galileo continued, as he knew all too well that one needed an "exquisite instrument" to replicate his observations.[42] However, he did not volunteer to send Clavius such an exquisite telescope but simply advised him to build a sturdy mount for whatever instrument he had because even the small shaking caused by the

37. In September 1610, Giuliano de' Medici informed Galileo of Kepler's observations and of his decision to publish the *Narratio* (*GO*, vol. X, p. 329).

38. *KGW*, vol. IV, p. 291.

39. Raffaello Gualterotti requested lenses on March 6, 1610 (*GO*, vol. X, p. 287) and Alessandro Sertini on March 27, 1610 (*GO*, vol. X, p. 306).

40. "[E]t m'ingegnerò d'adattare il tubo in forma della fiducia nel dorso dell'astrolabio per osservare anco i periodi; e scriverò a V.S. il tutto in lingua latina, acciò lo possi poi annettere nelle sue osservationi" (*GO*, vol. X, p. 317).

41. Magini was said to have received three large lenses from Santini, and that he thought to have a very good one among them. But he lacked good eyepieces and asked Galileo to send some (*GO*, vol. X, p. 446).

42. *GO*, vol. X, p. 431.

observer's pulse and breathing was enough to disrupt the observations.[43] He concluded that, in any case, he would show Clavius the "truth of the facts" during his forthcoming visit to Rome.[44] On October 9, Santini wrote Galileo that the Jesuits had not yet seen the satellites and added that "I think that these big shots, I mean in terms of reputation, are playing hard to get so that Your Lordship may feel obliged to send them an instrument."[45] Even then, however, Galileo did not send the Jesuits a telescope.

He was much more forthcoming with patrons and courtiers. In a January 7 letter he appears to have sent from Padua to a Florentine courtier, Galileo gave more useful tips about the telescope than in the *Nuncius* or in anything else he wrote that year. In that letter he explained how to minimize the shaking of the telescope caused by the observer's heartbeat and breathing, how to maintain the lenses, and how much movement one should allow the tubes carrying the two lenses so as to achieve proper focusing (as he probably was planning to send him two lenses but no casing).[46] More

43. *GO*, vol. X, p. 431.

44. *GO*, vol. X, p. 432.

45. "Io dubito che alcuni di questi pezzi più grossi, voglio dire di più riputassione, non stiano duri, acciò V.S. si metta di necessità di mandargli lei uno instrumento" (*GO*, vol. X, p. 445).

46. *GO*, vol. X, pp. 277–78. The editor of Galileo's *Opere*, Antonio Favaro, tentatively identifies the addressee of this letter as Don Antonio de' Medici—the cousin of Grand Duke Cosimo II. The identification is probably wrong because the titles used by Galileo to refer to the addressee of this letter are distinctly different from how he addressed Antonio de' Medici in a February 1609 letter (*GO*, vol. X, pp. 228–30). The addressee of this letter was of a lower social rank than Don Antonio's. What remains also unclear is whether Galileo was instructing his interlocutor on how to use the telescope in case the grand duke needed some help replicating Galileo's discoveries (if he decided to send him a telescope or a pair of lenses), or whether he was just giving this person tips on how to use his own telescope. If this letter was ever sent, it was probably addressed to a Florentine courtier Galileo was trying to use as an informal contact with the grand duke—someone who could inform him of what Galileo was observing and discovering. But Galileo's correspondence shows that he informed the Medici of his discoveries by writing directly to their secretary at the end of January, and that he mentioned no informal early contact through other people in that letter. Because of this and the fact that this letter—the January 7 letter—is never cited by anyone in any other letter preserved in Galileo's correspondence, I suggest that it might have never been sent. Galileo's secrecy may explain this. Most of the letter is about lunar observations, but toward the end Galileo mentions that he had observed certain fixed stars around Jupiter. When, a few days later, Galileo decided that these were not fixed stars but planets, he might have decided to shelve the letter so as not to give out hints about this last, more prized discovery.

importantly, he stressed that the objective lens needed to be stopped down with a diaphragm. As the lenses' shape was particularly irregular toward the edges, covering that part would significantly reduce aberrations.[47] Clavius was told by Galileo of this significant tip almost a year later (after he confirmed the satellites' observation in December 1610) and only because the Jesuit had asked Galileo why the telescopes he had sent to Rome (to cardinals) had stopped-down objectives.[48]

Predictably, Galileo did not loan his own instrument. He organized or participated in public observational séances in Venice, Padua, Bologna, Pisa, Florence, and later in Rome, but it appears that he never left the telescope in other hands, not even for a few hours.[49] These meetings were meant to provide demonstrations rather than to foster independent replications. Galileo would arrive, demonstrate, and depart. While people could look through the telescope, it appears that they did not have much of a chance to look inside it. During his visit to Bologna in April 1610, Martinus Horky had to sneak around Galileo's guard (probably while he was asleep) to make a cast of the telescope's objective lens.[50]

47. However, we do not have any evidence that Galileo actually sent the lenses or a telescope to Florence before his visit during the Easter vacation. The only evidence of a telescope in Florence is from April 20, 1610 (*GO,* vol. X, p. 341). The letter mentions a telescope kept in the Medici storage rooms, but does not say it is by Galileo. Also, because Galileo was in town at that time (and is actually mentioned in the letter), this could be an instrument he had brought with him from Padua.

48. Clavius' query is in *GO,* vol. X, p. 485. Galileo's response is at p. 501: "Hora, per rispondere interamente alla sua lettera, restami di dirgli come ho fatto alcuni vetri assai grandi, benchè poi ne ricuopra gran parte, et questo per 2 ragioni: l'una, per potergli lavorare più giusti, essendo che una superficie spaziosa si mantiene meglio nella debita figura che una piccola; l'altra è che volendo veder più grande spazio in un'occhiata, si può scoprire il vetro: ma bisogna presso l'occhio mettere un vetro meno acuto et scorciare il cannone, altramente si vedrebbero gli oggetti assai annebbiati. Che poi tale strumento sia incomodo da usarsi, un poco di pratica leva ogni incomodità; et io gli mostrerò come lo uso facilissimamente." On the relationship between diaphragms and performance in Galileo's telescopes, see Yaakov Zik, "Galileo and the Telescope," *Nuncius* 14 (1999): 31–67, p. 51, and "Galileo and Optical Aberrations," *Nuncius* 17 (2002): 455–65; Sven Dupré, "Galileo's Telescopes and Celestial Light," *Journal for the History of Astronomy* 34 (2003): 369–99.

49. Galileo to Kepler, August 19, 1610, *GO,* vol. X, p. 422. On meetings in Venice and Padua see Galileo to Vinta, March 19, 1610, *GO,* vol. X, p. 301.

50. *GO,* vol. X, p. 343.

Galileo's fears about the consequences of giving good telescopes to mathematicians or helping them construct their own were not unjustified. He knew from personal experience that after receiving an approximate verbal description of a telescope one could build a prototype in a single day, move from three-power to nine-power telescopes in a few weeks, and develop a twenty-power instrument in about four months and a thirty-power one in less than seven months.[51] Soon after, he witnessed a merchant like Santini build telescopes good enough to observe the satellites of Jupiter within two months from the publication of the *Nuncius* in March, and then supply lenses and entire telescopes to both Magini and the Roman Jesuits.[52] He also knew that increasingly powerful instruments were being constructed in Venice and elsewhere.[53] His friend Castelli reported that at the beginning of February (and thus before the publication of the *Nuncius*) a friar, Don Serafino da Quinzano, had shown him the Moon through a nine-power telescope that he had built on his own.[54] That was not quite the twenty-power instrument Galileo was using at the time, but it already had more than twice the power of the telescope that caused such a stir when it arrived in Venice about six months before.

51. On Galileo's quick progress, see van Helden, "Galileo and the Telescope," pp. 150–55. On the earlier developments of the telescope, see van Helden, "The Invention of the Telescope," *Transactions of the American Philosophical Society* 67 (1974), pt. 4, pp. 1–67.

52. Santini's lenses and telescopes to Magini are mentioned in *GO*, vol. X, pp. 378, 398, 437, 446, 451. On Santini's gifts of telescopes to the Jesuits, see *GO*, vol. XI, pp. 33–34.

53. Galileo's correspondence indicates that by mid-1610, low-power telescopes were common, their price had dropped, and the market was so saturated that some telescope makers were moving on to other (probably more provincial) cities. Then, Magini wrote him in October that Cardinal Giustiniani had managed to attract to Bologna a skilled glassmaker from Venice (Bortolo, the son of the emperor's glassmaker) who was quite good at grinding lenses for long (that is, high-power) telescopes and that Magini planned to use his services (*GO*, vol. X, p. 446). On October 15, 1609, Lorenzo Pignoria wrote from Padua that there were "most excellent telescopes," adding that they were produced by a few artisans, that is, not just by Galileo (*GO*, vol. X, p. 260). Hasdale wrote to Galileo from Prague that the emperor was getting increasingly better telescopes from Venice, one of which apparently had been produced by an artisan who worked for Galileo (*GO*, vol. X, pp. 401–2). On April 24, Gualterotti mentions a good telescope made by "Messer Giovambattista da Milano" whose quality Galileo appears to have praised (*GO*, vol. X, p. 341). Santini sent a new telescope to Florence (to the Venetian ambassador) on November 6, 1610, and asked Galileo to take a look at it (*GO*, vol. X, pp. 464–65). Galileo liked it (*GO*, vol. X, p. 479).

54. *GO*, vol. X, pp. 310–11.

His worries would only have increased had he known of the Jesuits' quick progress.[55] By the end of 1610 Grienberger had succesfully modified the eyepiece of a second telescope Clavius had received from Santini thereby turning it into a thirty-four-power instrument.[56] If Grienberger's information is correct, it means that the Jesuits had surpassed Galileo in terms of magnification power. (Another student of Clavius, Paul Guldin, wrote that Grienberger's instrument enlarged more than forty times).[57] Grienberger also claimed that the image quality of their high-power telescope was better than that of those that Galileo had sent to cardinals in Rome. The fact that making telescopes was a quickly spreading skill was evident not only to Galileo. At the end of September 1610, Santini wrote him:

> I do not understand, now that the telescope has become so common and easy, why the practitioners of the speculative sciences have not managed to clarify this matter [the existence of the satellites] and express their consensus.[58]

Such skills were spreading north of the Alps as well, either independently or through the networks of the Society of Jesus. As we will see in the next chapter, twenty-five-power telescopes were succesfully operated (and probably produced) in provincial Jesuit colleges like Ingolstadt within a year of Galileo's publication.[59] Already in August 1610, Kepler reported that when the elector of Cologne visited Prague, the elector showed him a telescope sent by Galileo, but argued that its resolution (though not its magnifica-

55. Since the summer of 1610, the Roman Jesuit Paolo Lembo had been producing increasingly good telescopes with which the mathematicians of the Collegio Romano were eventually able to see the satellites—though only when the sky was very clear. According to Grienberger, Lembo had developed his first telescopes on his own, without information or examples from the outside (*GO*, vol. XI, pp. 33–34).

56. *GO*, vol. XI, p. 34. I thank Albert van Helden for decoding this figure from Grienberger's letter.

57. Paul Guldin to Johann Lanz, February 13, 1611, reproduced in August Ziggelaar, "Jesuit Astronomy North of the Alps: Four Unpublished Jesuit Letters, 1611–1620," in Ugo Baldini, *Christoph Clavius e l'attività scientifica dei gesuiti nell'età di Galileo* (Rome: Bulzoni, 1995), p. 119.

58. "Io non so come, essendosi fatto tanto comune e facile questo uso del cannone, non sia da quelli che attendono alle specolative chiarito questa partita e dato l'assento" (*GO*, vol. X, p. 435).

59. As discussed in the next chapter, such an instrument was used by Christoph Scheiner to observe sunspots.

tion) was substantially worse than that of others he had purchased elsewhere.[60] The issue, then, was not whether people could develop powerful telescopes, but only how many weeks or months it would take them to move from three-power to twenty-power instruments, and beyond.[61] If we take Grienberger's and Kepler's testimonies at face value, it also appears that Galileo's superiority was instantiated only in very few of his instruments, not in those he sent to princes and cardinals. (Whether that was deliberate or accidental, we have no way to know).[62]

That Galileo worried about priority disputes rather than about the difficulties others might face in replicating his discoveries is confirmed by his statement that the *Nuncius* had been "written for the most part as the earlier sections were being printed" for fear that by delaying publication he would have "run the risk that someone else might make the same discovery and precede me [to print]."[63] Galileo's behavior in 1610 suggests he thought he had only a limited amount of time to discover whatever there was to be discovered with telescopes of that power range.[64] As he put it in the *Nuncius,* "Perhaps more excellent things will be discovered in time, ei-

60. "Mense Augusto Reverendissimus et Serenissimus Archiepiscopus Coloniensis Elector, et Bavariae Dux, Ernestus etc. Vienna Austriae redux instrumentum mihi commodavit, quod a Galilaeo sibi missum dicebat; quod ipse quidem aliis quibusdam, quae secum habebat, ex commoditate quam ipse inde videndo caperet, longe postposuit; questus stellas repraesentari quadrangulas." *KGW,* vol. IV, pp. 318–19. It appears that although Galileo's instrument was neither very good nor easy to use—"nec optimum nec commodissimum"—Kepler used it because of its greater magnification. On March 19, 1610, the elector of Cologne was listed by Galileo among the recipients of the first ten telescopes he deemed good enough to observe the satellites of Jupiter (*GO,* vol. X, pp. 298, 301).

61. Harriot, for instance, had a ten-power telescope by July 1610, a twenty-power by August, and a thirty-two-power by April 1611 (Roche, "Harriot, Galileo, and Jupiter's Satellites," p. 17).

62. Although it is quite reasonable to assume that, no matter what Galileo's intentions may have been, not all of his telescopes were equally good, it is also possible that he did not send his "A" telescopes to princes to reduce the risk that their mathematicians might discover new objects. Although not top-of-the-line, those telescopes could have sufficed at replicating Galileo's discoveries when used in conjunction with the *Nuncius.* For instance, although the telescope of the elector of Cologne used by Kepler showed the stars as bright squares, he was still able to observe the Medicean Stars.

63. *GO,* vol. X, p. 300.

64. The first phase of the race for telescopic discoveries was effectively over by 1612 with the discovery of sunspots. The second wave of discoveries started only in 1655 with Huygens (van Helden, "Galileo and the Telescope," p. 155).

ther by me or by others, with the help of a similar instrument."[65] Even the first available report about the use of the telescope indicates that those who managed to construct or have access to an instrument quickly pointed it to whatever celestial body they could spot.[66] How close the race must have been can be gathered from a January 1611 letter in which Grienberger mentioned to Galileo that even before the Jesuits had heard about his discovery of the phases of Venus at the end of December, they had independently observed them.[67]

While Copernican commitments may have played a role in setting the direction of further observations (as in the case of the discovery of the phases of Venus), most participants in this astronomical hunt seemed propelled by the desire to discover more novelties and, sometimes, get credit for them. Considerations of the possible pro-Copernican or anti-Ptolemaic significance of these discoveries were not on everyone's mind in the first half of 1610, but emerged more clearly after Galileo's claims had been widely accepted.[68]

Because of the speed with which others were learning how to build telescopes suitable for astronomical use, Galileo's uncooperative stance may have been the determining factor in achieving a monopoly over that first astronomical crop. He was first to discover the unusual appearance of Saturn (in the summer of 1610) and the phases of Venus (in the fall), and to determine the periods of the satellites—a result that both reinforced the epistemic status of the Medicean Stars and brought him more visibility.[69] His monopoly became almost self-sustaining. He managed to reclaim credit for the discovery of the sunspots from the Jesuits (although they and Johannes

65. *SN*, p. 36.

66. *Ambassades du Roy de Siam envoyé à l'Excellence du Prince Maurice, arrivé a la Haye le 10. Septemb. 1608* (The Hague, 1608), p. 11, reports that one of the very early telescopes had been aimed at the stars in the Netherlands as early as fall 1608.

67. *GO*, vol. XI, p. 34.

68. *GC*, pp. 94–96.

69. In the *Nuncius,* he exhorted other astronomers to find the satellites' periods (*SN*, p. 64). By this time he had only a figure for the outer satellite, which he put at about fifteen days (*GO*, vol. X, p. 289). That figure was corrected to more than sixteen days in the spring 1611 (*GO*, vol. XI, p. 114). There he also gave estimates for the period of the innermost at less than two days. He published his first full description of the satellites periods in 1612 in his *Discourse on Bodies in Water.* Those values were very close to modern ones. On Galileo's investigation of these periods, see Stillman Drake, "Galileo and Satellites Prediction," *Journal for the History of Astronomy* 10 (1979): 75–95.

Fabricius had been first to publish that discovery) and, years later, he succeeded in defending the referential status of his telescopes when other instrument makers, like Fontana in Naples, had produced more powerful ones.[70]

AN EMERGING FIELD, NOT A COMMUNITY

Galileo's noncooperative attitude and his focus on developing a Medici-based monopoly of telescopic astronomy reflected more than just his fears about being deprived of credit for future discoveries. Galileo and his readers did not belong to a professional community that could provide the kind of credit and rewards he sought. The lack of consensus about style of argumentation and standards of evidence as well as the scant interdependence among the members of this field hindered closure of the debate.[71] Additionally, the absence of established protocols for the evaluation of priority claims—some saying that verbal communication and third-party witnessing were sufficient, others linking priority to publication—fueled bitter disputes and added fragmentation to an already fragmented natural philosophical field. The very status and role of witnessing were also in flux. For instance, Galileo named many witnesses (many aristocratic, some not) in support of his priority claims, but did not name them in relation to epistemological matters, such as saying that a certain thing existed or that a certain claim was true.[72] Kepler, on the other hand, used and named witnesses as an integral part of the process of observation in his *Narratio*.[73] In

70. "From an early point, then, the authority of instruments was intertwined with personal authority. A strong argument can be made that after about 1612, Galileo's lead in telescope making had disappeared and that others had instruments of comparable quality. Yet Galileo ruled until his death as the undisputed master of telescopic astronomy" (van Helden, "Telescopes and Authority from Galileo to Cassini," *Osiris* 9 [1994]: 19). On related issues, see also Winkler and van Helden, "Representing the Heavens," pp. 214–16.

71. My analysis is broadly informed by Pierre Bourdieu's notion of "field," and especially by his discussion of how fields are established (Pierre Bourdieu, "The Social Space and the Genesis of Groups," *Theory and Society* 14 [1985]: 723–44; and "The Specificity of the Scientific Field," in Biagioli, *The Science Studies Reader,* pp. 31–50).

72. In the preface to the 1606 *Operations* Galileo mentioned the prince of Alsace and count of Oldenburg, the archduke of Austria, the Landgraf of Hesse, and the duke of Mantua to support his priority claims about the compass and its instructions. In the *Nuncius,* on the other hand, he mentioned no witness to his discoveries.

73. *KGW,* vol. IV, pp. 319–22.

the dispute on sunspots discussed in the next chapter, Scheiner felt comfortable naming "hearsay witnesses"—people who had not witnessed his discoveries but, upon hearing of them, found them acceptable.[74]

Galileo's correspondence shows that his discoveries were discussed in a field geographically dispersed over several courts and universities or punctuated by isolated individuals linked only through selective correspondence networks. It included few professional astronomers but many physicians, men of letters, diplomats, students, polymaths, and variously educated gentlemen. Political and religious boundaries mattered. A French or German mathematician did not have much incentive to engage, to credit, or even less to agree with the claims put forward by someone who operated on the other side of the Alps. Catholic mathematicians could dismiss the work of their "heretical" counterparts with relative impunity.[75] For instance, it would be interesting to know how much the fact that Kepler (a Protestant) worked at the court of Rudolph II (a Catholic) facilitated the interaction between Galileo and Kepler and, conversely, whether the Jesuits' eventual endorsement of Tycho was slowed down by the fact that he was a Protestant.

We should not take Kepler's endorsement of Galileo—one that crossed national and religious boundaries—to be just an example of good peer review at work.[76] It obviously reflected Kepler's recognition that the legitimation of Galileo's claims could provide him with further resources for his own Copernican program. But not all German Copernicans behaved like Kepler. Michael Maestlin, the Protestant Copernican from Tübingen, gleefully endorsed the most aggressive attack on Galileo's discoveries—Horky's *Peregrinatio contra nuncium sidereum*—and did so well after it had been disowned even by those early opponents of Galileo's (like Magini) who had probably more than a passing involvement with that text.[77] Maestlin even

74. *GO*, vol. V, p. 62.

75. Judging from the documents produced by the censurae librorum—the Jesuits' internal review of manuscripts for publications—Jesuit mathematicians were discouraged from citing Protestant mathematicians favorably. Ugo Baldini, *Legem Impone Subactis: Studi su filosofia e scienza dei gesuiti in Italia, 1540–1632* (Rome: Bulzoni, 1992), pp. 230–31.

76. In the *Dissertatio,* Kepler responded to those who accused him of having been too generous toward Galileo that "I do not think that I, a German, owe so much to Galileo, an Italian, that I have adulated him with prejudice for the truth and for my own deeply held position." (*KGW,* vol. IV, p. 287.)

77. *KGW,* vol. XVI, p. 334. Probably Maestlin was upset by finding no reference to his work on lunar appearance and especially lunar spots.

managed to read Kepler's glowing endorsement of Galileo in his *Dissertatio* as a timely sizing-down of the originality of the *Nuncius*' claims.[78]

Critical responses to the *Nuncius* (or even to rumors about Galileo's discoveries that circulated before the publication of the book) were presented first in private conversational settings and then communicated, often anonymously, through networks of scholarly and courtly correspondence and gossip. The remarkable metamorphoses that affected what went in and out of these channels did little to stabilize the debate. The proliferation of opinions was also fostered by the courtly format in which many of these views were presented and developed.[79] Upon receiving a copy of the *Nuncius,* a prince or his courtiers could ask court mathematicians and physicians for an opinion about the book. Critical responses were almost de rigueur in these contexts as they could generate lively and entertaining debates, but, by the same token, they did not tend to facilitate closure.[80] Sometimes the same people who dismissed the veracity of Galileo's claims proceeded to accuse him of having stolen those claims or instruments from others, apparently unfazed by the contradictory nature of their position.[81]

To modern ears, the tone of several of these early critiques appears harsh, perhaps libelous. This may have led some historians to overestimate the opposition to Galileo's discoveries (though in fact only one of these cri-

78. *KGW,* vol. XVI, p. 333.

79. *GC,* pp. 72–83.

80. These people were given little time to formulate their views and often were expected to respond on the spot, sometimes without having seen the book or tried a telescope. Because of this conversational format, commentators were only moderately accountable for their views and could modify or even reverse them at a moment's notice without much embarrassment or professional liability. For instance, a major astronomer like Magini could support (and largely share) Martinus Horky's vehement critique of Galileo (*GO,* vol. X, pp. 345, 365), but then turn around and write (or have others write) that he was a Bohemian madman as soon as Horky's attack on Galileo seemed to backfire (*GO,* vol. X, pp. 376–79, 384–85). In a differently structured republic of letters, the remarkable contradiction between Magini's public and private stances could have carried substantial costs. Moreover, some of these critiques seemed to be aimed not so much at Galileo's stars but rather at his sudden stardom. In September 1610, Magini remarked to Monsignor Benci that "in some universities, other mathematicians are paid better. For instance, recently Mr. Galilei has received 1,000 florins from the Venetians, and is currently retained by the grand duke with 1,200 scudi for life, although I know in my conscience that I am not at all inferior to but rather superior to him" (*GO,* vol. X, pp. 429–30).

81. Michael Maestlin is an example, but see also Georg Fugger's early response to the *Nuncius* in *GO,* vol. X, p. 316.

tiques was printed in 1610) or to read their tone as a sign of strong emotions stirred by cosmological incommensurabilities.[82] A more mundane explanation is that such a tone reflected the kind of discourse generated by controversial novelties (and by the sudden stardom of their producer) in a dispersed and marginally interdependent field that gave little incentive to produce more disciplined responses.[83] The same field that allowed Galileo to adopt an uncooperative stance toward other astronomers did not compel his critics to treat him respectfully either.[84]

As shown by its very title, the *Nuncius* was cast first of all as a report, an announcement. Its genre was closer to that of news reports than to the discipline-specific narratives one found in cosmology, natural philosophy, planetary astronomy, natural history, etc. The choice of such a nondisciplinary literary genre may signify Galileo's assessment of the cross-disciplinarity of his audience. It may also reflect Galileo's desire to showcase the Medicean Stars as widely as possible to please his patrons and—in case his bid for a position at the Medici court did not pan out—to find other princely patrons.[85]

PERIODIC EVIDENCE VS. INSTANTANEOUS PERCEPTION

An interesting range of narrative and pictorial tactics followed from Galileo's decision to address his claims to a broad, unspecific audience, and from his interest to keep his readers from becoming his competitors. At first

82. Martinus Horky, *Brevissima peregrinatio contra nuncium sidereum* (Modena: Cassiani, 1610; reproduced in *GO*, vol. III, pt. 1, pp. 129–45). A second critique, Francesco Sizi's *Dianoia astronomica, optica, physica* (Venice: Bertani, 1611; reproduced in *GO*, vol. III, pt. 1, pp. 203–50), was written in 1610 but was published only in 1611, after Galileo's claims had been widely accepted. It had little or no noticeable impact on the debate.

83. On the transition from this kind of sociability to more interdependent ones, see Biagioli, "Etiquette, Interdependence, and Sociability in Seventeenth-Century Science," pp. 193–238.

84. *GC*, pp. 60–73.

85. In this sense, the function of the *Nuncius* could be compared to that of Tycho Brahe's 1598 *Astronomiae instauratae mechanica*—the luxuriously produced book in which Tycho described his famous instruments, the astronomical palace he constructed at Hven, and the services he could provide for a suitably royal or imperial patron (John Robert Christianson, *On Tycho's Island* [Cambridge: Cambridge University Press, 2000], pp. 219–20, 223–24). That Tycho disclosed so much about his instruments is an indication of his dire patronage needs, and of the fact that, unlike Galileo's telescope, his large instruments were not easily reproducible.

glance, the *Nuncius* appears to present a straightforward account of sequential observations. But the narrative structure that wove these observations into physical claims displayed a peculiar kind of demonstrative logic—a logic that well matched Galileo's epistemological and social predicament at the beginning of 1610.

Galileo's first goal was to gain assent for his claims, minimize the risk of losing priority over future discoveries, and cast his reluctance to provide information about the telescope as inconsequential to the acceptance of his discoveries. Second, he could not present himself as someone whose claims could be accepted on the grounds of his personal credibility. By the time the *Nuncius* was published, few readers knew of its author. Narratives that deemphasized the author's personal qualities while stressing their internal logic helped Galileo bypass the problems posed by his modest professional and social status. Third, narratives whose acceptance did not appear to hinge on their author's adherence to specific disciplinary conventions had a better chance to be understood and accepted by Galileo's diverse audiences.

The logic of Galileo's narratives rested on the specificity of his observational protocols.[86] The production and reproduction of his observations was a time-consuming process not only in the obvious sense that much labor and effort went into it but, more importantly, in the sense that the evidence behind those discoveries was inherently historical. Like other astronomical phenomena, the satellites of Jupiter were observed as a process (and, I argue, were probably observable only as a process). The time dimension of those phenomena was as important as their spatial location.

One does not see the precession of the equinoxes by looking in the direction of the celestial pole for a few hours but detects it by comparing and interpolating the observations of the motion of the celestial pole through the stars over centuries. Similarly, one did not see the satellites of Jupiter just by pointing the telescope toward that planet for a few minutes. That would have shown, at best, a few bright dots. What enabled their discovery was not a specific gestalt that immediately turned those dots into satellites, but a commitment to produce the suitable apparatus and conduct observations over several days so as to record the periodic motions of the

86. These protocols have been discussed, in various degrees of depth, in Drake, "Galileo and Satellites Prediction," pp. 75–95; van Helden in *SN*, pp. 10–16; Alan Chalmers, *Science and Its Fabrication* (Minneapolis: University of Minnesota Press, 1990), pp. 54–55; and Dear, *Discipline and Experience*, pp. 107–11. What, in my view, has not been previously addressed is how Galileo's observational practices dovetailed with his concerns about minimizing disclosure and maximizing credit.

satellites and differentiate them from other visual patterns (be they fixed stars or optical artifacts produced by the instrument). Because of the features of early telescopes (narrow field of vision, double images, color fringes, and blurred images especially toward the periphery), people who looked through a telescope for only a few minutes could legitimately believe that Galileo's claims were artifactual, as numerous spurious objects could be seen through a telescope's eyepiece at any given time.

This view of Galileo's process of discovery is no a posteriori reconstruction but conforms to his log entries, to the *Nuncius,* and to a letter written immediately after his first observation of Jupiter. When he observed Jupiter for the first time on January 7, 1610, Galileo wrote that he had seen three fixed stars near the planet, two to the east and one to the west.[87] In the *Nuncius* he added that these stars seemed "brighter than others of equal size" and "appeared to be arranged exactly along a straight line and parallel to the ecliptic" but, in and of itself, their peculiar appearance and arrangement did not cause him to doubt that they were fixed stars.[88]

At first, he "was not in the least concerned with their distances from Jupiter," but on the following night he noticed that while the three stars had remained close to the planet, they had all moved to the west. Even then, Galileo did not think that the stars had shifted. Instead, he assumed that Jupiter must have moved (though he was puzzled that, according to his tables, it should have gone in the opposite direction). Clouds prevented him from observing on the following night. On January 10, however, he was surprised to see that only two stars were visible and that they had again switched sides, this time from the west to the east. He could make sense of the missing star by thinking that it must have been hidden by Jupiter, but could not believe that Jupiter had moved around again. On January 11, there were still only two stars to the east of Jupiter, but they had moved much farther to the east of the planet, were closer to each other, and one of them appeared much larger (though on the previous night they had appeared to be of equal size).[89]

Only at that point did he conclude that what he had observed were not

87. *GO,* vol. X, p. 277.

88. *SN,* p. 64. In the January letter he had already remarked that planets appeared well demarcated ("like small full moons") when observed through the telescope, but that fixed stars remained so shimmering that their shape could not be detected. It seems, therefore, that the three "fixed stars" around Jupiter had struck him as being of the size of stars while looking more like planets.

89. *SN,* pp. 65–66.

fixed stars but planets (wandering stars).[90] Both the *Nuncius* and his log show that from that night on Galileo began to record the changing distances between the new planets and Jupiter, having probably decided that the robustness of his claims rested on the determination of their motions.[91] On January 13, after having sturdied the telescope's mount, he observed a fourth satellite.[92] Since January 15, all his log entries were made in Latin, suggesting that on that date he decided to publish the *Nuncius* and to include the daily positions of the satellites in it.[93]

To Galileo, then, the evidence that counted was not a snapshot of individual luminous dots around Jupiter, but the "movie" of their motions.[94] It was a chronological perspective that linked his string of observations and turned the luminous bodies near Jupiter into satellites, not fixed stars. On the title page of the *Nuncius,* in fact, Galileo identified the satellites with their motions—"four planets flying around the star of Jupiter at unequal intervals and periods with wonderful swiftness"—a characterization that was then repeated in the text.[95]

The *Nuncius*' mapping of the satellites' motions did not stop on January 13, but continued with painstaking descriptions of more than sixty configurations (which he also represented as diagrams) of the four satellites over forty-four almost consecutive nights (fig. 2). The textual and diagrammatic description of their movements occupies a large portion (about 40 percent) of Galileo's text. Taken at face value, this section may appear tedious (Drake's English translation edited out most of it) as it does not present complex arguments or exciting evidence.[96] And yet, Galileo included it and continued to observe the satellites for several more weeks de-

90. *GO,* vol. III, pt. 2, p. 427; *SN,* p. 66.

91. *GO,* vol. III, pt. 2, p. 427.

92. "Havendo benissimo fermato lo strumento" (*GO,* vol. III, pt. 2, p. 427; *SN,* p. 67).

93. *GO,* vol. III, pt. 2, p. 427.

94. *SN,* pp. 67–83. My reference to Galileo's observation as a kind of movie is not meant metaphorically. While Galileo's visual narrative is articulated on the printed page rather than on film, its logic is distinctly cinematic. As Jimena Canales has recently shown, astronomy's imaging techniques played a direct role in the history of film technology (Canales, "Photogenic Venus: The 'Cinematographic Turn' in Science and Its Alternatives," *Isis* 93 [2002]: 585–613).

95. *SN,* pp. 26, 36, 64.

96. The translation included in Galileo, *Discoveries and Opinions of Galileo,* trans. Stillman Drake (New York: Doubleday, 1957), excluded all the observations from January 14 to February 25.

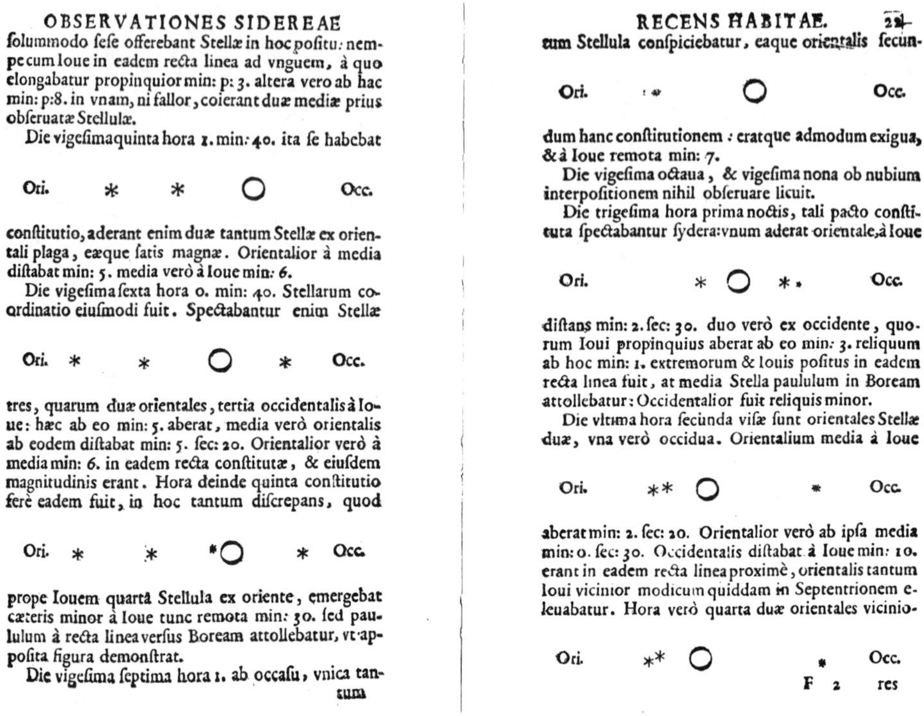

OBSERVATIONES SIDEREAE

ſolummodo ſeſe offerebant Stellæ in hoc poſitu: nempe cum Ioue in eadem recta linea ad vnguem, à quo elongabatur propinquior min: p: 3. altera vero ab hac min: p:8. in vnam, ni fallor, coierant duæ mediæ prius obſeruatæ Stellulæ.

Die vigeſimaquinta hora 1. min: 40. ita ſe habebat

Ori. * * ○ Occ.

conſtitutio, aderant enim duæ tantum Stellæ ex orientali plaga, eæque ſatis magnæ. Orientalior à media diſtabat min: 5. media verò à Ioue min: 6.

Die vigeſima ſexta hora 0. min: 40. Stellarum coordinatio eiuſmodi fuit. Spectabantur enim Stellæ

Ori. * * ○ * Occ.

tres, quarum duæ orientales, tertia occidentalis à Ioue: hæc ab eo min: 5. aberat, media verò orientalis ab eodem diſtabat min: 5. ſec: 20. Orientalior verò à media min: 6. in eadem recta conſtitutæ, & eiuſdem magnitudinis erant. Hora deinde quinta conſtitutio ferè eadem fuit, in hoc tantum diſcrepans, quod

Ori. * * *○ * Occ.

prope Iouem quarta Stellula ex oriente, emergebat cæteris minor à Ioue tunc remota min: 30. ſed paululum à recta linea verſus Boream attollebatur, vt appoſita figura demonſtrat.

Die vigeſima ſeptima hora 1. ab occaſu, vnica tantum

RECENS HABITAE. 21

tum Stellula conſpiciebatur, eaque orientalis ſecun-

Ori. * ○ Occ.

dum hanc conſtitutionem: eratque admodum exigua, & à Ioue remota min: 7.

Die vigeſima octaua, & vigeſima nona ob nubium interpoſitionem nihil obſeruare licuit.

Die trigeſima hora prima noctis, tali pacto conſtituta ſpectabantur ſydera: vnum aderat orientale, à Ioue

Ori. * ○ * . Occ.

diſtans min: 2. ſec: 30. duo verò ex occidente, quorum Ioui propinquius aberat ab eo min: 3. reliquum ab hoc min: 1. extremorum & Iouis poſitus in eadem recta linea fuit, at media Stella paululum in Boream attollebatur: Occidentalior fuit reliquis minor.

Die vltima hora ſecunda viſæ ſunt orientales Stellæ duæ, vna verò occidua. Orientalium media à Ioue

Ori. ** ○ * Occ.

aberat min: 2. ſec: 20. Orientalior verò ab ipſa media min: 0. ſec: 30. Occidentalis diſtabat à Ioue min: 10. erant in eadem recta lineaproximè, orientalis tantum Ioui vicinior modicum quiddam in Septentrionem eleuabatur. Hora verò quarta duæ orientales vicinio-

Ori. ** ○ * Occ.

F 2 res

FIGURE 2. Example of Galileo's diagrams of the motions of the Medicean Stars in *Sidereus nuncius* (1610).

spite being already certain of his claims and despite his fear that any delay in publication could deprive him of priority. His actions clearly indicate the importance he placed on this section—a section he then planned to expand in a revised edition of the *Nuncius*.[97] Although after the publication of the *Nuncius* he refrained from giving telescopes to other astronomers, he sent

97. *GO*, vol. X, p. 373. Such an edition, however, never materialized. The last observation of the satellites reported in the *Nuncius* is from March 2, and the book was off the press on March 13. The lunar observations included in the *Nuncius* dated from much earlier. According to Ewen Whitaker's reconstruction of the dating of Galileo's lunar observations and drawing, all but one were done by December 18 (Ewen Whitaker, "Galileo's Lunar Observations and the Dating of the Composition of the *Sidereus nuncius*," *Journal for the History of Astronomy* 9 [1978]: 155–69). His essay also recapitulates the previous debate about the dating of such observations. This shows that from January 7 to March 2 Galileo dedicated himself almost exclusively to observing the satellites to substantiate a claim he was already sure of by January 11.

them records of his continued observations.[98] In fact, if one trusted Galileo's description of their rapid movements (and that such movements seemed to lie within a specific plane), it would have been difficult to claim that the satellites were optical artifacts produced by the telescope. The determination of the satellites' periods would have provided even stronger evidence for their existence, but a preliminary mapping of the luminous dots' regular motions along a plane already cast them as strong candidates for physical phenomena. Galileo did not first establish his new objects and then proceed to measure their periods. In this case, the object was its periods. The finer the determination of the period, the more stable the object, and vice versa.

The only feature of the telescope Galileo discussed at some length at the beginning of the *Nuncius* was not its construction and optical principles, but its use for measuring angular distances, that is, for tracking the movements of the satellites and detecting their periods.[99] And Galileo's exhortation to his colleagues to go beyond what he had done only concerned the periods of objects he had already detected, not new discoveries. Read in the context of his monopolistic ambitions, his saying "I call on all astronomers to devote themselves to investigating and determining their periods" does not sound like an attempt to encourage others to take up telescopic astronomy, but rather an attempt to channel his competitors' drive in directions useful to him. Even if they preceded Galileo at determining the periods, their confirmation of the physical reality of the satellites would have still helped him.

Galileo used the same "historical" logic of observation to argue that, contrary to received views, the lunar surface was not smooth but rugged like the Earth's.[100] As I show in the next chapter, he adopted the same approach in his 1613 book on sunspots. In the *Nuncius,* he opened the more detailed discussion of the Moon's appearance with the observation that

> when the Moon displays herself to us with brilliant horns, the boundary dividing the bright from the dark part does not form a uniformly oval line, as it would happen in a perfectly spherical solid, but is marked by an uneven, rough, and sinuous line, as the figure shows.[101]

98. *GO,* vol. X, p. 403.

99. *SN,* pp. 38–39.

100. *SN,* p. 40. See also his description of the lunar surface in the January 7 letter in *GO,* vol. X, pp. 273–77.

101. *SN,* p. 40.

On the opposite page, he inserted his first engraving of the Moon (fig. 3). As with the satellites of Jupiter, Galileo's problem was to show that physical objects (valleys and ridges) were behind the irregular visual appearance of the terminator. And, as with the satellites of Jupiter, his argument did not stop at one snapshot of the irregular pattern of bright and dark spots on the lunar surface but continued with a discussion of how that visual pattern *changed* in time:

> Not only are the boundaries between light and dark on the Moon perceived to be uneven and sinuous, but, what causes even greater wonder is that very many bright points appear within the dark part of the Moon, entirely separated and removed from the illuminated region and located no small distance from it. Gradually, after a small period of time, these are increased in size and brightness. Indeed, after two or three hours they are joined with the rest of the bright part, which has now become larger. In the meantime, more and more bright points light up, as if they were sprouting, in the dark part, grow, and are connected at length with that bright surface as it extends farther in this direction.[102]

He then repeated this same kind of "historical" analysis for particularly conspicuous dark and bright spots, showing how their changing appearances were consistently connected to the changing angle at which sunlight struck the lunar surface as the Moon went through its phases:

> I would by no means be silent about something deserving notice, observed by me while the Moon was rushing toward first quadrature, the appearance of which is also shown in the above figure [fig. 3]. For toward the lower horn a vast dark gulf projected into the bright part. As I observed this for a long time, I saw it very dark. Finally, about after two hours, a bit below the middle of this cavity a certain bright peak began to rise and, gradually growing, it assumed a triangular shape. . . . Presently three other small points began to shine around it until, as the Moon was about to set, this enlarged triangular shape, now made larger, joined together with the rest of the bright part [. . .] it broke out into the dark gulf.[103]

He also set up two complementary observations by showing the Moon at first and second quadrature, that is, when the Moon is half full but its bright and dark sides are switched around (fig. 4). By doing so, he tried to show how the irregular patterns of lights and shadows are inverted in the two cases and that, therefore, they constituted the negative and positive picture of the same physical features of the lunar surface.

102. *SN*, p. 42.

103. *SN*, pp. 42–43.

As in his discussion of the movements of the satellites of Jupiter, Galileo used pictorial representations to guide his readers through the changing patterns of lunar lights and shadows. In this case, however, the movie had only four consecutive frames, one of which was reproduced on two different pages to lend more continuity to the visual narrative. Though these few pictures were strategically chosen and placed to maximize their narrative potential (figs. 4 and 5), it was mostly the detailed verbal narrative that created an intertextual, cinematic effect.[104] Galileo was probably aware of the jumpy nature of his lunar movie if he wanted to include engravings covering all the phases of the Moon in a future reissue of the *Nuncius*.[105]

Despite the different resolution of the visual narratives about the Moon and the satellites of Jupiter, the argument's logic was the same in both cases. The existence of lunar valleys and mountains did not hinge on a few disjointed observations, but on the pattern traced by dark and bright spots as they changed through several interrelated observations—observations that would yield the same pattern if repeated over different Moon cycles: "day by day these [spots] are altered, increased, diminished, and destroyed, since they only derive from the shadows of rising prominences."[106] Being consistently connected to the phases of the Moon, these changing visual patterns could not be easily dismissed as optical artifacts produced by the telescope.[107] Therefore, while having the appearance of "natural histories"

104. For instance, each of the two pairs of images reproduced in fig. 4 and fig. 5 marks two stages of a temporal process, but they also work in juxtaposition with each other. In one case (fig. 5), that relationship is remarked upon by Galileo himself: "The following figures clearly demonstrate this double appearance" (*SN*, p. 45). The relationship between Galileo's somewhat crude pictures of the Moon and his more detailed narrative has been discussed in Winkler and van Helden, "Representing the Heavens," pp. 207–9.

105. "I want to draw the phases of the Moon for a whole period with the utmost diligence, and imitate them in minute detail [. . .] and I want to have them engraved in copper by an excellent artist" (*GO*, vol. X, p. 300). Galileo was aware then that copperplates used in the *Nuncius* were not as good as they could be and planned to include better illustrations of the Moon in a revised edition that, however, never appeared (*GO*, vol. X, p. 373).

106. *SN*, p. 48. Galileo was also able to use the fact that the appearance of some spots did not change with the changing angle of illumination to argue that the relative darkness and brightness of these (flat) spots had to be linked to material rather than topographical differences (*SN*, p. 48).

107. A few years later, during the debate with the Jesuit mathematician Christoph Scheiner on the discovery and nature of sunspots, Galileo resorted again to periodic evidence, not snapshots. His claims about the status of sunspots as objects were inseparable from the description of their periodical movements and of how their shape changed in time.

of satellites or lunar peaks and valleys, these pictures functioned like visual arguments.

However, even those willing to accept that such visual patterns were not artifactual did not need to agree that they were about ridges and valleys.

OBSERVAT. SIDEREAE

vergit primo loco dicamus, quam facilioris intelligentiæ gratia in duas partes diſtinguo, alteram nempè clariorem, obſcuriorem alteram: clarior videtur totum Emiſphærium ambire, atque perfundere; obſcurior verò veluti nubes quædam faciem ipſam inficit, maculoſamque reddit; iſtæ autem maculæ ſuboſcuræ, & ſatis amplæ vnicuique ſunt obuiæ, illaſque æuum omne conſpexit; quapropter magnas, ſeu antiquas eas appellabimus, ad differentiam aliarum macularum amplitudine minorum, at frequentia ita conſitarum, vt totam Lunarem ſuperficiem, præſertim verò lucidiorem partem conſpergant; hæ verò à nemine ante nos obſeruatæ fuerunt; ex ipſarum autem ſæpius iteratis inſpectionibus, in eam deducti ſumus ſententiam, vt certò intelligamus, Lunæ ſuperficiem, non perpolitam, æquabilem, exactiſſimæque ſphæricitatis exiſtere, vt magna Philoſophorum coors de ipſa, deque reliquis corporibus cœleſtibus opinata eſt, ſed contra inæqualem, aſperam, cauitatibus, tumoribuſque confertam, non ſecus, ac ipſiusmet Telluris facies, quæ montium iugis, valliumque profunditatibus hincindè diſtinguitur. Apparentiæ verò ex quibus hæc colligere licuit eiuſmodi ſunt.

Quarta aut quinta poſt coniunctionem die, cum ſplendidis Luna ſeſe nobis cornibus offert, iam terminus, partem obſcuram à luminoſa diſſidens, non æquabiliter ſecundum oualem lineam extenditur, veluti in ſolido perfectè ſphærico accideret; ſed inæquabili, aſpera, & admodum ſinuoſa linea deſignatur, veluti appoſita figura repræſentat. complures enim veluti excreſcentiæ lucidæ vltra lucis tenebrarumque confinia in partem obſcuram extenduntur, & contra tenebricoſæ particulæ intra lumen ingrediuntur. Quinimo, & magna nigricantium macularum

RECENS HABITAE. 8

cularum exiguarum copia, omnino à tenebroſa parte ſeparatarum, totam ferè plagam iam Solis lumine perfuſam vndiquaque conſpergit, illa ſaltem excepta parte quæ magnis, & antiquis maculis eſt affecta. Adnotauimus autem, modo dictas exiguas maculas in hoc ſemper, & omnes conuenire, vt partem habeant nigricantem locum Solis reſpicientem; ex aduerſo autem Solis lucidioribus terminis, quaſi candentibus iugis coronentur. At conſimilem penitus aſpectum habemus in Terra circa Solis exortum, dum valles nondum lumine perfuſas, montes verò illas ex aduerſo Solis circundantes iam iam ſplendore fulgentes intuemur: ac veluti terreſtrium cauitatum vmbræ Sole ſublimiora petente imminuuntur, ita & Lunares iſtæ maculæ, creſcente parte luminoſa tenebras amittunt.

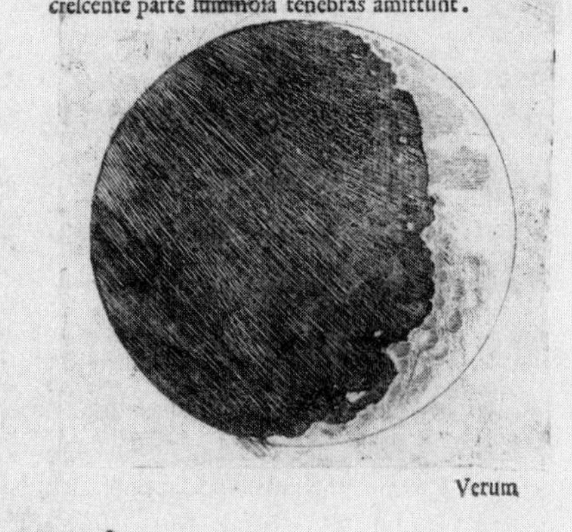

Verum

FIGURE 3. First illustrations of the lunar surface in *Sidereus nuncius* (1610). (Reproduction courtesy of Houghton Library, Harvard University.)

FIGURE 4. (*facing*) Second and third illustrations of the lunar surface in *Sidereus nuncius* (1610). (Reproduction courtesy of Houghton Library, Harvard University.)

FIGURE 5. (*facing*) Last two illustrations of the lunar surface in *Sidereus nuncius* (1610). The picture at the bottom is a repeat from the previous page. (Reproduction courtesy of Houghton Library, Harvard University.)

OBSERVAT. SIDEREAE

ctum daturam. Depreſſiores inſuper in Luna cernuntur magnæ maculæ, quàm clariores plagæ; in illa enim tam creſcente, quam decreſcente ſemper in lucis tenebrarumque confinio, prominente hincindè circa ipſas magnas maculas contermini partis lucidioris; veluti in deſcribendis figuris obſeruauimus; neque depreſſiores tantummodò ſunt dictarum macularum termini, ſed æquabiliores, nec rugis, aut aſperitatibus interrupti. Lucidior verò pars maximè propè maculas eminet; adeò vt, & ante quadraturam primam, & in ipſa fermè ſecunda circa maculam quandam, ſuperiorem, borealem nempè Lunę plagam occupantem valdè attollantur tam ſupra illam, quàm infra ingentes quædam eminentiæ, veluti appoſitæ præſeferunt delineationes.

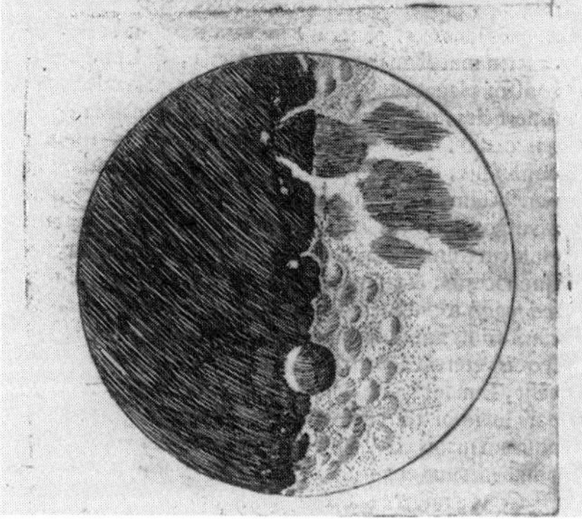

RECENS HABITAE. 10

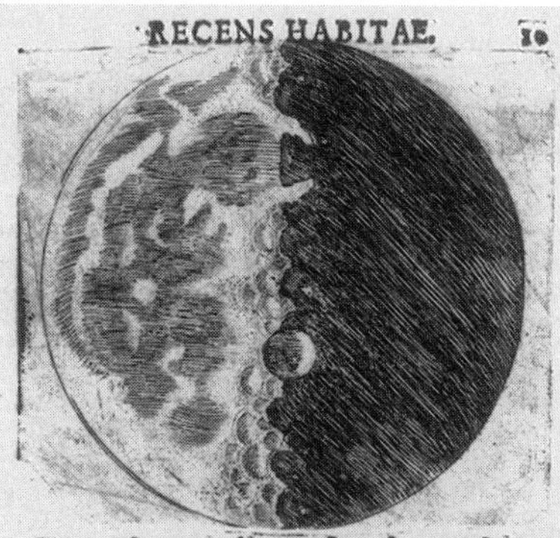

Hæc eadem macula ante ſecundam quadraturam nigrioribus quibuſdam terminis circumuallata conſpicitur; qui tanquam altiſſima montium iuga ex parte Soli auerſa obſcuriores apparent, quà verò Solem reſpiciunt lucidiores extant; cuius oppoſitum in cauitatibus accidit, quarum pars Soli auerſa ſplendens apparet, obſcura verò, ac vmbroſa, quæ ex parte Solis ſita eſt. Imminuta deinde luminoſa ſuperficie, cum primum tota fermè dicta macula tenebris eſt obducta, clariora mōtium dorſa eminenter tenebras ſcandunt. Hanc duplicem apparentiam ſequentes figuræ commoſtrant.

FIGURE 4.

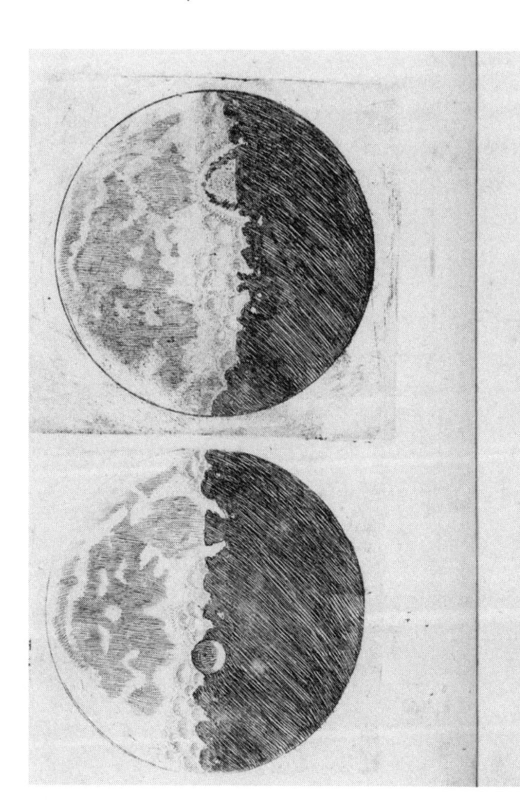

RECENS HABITAE. 11

Vnum quoque obliuioni minimè tradam, quod nō niſi aliqua cum admiratione adnotaui: medium quaſi Lunæ locum à cauitate quadam occupatum eſſe reliquis omnibus maiori, ac figura perfectæ rotunditatis; hanc prope quadraturas ambas conſpexi eandemque in ſecundis ſupra poſitis figuris quantum licuit imitatus ſum. Eundem quo ad obumbrationem, & illuminationem facit aſpectum, ac faceret in terris regio conſimilis Boemiæ, ſi montibus altiſſimis, inque peripheriam perfecti circuli diſpoſitis occluderetur vndique: in Luna enim adeò elatis iugis vallatur, vt extrema hora tenebroſæ Lunæ parti contermina Solis lumine perfuſa ſpectetur, priuſquàm lucis vmbræque terminus ad mediam ipſius figuræ diametrum pertingat. De more autem reliquarum macularum, vmbroſa illius pars Solem reſpicit, luminoſa verò verſus tenebras Lunæ conſtituitur; quod tertio libenter obſeruandum admoneo, tanquam firmiſſimum argumentum, aſperitatum, inæqualitatumque per totam Lunæ clariorem plagam diſperſarum; quarum quidem macularum ſemper nigriores ſunt illæ, quæ confinio luminis, & tenebrarum conterminæ ſunt; remotiores verò tum minores, tum obſcuræ minus apparent, ita vt tandem cum Luna in oppoſitione totum impleuerit orbem, modico, admodumque tenui diſcrimine, cauitatum opacitas ab eminentiarum candore diſcrepet.

Hæc quæ recenſuimus in clarioribus Lunæ regionibus obſeruantur, verum in magnis maculis talis nō conſpicitur lacunarum, eminentiarumque differentia, qualem neceſſariò conſtituere cogimur in parte lucidiori, ob mutationem figurarum ex alia, atque alia illuminatione radiorum Solis, prout multiplici poſitu Lunam reſpicit; at in magnis maculis exiſtunt quidem

areolæ

FIGURE 5.

They may have been a movie, but a movie about what? Throughout the discussion of the changing visual appearance of the Moon during its phases, Galileo made repeated analogies to how terrestrial mountains and valleys are variously illuminated and cast shadows of different length during the day. The analogy may be read as an anti-Aristotelian argument because it simultaneously undermined the unique status of the Earth while claiming that the Moon was not as pristine as the philosophers expected it to be. Galileo might not have opposed such a reading. There was, however, a more specific, local role for the Earth-Moon analogy in the *Nuncius*.

Galileo argued that the satellites of Jupiter were real because they had periodical motions, but he did not need to convince anyone that planets (the category in which he placed the Medicean Stars) had periods. The case of the lunar valleys and mountains was different. Galileo needed to hinge the physical status of these topographical features on the periodicity of their appearances, but in this case he did not have an astronomical exemplar for that kind of movie. As a result, he compared those changing patterns of lights and shadows to those cast by terrestrial mountains over valleys at different times of the day. It has been often remarked, quite correctly, that the ruggedness of the Moon supported the anti-Aristotelian claim that the Earth was not unique in its topographical features. This argument, however, worked only after one had accepted Galileo's claims about the ruggedness of the Moon. To get to the point where one could drew an analogy between the Moon and the Earth, Galileo drew an analogy between the periodic appearances of the topography of the Earth and that of the Moon. Before he could use the Moon to end the cosmological uniqueness of the Earth, he needed a messy Earth to show that he was telling the truth about the messy Moon.

I do not argue that everyone should have felt compelled to accept Galileo's cinematic logic. At the same time, there was nothing revolutionary about the protocols and inferences he asked his readers to follow. The way he processed telescopic evidence to argue for the existence of the satellites of Jupiter or for the irregularity of the lunar surface was the same used by traditional astronomers to detect the precession of the equinoxes or other time-based phenomena (with the important difference that, in this case, the periods involved were on the order of days, not centuries).

The *Nuncius'* crucial novelty as a narrative was that it translated these practices from a series of numerical observations into a form that could be appealing to the philosophically curious, not only to professional astronomers. Although Galileo made his inferences from geometrical entities (the angular distances of the satellites from Jupiter, the relation between lunar shadows and the height of lunar mountains, etc.) he presented his claims in visual terms—as movies about satellites and shadows. Judging from how few people rejected the *Nuncius,* it appears that its narratives succeeded at least in casting Galileo's claims as plausible.

TIME AND ITS MARKETS

The few practitioners who, after reading the *Nuncius,* went on to observe the satellites of Jupiter did adopt the observational practices Galileo had described in his text. For example, Kepler and the Roman Jesuits confirmed Galileo's claims after conducting a series of interrelated observations of the Medicean Stars.[108] The Jesuits remained doubtful about the reality of the satellites after observing them for a few nights, but their skepticism gave way after conducting daily observations over two weeks and noticing their revolutions.[109] As Clavius wrote on December 17,

> Here in Rome we have seen them [the Medicean Stars]. I will attach some diagrams at the end of this letter from which one can see most clearly that they are not fixed stars, but errant ones, as they change their position in relation to Jupiter.[110]

Analogously, on October 9, Santini wrote Galileo that he had seen the satellites again, "several times, in different positions, so that I have no doubt [about their existence]."[111] In May 1611, Luca Valerio, a Roman mathematician, added a more explicit epistemological commentary to the practice of consecutive observations:

> It has never crossed my mind that the same glass [always] aimed in the same fashion toward the same star [Jupiter] could make it appear in the same place, surrounded by four stars that always accompany it . . . in a fashion that one evening they might appear, as I have seen them, three to the west

108. The structure of Kepler's *Narratio* resembles that of the *Nuncius.* In it, Kepler listed daily observations of the satellites of Jupiter from August 30 to September 9 (*KGW,* vol. IV, pp. 319–22). Right after the last entry Kepler simply wrote that these observations confirmed Galileo's claims and that he returned the telescope to the elector of Cologne.

109. The Jesuits' observational log shows that they had been recording the daily positions of the satellites since November 28 (*GO,* vol. III, pt. 2, p. 863).

110. *GO,* vol. X, p. 484.

111. *GO,* vol. X, p. 445.

> and one to the east [of Jupiter], and other times in very different positions, because the principles of logic do not allow for a specific, finite cause [the telescope] to produce different effects when [the cause] does not change but remains the same and maintains the same location and orientation.[112]

As we will see in the next chapter, the Jesuit Christoph Scheiner deployed a similar argument to convince his readers that the dark spots he observed on the Sun in 1611 were not telescopic artifacts.[113]

While we have much evidence about the importance of time in the corroboration of Galileo's observations, we have no indications that a tacit and cosmology-informed perceptual gestalt played a role in that process. Both Galileo and Kepler were Copernicans, but the Jesuits were not (though they were growing increasingly skeptical about the Ptolemaic system). There is no clear evidence about Santini's cosmological beliefs, but none of his letters addressed those issues, thus suggesting that he was not particularly concerned with the discoveries' cosmological implications. Cosmological beliefs, it seems, motivated the observers' behavior but did not frame their perceptions.[114] Those who observed the satellites had to invest weeks and months in the project, and did so because they had something to gain (or at least nothing to lose) from corroborating Galileo's claims.

Symmetrically, the rejection of these discoveries did not result from cosmological or perceptual incommensurabilities or from the lack of a satisfactory description of the telescope's workings. Simply, those who opposed Galileo's claims did not take sufficient time to conduct long-term observations. By observing for only a short time, they could plausibly argue that the evidence available to them was, at best, insufficient. And because of the

112. Antonio Favaro, *Amici e corrispondenti di Galileo,* ed. Paolo Galluzzi (Florence: Salimbeni, 1983), vol. I, p. 573. A similar point is made, in a more humorous fashion, by Galileo in a May 21, 1611, letter to Piero Dini in which he promises 10,000 scudi to whoever can construct a telescope that shows satellites around one planet but not others (*GO*, vol. XI, p. 107).

113. Chapter 3, this volume, pp. 162, 180.

114. Valerio's case is more ambiguous because a few years after he wrote the letter I cited, his membership in the Accademia dei Lincei was suspended when he declined to endorse the academy's full support of Galileo's pro-Copernican position in the "Letter to the Grand Duchess." However, Valerio's stance in 1615 was not informed by direct geocentric commitments, but rather from the desire to stay out of dangerous cosmological debates. On this dispute, see Mario Biagioli, "Knowledge, Freedom, and Brotherly Love: Homosociality and the Accademia dei Lincei," *Configurations* 3 (1995): 139–66.

structure of the field, there were no shared professional norms that compelled Galileo's opponents to abide by his rules and invest time and resources to engage in the long-term observations needed to test his assertions, or to require Galileo to give them telescope time, telescopes, or instructions about how to build them. Galileo's monopolistic attitudes were as ethical or unethical as his critics' allegedly stubborn or obscurantist dismissals.

In the case of the philosopher Cremonini, geocentric beliefs translated into an absolute refusal to observe. It was reported that he did not want to look through the telescope for fear it would give him a headache.[115] However, it is not that Cremonini was unable to see the satellites of Jupiter because he was an Aristotelian, but simply that such an observation would have been a very unwise investment of time and resources for someone of his disciplinary affiliation and professional identity.

Unlike Cremonini, other critics did look through the telescope, though only for a short time. Because of the brevity of their observations, they remained vocally skeptical about Galileo's reading of those changing patterns of bright spots as satellites. In a letter sent to Kepler right after Galileo's visit to Bologna, Horky wrote that the instrument worked wonderfully when aimed at terrestrial objects but performed poorly when pointed at the sky. Horky was probably correct in saying that fixed stars appeared double, a fact that may have made him justifiably skeptical about Galileo's other claims.[116]

But while he did not share Galileo's perception of the significance of those spots, he did see them nevertheless. A few weeks later, in the *Peregrinatio contra nuncium sidereum,* Horky added that when he tried to observe Jupiter he saw "two globes or rather two very minute spots" near Jupiter on April 24, and detected "all four very small spots" on April 25.[117] He did not believe that those spots were satellites and yet the fact that Ga-

115. On Cremonini's refusal to confront Galileo's discoveries see *GO,* vol. XI, p. 100, and esp. p. 165.

116. *GO,* vol. X, p. 343.

117. "24 Aprilis nocte sequente vidi duos solummodo globulos aut potius maculas minutissimas." When Horky asked Galileo why the two other stars were not visible despite the fact that the night was clear, he allegedly received no answer (*GO,* vol. III, pt. 1, p. 140). The next night, "Iupiter occidentalem exhibuerat, cum omnibus suis novis quator famulis supra nostrum Bononiensem Horizontem apparuit. Vidi omnes quator maculas minutissimas a Iove presilientes cum ipsius Galileo perspicillo, cum quo illas se invenisse gloriatur" (*GO,* vol. III, pt. 1, p. 141). Interestingly, Horky did not admit to having seen any of these "spots" in the April 27 letter to Kepler.

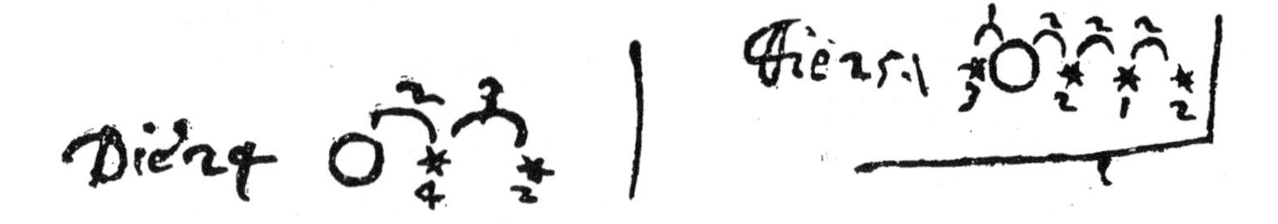

FIGURE 6. Entries for April 24 and 25 in Galileo's observational log. From *GO*, vol. III, pt. 2, p. 436.

lileo's own records for those two nights report exactly the same configurations shows that Horky's cosmological beliefs did not prevent him from registering the phenomena as Galileo saw them (fig. 6).[118]

Giovanni Magini was another of Galileo's early opponents. A professor of mathematics at Bologna, a supporter of geocentric astronomy, and Horky's employer, Magini was among those who spent two nights observing with Galileo. He did not publish a critique of his claims but worked hard at undermining Galileo's credibility through letters describing his fiasco.[119] However, when he described those events to Kepler a few weeks later, Magini adopted a much more accommodating stance, simply saying that those who observed with Galileo at Bologna were unable to see the satellites perfectly.[120]

Magini had a point. He and Horky had reasonable grounds for skepticism and little incentive to take time to observe. Furthermore, Galileo did little to change their minds. He did not visit Bologna on the way to Florence, but only on the way back to Padua, after he had shown the Stars to the grand duke and his family.[121] Eager to reach Florence as soon as possible, he actually changed his travel plans and skipped a stopover he had previously planned in Bologna.[122] Furthermore, his one visit on the way

118. Galileo's manuscript log in *GO*, vol. III, pt. 2, p. 436, last line. Kepler too spotted the congruence between Horky's report and the configurations of the satellites he had received from Galileo (*GO*, vol. X, p. 416).

119. *GO*, vol. X, pp. 345, 365.

120. *GO*, vol. X, p. 359.

121. That Galileo stopped to observe in Bologna on his way to Florence is a claim commonly found in the secondary literature, but is not supported by any evidence contained in Galileo's correspondence or observational log.

122. Galileo's observational log shows that on the night of April 2 he was already close to Bologna (he observed in Firenzuola), suggesting he may have left Padua on April 1. On

back to Padua was very short and yielded only two observational sessions. A few more nights could have made the satellites' periodic behavior more evident.

But on April 26, a few hours after the end of the second session, Galileo left. Horky assumed that Galileo, demoralized by his failure, had left early in the morning to avoid further confrontations with his critics.[123] More likely, he simply needed to rush back to Padua to teach. In a March 13 letter to the Medici secretary, Galileo stated that the Easter recess at Padua lasted about twenty-three or twenty-four days and that he could leave only on April 2 (probably at the very beginning of the vacation).[124] This suggests that the recess ended around Monday, April 26, the day Galileo left Bologna for Padua. He was cutting it quite close. But if he had strong reasons not to delay the departure any further, there is no evidence that he could not have arrived in Bologna a few days earlier.[125] He spent almost three weeks in Tuscany, but dedicated only two days to Bologna. Although he may have regretted his rush later on after realizing the harm done by Horky's and Magini's opposition, at that point Galileo seemed content with having shown the satellites to the grand duke and treated the assent of his "colleagues" in Bologna as a side dish for which it was not worth shortening his Tuscan stay by a few days.

April 3 he was already observing in Florence, indicating that he did not stop in Bologna and did not catch the Medici carriage that was supposed to pick him up on April 5. On that day, in fact, he was already at San Romano, on the way to Pisa to meet the grand duke (*GO*, vol. III, pt. 2, p. 436). On March 13, Galileo asked Vinta to send a carriage to Bologna "on the Monday of the week of Passion," that is, the week leading to Easter (*GO*, vol. X, p. 289). Other letters confirm the appointment (*GO*, vol. X, pp. 303, 307). In the Gregorian calendar, Easter fell on April 11, 1610. This means that Monday of Easter week was April 5. I thank Owen Gingerich for providing the date of Easter 1610.

123. "Galileo became silent, and on the twenty-sixth, a Monday, dejected, he took his leave from Mr. Magini very early in the morning. And he gave us no thanks for the favors and the many thoughts, because, full of himself, he hawked a fable. Mr. Magini provided Galileo with distinguished company, both splendid and delightful. Thus the wretched Galileo left Bologna with his spyglass on the twenty-sixth" (Horky to Kepler, April 27, 1610, *GO*, vol. X, p. 343). English translation by Albert van Helden in *SN*, p. 93.

124. *GO*, vol. X, p. 289.

125. He probably arrived in Bologna on either April 23 or 24, as he was still in Florence on April 20, but was gone by April 24 (*GO*, vol. X, p. 341).

TELESCOPES AND BLACKBOXES

If Galileo's observations did not require revolutionary gestalt switches, neither did they depend on his tacit instrument-making skills. Others were able to develop those skills in a matter of months without written or hands-on instruction from Galileo, thus suggesting that the "secret" of his telescope was little more than a trade secret. If one shared Valerio's conclusion (as Clavius, Santini, and Kepler did) that the satellites of Jupiter could not be dismissed as optical artifacts (because, under ceteris paribus conditions, one would expect telescopic artifacts to have a rather stable appearance, not orderly motions), then one could consider the telescope's status as relatively unproblematic despite the fact that no one, including Galileo, seemed able to provide a comprehensive explanation of how it worked.

Many studies of experimental replications use the notion of "blackbox" or similar ones to describe the process through which an instrument comes to be seen as a reliable producer of claims about nature.[126] Typically, blackboxing takes place within a community of users who exchange textual information and bodily skills about the construction, use, and calibration of a given instrument. In the early stages of blackboxing, when neither instrument nor claim is stabilized, the relationship between the two is necessarily circular—what Collins calls the experimenter's regress. Through a range of negotiations that are inherently social in nature, the community of experimenters breaks the regress and brings the dispute to closure, that is, it certifies a true claim, a winner, and the standard instrument to be used to calibrate other instruments. Knowledge that was initially tacit, private, contested, and body-bound is thus transformed into something that can travel and be standardized and unquestioned (if only temporarily).[127]

126. As far as I know, the concept of blackbox was originally applied to contexts of experimental replication in Trevor Pinch, *Confronting Nature: The Sociology of Solar-Neutrino Detection* (Dordrecht: Reidel, 1986), pp. 212–14. My references here are to the notion of blackbox described in Pinch and to the discussion of the experimental replication in Collins, *Changing Order.*

127. Bruno Latour uses a slightly different image of the blackbox. The main differences are that he does not consider the role of bodily skill and tacit knowledge and its transformations during the process of blackboxing, and that he stresses the instability of the blackbox more than the practitioners of SSK do. As he puts it, the blackbox has no inertia. When it becomes disconnected from the network of humans who sustain it, it ceases to exist as a blackbox (Bruno Latour, *Science in Action* [Cambridge: Harvard University Press, 1987], p. 137). My discussion here refers to Pinch's and SSK's notion of blackbox, not Latour's.

As useful as it has been in analyzing disputes within experimental communities, the blackbox does not capture the process through which Galileo's instrument and evidence were accepted and replicated. For instance, the corroboration of the discoveries presented in the *Nuncius* did not rest on a transfer of textual or tacit knowledge about lens grinding or observational techniques from Galileo to other astronomers—a transfer he did his best to prevent. Although I have argued that Galileo developed a monopoly on early telescopic astronomy and acted as if his instruments were superior to those of his competitors, those moves were not connected to a blackboxing process. His monopoly over telescopic astronomy may bear a family resemblance with the canonization of claims, practitioners, and instruments resulting from blackboxing, but Galileo pursued that monopoly to maximize his credit, not to break an experimenters' regress that may have prevented the stabilization of his claims. He did not worry that people who build their own telescopes may end up with claims that contradicted and destabilized his own, but that people could be so fast at corroborating his claims that they would not stop there. While implying that his telescopes were the best around, Galileo did not invite people to copy them. He was simply trying to deter people from entering the game by raising the cost of entering that game.

Similarly, the descriptions of telescopic observations the *Nuncius* offered to those readers who did not have access to telescopes was part of a strategy of control, not the kind of community building that, a few decades earlier, Boyle tried to foster through the detailed description of his experiments and his apparatus.[128] Like Boyle, Galileo was trying to satisfy his readers with narrative simulations of his own experience so that they would not feel the need to pursue it on their own, but their motives for pursuing these literary strategies were quite different. Unlike Boyle, who provided those narratives because he knew that replication was difficult and probably out of the reach of most of his readers, Galileo crafted his reports so as to minimize the chance that his readers would try to replicate his claims and then turn into his competitors.[129] (A seventeenth-century observer whose

128. Steven Shapin, "Pump and Circumstance: Robert Boyle's Literary Technology," *Social Studies of Science,* 14 (1984): 481–520, esp. 487–97, and Steven Shapin and Simon Schaffer, *Leviathan and the Air Pump* (Princeton: Princeton University Press, 1985), pp. 59–65.

129. This is the motivation Shapin and Schaffer find behind Boyle's development, in the mid-1670s, of more detailed narrative reports of experiments (Shapin, "Pump and Circumstance," pp. 490–91; Shapin and Schaffer, *Leviathan and the Air Pump,* pp. 59–60).

tactics bear some resemblance to Galileo's is Leeuwenhoek—another practitioner steeped in the artisanal tradition).[130] It is also possible that the absence of any mention of independent witnessing in the *Nuncius* may reflect Galileo's concern with jeopardizing his priority claims by sharing his findings with witnesses before their publication. In sum, Galileo cast his readers not as colleagues in an emerging philosophical community, but as remote, credit-giving consumers.

Similar considerations apply to the status of Galileo the observer. The narrative logic of the *Nuncius* not only reduced the pressure on Galileo to disclose the workings and manufacture of the telescope, but it also cast his personal trustworthiness as something of a nonquestion. Although readers of the *Nuncius* were asked to believe Galileo's claims about spending several nights on the roof of his house observing the changing positions of the satellites of Jupiter or the changing appearances of the Moon, they were not required to trust the accuracy of all the specific observations he reported. Because Galileo's claims were about the recursiveness of certain patterns, their robustness did not rely on one crucial observation or experiment, nor did it hinge primarily on his personal qualifications as a trustworthy observer.[131]

130. Like Galileo, Leeuwenhoek provided long and detailed narratives of his observations in his letters to the Royal Society, many of which were printed in the *Philosophical Transactions.* But he did not disclose information about the construction of his microscopes, nor did he give access to the best of his instruments (Thomas Birch, *The History of the Royal Society of London* [London: Millar, 1757], vol. IV, p. 365). There were, however, at least two major differences between Galileo and Leeuwenhoek. Unlike Galileo, the Dutch microscopist was not a good draftsman. Unwilling to share his work with artists, he did the drawing of the illustration himself, usually with modest results (Philippe Hamou, *La mutation du visible: Microscopes et Télescopes en Angleterre de Bacon à Hooke* [Villeneuve d'Ascq: Presses Universitaires du Septentrion, 2001], p. 161). The other is that Leeuwenhoek's distinctive skills were not limited to the construction of the microscope but included the preparation of the specimens. He could share examples of the latter without disclosing the former (*LE*, vol. I, p. 119).

131. Unlike later seminal texts of experimental philosophy, the *Nuncius* was not cast as the exemplar of a philosophical "form of life." The kind of observational narratives presented in it are not cast as examples of how any telescopist should write his findings. They are simply the way Galileo chose to convince his readers in that instance, concerning a specific set of phenomena. More generally, he tried neither to blackbox the telescope nor to stabilize the community of its users. He adopted the customary protocols of long-term observational astronomy, but did not treat other astronomers as colleagues. The *Nuncius* tried to get credit from whatever constituency it could reach, and it did so by minimizing (not maximizing) its reliance on the social conventions and values of any given community.

INVENTIONS AND DISCLOSURE

Before 1610, Galileo participated in various professional and social groups that, in different ways, accustomed him to the value of limited disclosure and to the appreciation of economies of reward based on local patronage. Though by no means a reclusive scholar, Galileo seemed content to limit his audience to small groups of Paduan academics, Venetian patricians, and Florentine courtiers with whom he discussed philosophy, music, mathematics, and literature.[132] He also interacted with "low-culture" practitioners: artists, artisans, and engineers. Until 1610, the only mathematical or philosophical publication under his name was the short instruction manual of the geometrical and military compass.[133] Like the *Nuncius,* the compass' manual did not describe the instrument's construction.

Placed in this context, the monopolistic tactics Galileo displayed in the *Nuncius* and his carefully controlled distribution of telescopes were not just the actions of an author who "held back"; they could be seen also as the behavior of someone who knew little about what to expect from larger audiences. Being new at writing a "best seller" like the *Nuncius,* he continued to align his discourse to the local and noncooperative credit systems he was familiar with, extending them to cover much wider audiences but without fully recasting them into a cooperative framework—a framework he had few exemplars for and from which, in any case, he had little to gain.

Some of Galileo's tactics came from his astronomical background, but others came from the world of inventors and instrument makers. He had been designing and producing instruments and machines prior to 1609, and his career as an inventor peaked precisely with the development of the telescope in the nine months leading to the publication of the *Nuncius.* Before he realized he could gain more credit for his discoveries than for his instrument, Galileo focused on the telescope as his ticket to success. By March

132. Although he was forty-six by the time he wrote the *Nuncius,* Galileo had made no prior attempts to reach broader readerships through his publications, and his correspondence had been modest in volume and geographically limited to Italy. Even Kepler's 1597 invitation to engage in an epistolary dialogue about Copernicanism did not move him (*GO,* vol. X, pp. 69–71).

133. He was also a possible author or coauthor of a short 1605 pseudonymous satirical publication on the new star observed in 1604. This publication, written in Paduan dialect, was meant for local consumption. In 1607, Galileo published an attack on Capra's plagiarism of his book on the geometrical and military compass, but that was not a philosophical or mathematical text.

1610 he had fashioned himself as the discoverer of the Medicean Stars, but just a few months before he was still casting himself as the inventor of the first high-power telescope—an instrument he marketed for its military (not astronomical) applications. Galileo the inventor turned quickly into Galileo the discoverer, but the metamorphosis was never complete.

Many readers seemed to recognize the inventor's "voice" in the *Nuncius,* as neither supporters nor critics objected to his secretive attitudes. Some wished he had given out telescopes or information how to build them, but did not expect him to do so—at least not for free. The elector of Fraising, for instance, read the *Nuncius* and, disappointed with how little Galileo had shared with his readers about the construction of high-power telescopes, offered him a reward if he communicated his secret to him and promised not to divulge it to others.[134] Galileo, then, was treated as an artisan entitled to have proprietary attitudes about the "secret" of his device.[135] The word *secreto* appeared often in correspondence discussing early telescopes, thus confirming that most of Galileo's contemporaries had a clear sense of the economy in which these instruments circulated.

The protection of inventions depended on local legal and administrative practices, and was necessarily limited to the state that issued it.[136] Typically, the inventor was expected to show the appropriate officials a working example of the device for which he sought a temporary monopoly within that

134. On April 14, 1610, Galileo's brother, Michelangelo Galilei, reported that according to the elector, "non havendo voi, in questo vostro primo libro, insegnato chiaramente tal fabbrica, il pare che sia di mancamento; et dice, se metterete in esecuzione quello che scrivete, che vi farete immortale; et vi prega, non volendo voi insegnare a altri detta fabbrica, al manco siate contento di volerne compiacere S.A., che vi si dimostrerà quel principe che egli è" (*GO,* vol. X, p. 313).

135. Later, on January 7, 1611, Mark Welser wrote Galileo that "I can tell you that information about how to build [telescopes] is much desired here [in Germany]," but did not intimate that Galileo's secrecy was seen as unethical (*GO,* vol. XI, p. 14).

136. According to Christine MacLeod, Italian states had been at the forefront of the development of property rights for technical achievements, and these legal and administrative models were then transferred to northern Europe and England. In particular, Venice "was the first to regularize in law the award of monopoly patents, the Senate ruling of 1474 that inventions should be registered when perfected: the inventor thereby secured sole benefit for ten years, with a penalty of 100 ducats for infringement, while the government reserved the right to appropriate registered inventions." In order to expand the geographical coverage of their patents, inventors registered them in other states, provided they were deemed interesting enough to deserve that treatment (Christine MacLeod, *Inventing the Industrial Revolution* [Cambridge: Cambridge University Press, 1988], p. 11).

state's jurisdiction, but did not need to provide a description of that device.[137] For instance, a north European artisan approached the Venetian senate in August 1609 asking for one thousand ducats for a low-power telescope, but did not want the Venetian authorities to examine the instrument, but only to look through it.[138] Paolo Sarpi, acting as the senate's advisor, did oppose the offer but not because he thought that the inventor's position was unethical.[139] What he objected to was the high price demanded for an instrument whose "secret" was proving to be remarkably short-lived. Similarly, disclosure was not mentioned in any of the documents related to Hans Lipperhey's October 1608 application for a patent for the telescope he filed in the Netherlands.[140]

The definition of inventor was a local matter.[141] If the authorities

137. In England, for instance, the legal demand for written specifications emerged only in the early eighteenth century, and such specifications were made public only towards the end of the century. Before then, "[i]t was rare to demand anything of the patentee." MacLeod, *Inventing the Industrial Revolution,* pp. 11–13.

138. On August 22, 1609, Giovanni Bartoli, the Medici agent in Venice, wrote to Florence about a foreign artisan's offer of a telescope to the Senate, and that the instrument was tried out from St. Mark's bell tower (like Galileo's a few days later), but that many thought that its "secret" was well known in France and elsewhere, and that similar instruments were quite cheap outside Venice (*GO,* vol. X, p. 250). The reference to the foreign artisan not allowing any internal inspection of the telescope is found in Micanzo's biography of Sarpi: "L'occhiale, detto in Italia del Galileo, trovato in Olanda, fu da lui penetrato l'artifizio quando, presentatone uno alla serenissima signoria con dimanda di mille zecchini, fu al padre dato carico di far le prove a che potesse servire e dirne il suo giudizio; e perchè non gl'era lecito aprirlo e vedere, imaginò ciò che potesse" (Fulgenzio Micanzo, *Vita del Padre Paolo,* reprinted in Paolo Sarpi, *Istoria del Concilio Tridentino* [Turin: Einaudi, 1974], vol. II, pp. 1372–73). MacLeod argues that in sixteenth-century England, an inventor was not required to share his secret if the technology he was bringing into the country helped the "furtherance of trade" (MacLeod, *Inventing the Industrial Revolution,* p. 13).

139. Sarpi may have done some "technology transfer" here. On August 29, 1609, Bartoli wrote to the Medici secretary that Sarpi told Galileo about "the secret he had seen [the foreigner's telescope]" and that Galileo, moving from that tip, was able to produce a better instrument (*GO,* vol. X, p. 255). Micanzo reports the same in his biography of Sarpi (Micanzo, *Vita del Padre Paolo,* p. 1373). Given Sarpi's role in the Venetian government, it would have been quite ethical for him to facilitate Galileo's successful development of the telescope by feeding him information that could lead to a better instrument (and then to see it accepted and rewarded by the senate).

140. Van Helden, "The Invention of the Telescope," pp. 36–44.

141. Such a definition of inventor makes sense in a context in which many inventors were itinerant artisans making a living out of spreading a country's technology into another

deemed a certain device useful or protectable (or both), they might issue a privilege (through a "letter patent") to a person who was not necessarily the original inventor but simply the one who made available or perfected that technology within the jurisdiction of the privilege-granting authorities. One could obtain a privilege for the exclusive use of the printing press in Venice for a certain amount of time despite the fact that his name was Speyer, not Gutenberg.[142] Galileo's gift of the telescope to the Venetian senate in 1609 did not amount to a proper patent application (most likely because he knew that such a privilege would have been unenforceable given how widespread telescope-making skills had become).[143] However, some of his interactions with the Venetian senate (such as the monopoly he offered them for the use and production of the instrument, and the higher salary and tenure at Padua he received as a counter-gift from the senate) conformed to artisanal and legal practices according to which he was the inventor of that kind of telescope within the jurisdiction of the Republic of Venice.[144] Some inventors did not request patents but donated their devices

(very much like the foreigner who first brought the telescope to Venice). In sixteenth-century England, "the rights of the first inventor were understood to derive from those of the first importer of the invention" (MacLeod, *Inventing the Industrial Revolution,* p. 13). Such a definition was still held in the early seventeenth century (ibid., p. 18).

142. On September 18, 1469, Johannes of Speyer received a privilege from the Collegio of the Signoria for printing in Venice and his dominion for five years (Leonardas Gerulaitis, *Printing and Publishing in Fifteenth-Century Venice* [London: Mansell, 1976], p. 21).

143. This problem had already caused Lipperhey to have his patent application denied in the Netherlands. On October 14, 1608, the States-General commented that "we believe there are others [other inventors] as well and that the art cannot remain secret at any rate, because after it is known that the art exists, attempts will be made to duplicate it, especially after the shape of the tube has been seen, and from it has been surmised to some extent how to go about finding the art with the use of lenses" (van Helden, "The Invention of the Telescope," pp. 38–39). Lipperhey, like Galileo after him, stressed the military application of the telescope.

144. Galileo's dedication of the telescope to the Venetian doge is not a simple letter, but a formal document that was officially debated and discussed by the senate. If it did not ask for specific quid pro quo, that was for politeness' sake. The senate did understand that Galileo was offering them a device (whose military applications were clearly laid out in the letter of presentation) in exchange for a better salary at Padua (which they did give him, together with tenure). Galileo's letter of presentation is in *GO*, vol. X, pp. 250–51. On the debate on whether Galileo was or was not the inventor of the telescope, see Edward Rosen, "Did Galileo Claim He Invented the Telescope?" *Proceedings of the American Philosophical Society* 98 (1954): 304–12.

to their rulers in exchange for a job or a pension.[145] Even the patent application for the telescope filed by Lipperhey at The Hague in 1608 asked for an annual pension in case the patent itself were to be denied.[146] Galileo's gift of the telescope to the senate in exchange for tenure and a salary raise fits squarely in this tradition.[147]

The workings of the early privilege system explain not only Galileo's secrecy but also his sense of to whom disclosure was due. As the Venetian senate was the institutional patron to reward Galileo for his gift of the tel-

145. Inventors who had developed something directly useful to the state itself (as distinct from a technology that could foster a state's industry and trade) would not usually apply for a patent, especially knowing that states could take over their patents if they wished to do so. On the issue of state appropriation of patents, see the Venetian Senate ruling of 1474 in MacLeod, *Inventing the Industrial Revolution,* p. 11.

146. "On the request of Hans Lipperhey, born in Wesel, living in Middelburg, spectacle-maker, having discovered a certain instrument for seeing far, as has been shown to the Gentlemen of the States, requesting that, since the instrument ought not to be made generally known, he be granted a patent for thirty years under which everyone would be forbidden to imitate the instrument, or otherwise, that he be granted a yearly pension for making the said instrument solely for the use of the land, without being allowed to sell it to any foreign kings, monarchs, or potentates; it has been approved that a committee consisting of several men of this assembly will be appointed in order to communicate with the petitioner about his invention, and to ascertain from the same whether he could improve it so that one could look through it with both eyes, and to ascertain from the same with what he will be content, and, upon having heard the answers to these questions, to advise [this body], at which time it will be decided whether the petitioner will be granted a salary or the requested patent" (van Helden, "The Invention of the Telescope," p. 36). Notice that the section about the pension matches quite closely what Galileo requested two years later in Venice.

147. Additionally, offering a device to a prince in exchange for a job or a pension made particular sense when such a device had no great commercial potential. In fact, in the summer of 1609, the telescope had only two financially rewarding applications: as gadgets for rich gentleman, and for military intelligence. Galileo already knew that the market for "play" telescopes was becoming quickly saturated and that prices were dropping fast. Also, the more telescopes one produced and sold, the more likely it was that his "secret" would be copied. The military market was more appealing. The telescope did have military applications, but one can also speculate that Galileo had plenty of reasons to amplify the range and importance of such applications, as he did in the formal presentation of the instrument to the Venetians. Selling the telescope to a prince was the best deal he could think of under those circumstances. After buying Galileo's device, the Venetians would have had all the interest to keep its secret for as long as possible, thereby lengthening Galileo's leverage. And Galileo could still enjoy tenure and a higher salary after the secret was gone.

escope, it was the senate (not the readers of the *Nuncius* or fellow astronomers) that Galileo felt obliged to share the secret of the telescope with.[148] The only substantial description of Galileo's instrument (including the focal length of the objective lens, angle of view, and overall size of the instrument) is found in one of Sarpi's private letters.[149] Probably Sarpi had access to that information not by virtue of being one of Galileo's friends, but by having examined his instrument on behalf of the senate.[150] Galileo had the same sense of obligation toward his next patron, the grand duke of Tuscany. In 1610, two months after the publication of the *Nuncius,* Galileo told Belisario Vinta, the grand duke's secretary: "I do not wish to be forced to show to others the true process for producing [telescopes], except to some granducal artisan."[151] There was nothing unusual in Galileo's conduct. When Evangelista Torricelli—Galileo's pupil and successor at the Medici court—was on his deathbed in 1647, he ordered that his "secret for manufacturing lenses for the spyglass or telescope" together with his lens-grinding equipment, polishing materials, templates, etc. be put in a large padlocked crate at the grand duke's disposal.[152] The crate was delivered

148. In his "Racconto istorico della vita del Sig. Galileo Galilei," Vincenzio Viviani writes that, together with an instrument, he gave the Venetian doge and senate "a text in which he explained the fabrication, use, and applications—both on land and at sea—that were to be had" (*GO,* vol. XIX, p. 609).

149. "Constat, ut scis, instrumentum illud duobus perspicillis (lunettes vos vocatis), sphaeric ambobus, altero superficiei convexae, altero concavae. Convexus accepimus ex sphaera, cuis diameter 6 pedum; concavum, ex alia, cuius diameter latitudine digiti minor. Ex his componitur instrumentum circiter 4 pedum longitudinis, per quod videtur tanta pars objecti, quae, si recta visione inspiceretur, subtenderet scrupula l.a 6; applicato vero instrumento, videtur sub angulo maiori quam 3 graduum" (Sarpi to Leschassier, March 16, 1610, *GO,* vol. X, p. 290). Why Sarpi felt free to share this information with his Parisian friend remains an open question.

150. Another, more approximate description of the telescope (of its length and the diameter of the objective lens) is found in the diary of one of the Venetian officers, Antonio Priuli, who observed with Galileo on August 21, 1609 (*GO,* vol. XIX, p. 587).

151. *GO,* vol. X, p. 350.

152. The most detailed description of Torricelli's transfer of his "secret" is in a letter from Lodovico Serenai to Raffaello Magiotti (December 21, 1647), in Giuseppe Rossini (ed.), *Lettere e documenti riguardanti Evangelista Torricelli* (Faenza: Lega, 1956), pp. 40–41. Additional details are in other letters by Serenai at pp. 2, 5, 6, 22, 18, 40, in the draft of Torricelli's will at pp. 133–34, and in the postmortem inventory of his possessions at pp. 140, 153. See also Paolo Galluzzi, "Evangelista Torricelli: concezione della matematica e segreto degli occhiali," *Annali dell'Istituto e Museo di Storia della Scienza di Firenze* 1 (1976): 71–95, esp. pp. 85–90.

personally to the grand duke two days after Torricelli's death.[153] Disclosure was given to the source of credit and, in Galileo's and Torricelli's cases, credit came from patrons, not "colleagues."

There is some irony here. Much of Galileo's career plans after 1610 focused on gaining recognition as a philosopher, not as a mathematician or an instrument maker. However, his career as a philosopher hinged on the fact that at that time he was not perceived as a philosopher but as a remarkable instrument maker and that, as such, he was entitled to keep his secrets. Such a socially sanctioned right to secrecy allowed him to develop a monopoly on observational astronomy and obtain the title of philosopher he desired so much, despite the fact that secrecy was not exactly a customary value among philosophers.

One could even say that Galileo's secrecy was not an entitlement but a duty. Having been rewarded by the senate for his telescope, he was obliged not to divulge its secret or to sell his instruments to anyone other than his employers. Giovanni Bartoli, the Medici representative in Venice, wrote to the Florentine court in October 1609 that Galileo's instruments were considered the best and that he was building twelve of them for the senate. However, he continued, Galileo could not teach anyone how to build them because he had been ordered by the senate not to divulge the secret.[154] During his flirtation with the Medici, then, Galileo was treading on delicate grounds as he was enticing a new patron with an instrument for which he had already been rewarded by another patron.[155] Because the senate had rewarded the telescope for its military applications, sending an instrument to the Medici might have been construed as treason.[156] That Galileo kept

153. Rossini, *Lettere e documenti,* p. 153.

154. *GO,* vol. X, p. 260. Bartoli repeated the point a few days later (*GO,* vol. X, p. 260). Contrary to what I have written elsewhere (*GC,* pp. 45–47) the fact that Galileo never sold his instruments was not just a matter of social self-fashioning but also of legal obligations. If Galileo could not divulge the secret, one can assume he could not sell his instruments either.

155. On June 5, 1610, the Medici promised Galileo that "in the meantime [your appointment] will be kept as secret as possible" (*GO,* vol. X, p. 369). The Medici resident in Venice wrote to the Florentine court on June 26 that "I have been asked if it is true that Dr. Galilei is going to serve the grand duke with a great salary. I answered I didn't know anything about it. If what they say is true, and is found out, it could give him trouble here" (*GO,* vol. X, p. 384).

156. One way to interpret Galileo's behavior is that by the beginning of 1610 everyone understood that the "secret" of the telescope was hopelessly public, and that his contract with the Venetians was therefore more nominal than actual. In any case, it is interesting

promising to send the grand duke a good telescope but only took one to Florence himself at Easter time suggests that he probably felt he could not send an instrument to the Medici without upsetting the Venetians.[157] Had there been an international patent law, Galileo could have been in serious trouble. Instead, he could simply cross the River Po and start a new professional life.[158]

The conventions of early patents may also explain why Galileo never provided a description of the optical processes of image formation through a telescope. Unlike the historians and philosophers of science who have seen this alleged failure as potentially damaging to the epistemological status of the telescope, Galileo seemed unfazed by it.[159] He did not seem to be familiar with the most relevant optical literature (Kepler's 1604 *Ad Vitellionem paralipomena*) nor did he seem anxious to fill his knowledge gap. Years later, he still did not think that Kepler's *Dioptrice* (a text in which the

---

that Galileo's disclosure of the telescope to the Medici, his proposal to send several of them to European princes, and his acceptance of a contribution toward the cost of producing those instruments came only after both Galileo and the Medici understood that a position for him at the Florentine court was a serious possibility.

157. The Medici seemed to understand that they were in a peculiar position. Around April 1610, Alfonso Fontanelli, a diplomat from Modena who had observed with Galileo's telescopes (probably at Pisa) jokingly told the grand duchess that as soon as other nobles heard of the quality of the Medici's instruments, they would flood them with requests. To this, the grand duke and the grand duchess replied that the telescopes they had did in fact belong to the Venetians and that the Medici could not give them to anyone else (*GO*, vol. X, p. 347). The grand duchess might have used this argument as a way to deflect requests for telescopes, or she might have actually stated a common view about the ownership of the telescopes at that time. Things are a bit murkier because at this time the Medici were also evaluating Galileo's request to distribute telescopes to European princes through their diplomatic networks.

158. We know from Sagredo that Galileo's departure from Venice (and especially the modalities of such departure) had upset several people there. I had previously thought that the Venetians' indignation had to do with what they must have perceived as Galileo's ingratitude (*GC*, pp. 44–45). In light of this new evidence, the Venetians probably thought that Galileo had behaved unethically, perhaps even illegally. As a thought experiment, it may be interesting to consider what could have happened to Galileo had his new patron been a prince who, unlike the Medici, did not have friendly relations with the Republic of Venice.

159. See, for example, Peter Machamer, "Feyerabend and Galileo: The Interaction of Theories, and the Reinterpretation of Experience," *Studies in History and Philosophy of Science* 4 (1973): 1–46, esp. 13–27.

German mathematician discussed the process of telescopic image formation) actually shed much light on the workings of the telescope and told Jean Tarde that Kepler's book was so obscure that maybe not even its author had understood it.[160] Be that as it may, the legal and cultural customs surrounding inventions did not require Galileo to produce any explanation of the optical workings of the telescope.

INVENTIONS, DISCOVERIES, AND NATURAL MONUMENTS

Despite all the ties between Galileo's telescope and the culture of inventors, the *Nuncius* was no patent application. Written in Latin and printed in 550 copies, it addressed a European audience, not a local political authority. It did open with a brief discussion of the telescope, but its stated purpose was to report discoveries. These discoveries, however, were still dedicated and tailored to a patron, the Medici, who was as local as the Venetian senate to which Galileo had previously offered the telescope. And while the *Nuncius* made the discoveries public, it maintained artisan secrecy about the instrument that made them possible. So what kind of genre did the *Nuncius* belong to? The short answer is that both the *Nuncius* and the Medicean Stars belonged to a new, unstable economy Galileo tried to develop by borrowing ingredients from the economies of inventions, discoveries, and artworks.

The Stars were not discoveries in the modern sense of the term. Unlike late-seventeenth-century natural philosophers, Galileo did not place his discoveries in the public domain in exchange for nonmonetary credit that accrued on his name. Like inventors, Galileo did receive financial rewards for his work from his local patron. But, unlike other inventions, the Medicean Stars could not be used locally and kept secret within that jurisdiction. Their value was predicated on widespread visibility, not secrecy. And while novelty was not an issue in the economy of early inventions, the Medici did care a great deal about the fact that the Stars were a new kind of object and that Galileo was their original discoverer. Both their value and the exchanges of gifts and counter-gifts between Galileo and the Me-

160. Galileo requested Kepler's *Ad Vitellionem paralipomena* from Giuliano de' Medici on October I, 1610 (*GO*, vol. X, p. 441). By December 1612 he also had a copy of Kepler's *Dioptrice (GO*, vol. XI, p. 448). Tarde's remarks are in "Dal Diario del Viaggio di Giovanni Tarde in Italia" (*GO*, vol. XIX, p. 590).

dici were made possible by their novelty and by Galileo's status as their first discoverer.

The Medici did not own the Stars the way they could own an invention, and yet the satellites of Jupiter needed to "belong" to the Medici for Galileo to be rewarded. This tension could be read in the dedication: Galileo simultaneously presented the Stars as natural entities (remote beyond the possibility of ownership) and as monuments he was dedicating to the Medici—the most permanent monuments anyone could give them. Although these were objects he had carved out of nature for his patrons, they were not presented as artifacts like a statue chiseled out of a block of marble. The Medicean Stars were natural monuments. They were tied to the Medici through their name, and yet they were not objects they could keep and display in their galleries. Perhaps one could think of them as a peculiar artwork displayed in a celestial museum, globally visible because the Medici had "loaned" it to all viewers at once—a paradoxical artwork that needed to be simultaneously natural and artifactual, local and global.

The peculiar economy of the Medicean Stars matched the peculiar kind of credit Galileo received from them. He was neither a modern scientific author who receives reputation in exchange for the discoveries s/he makes public, nor an inventor or artist who owns (and therefore can sell) his/her work. Galileo dedicated his discoveries to the Medici, but did not really sell them because they were not something he could truly sell. The Medici gave him both financial rewards (of the kind given to artists or inventors), but also more symbolic rewards (such as the title of philosopher) because what he dedicated to them was not a piece of property exchangeable through monetary transactions.[161] Galileo's author function was equally hybrid: he was the discoverer of his patron's Stars. Like a court artist who could be very famous and yet remain someone's artist, Galileo could only be the personal philosopher of the grand duke of Tuscany. Despite the global visibility of his discoveries, Galileo was a philosopher in a very local sense.

Comparable hybridity is found in the parameters of evaluation and re-

161. Merton has argued that eponymy reflects the fact that scientists do not receive direct monetary credit from their discoveries as they could from their inventions (Robert Merton, "Priorities in Scientific Discoveries," in *The Sociology of Science: Theoretical and Empirical Investigations* [Chicago: University of Chicago Press, 1973], pp. 286–324). Because they cannot claim their findings as real property, they may attach their names to them as a gesture of symbolic ownership for their work. In this case, however, eponymy was tied to the patron's (not the discoverer's) name.

ward of the Medicean Stars. Inventions were evaluated locally and their reward involved little or no disclosure. The evaluation of later scientific discoveries depended, by contrast, on the judgment of a nonlocal community based on the information disclosed by the author. The Medicean Stars fell somewhere in between. They were not evaluated only by either a local patron or a dispersed community of peers, but through a process in which a few external reports were brought to the local patron, who integrated them with his own assessment of Galileo's claims.

Unlike discoveries, inventions were rewarded for their local utility, not for their nonlocal truth status. Galileo was rewarded for dedicating to the Medici a discovery that they came to accept as true, but also for giving them "natural monuments"—something whose utility was symbolic and nonlocal rather than practical and local. Consequently, the process through which the Stars were evaluated and rewarded was predicated on the Medici's perception that the Stars had some kind of utility that offset the risks they would have taken by rewarding and tying their family name to them. However, the meaning of "utility" and "risk" was much less financial, material, and local than it was in the economy of invention.[162]

FROM EXPERTISE TO LOCATION

A comparison between Venice and Florence may shed some light on these differences. In Venice Galileo needed only to convince elderly senators to climb up the many steps to St. Mark's tower. Once up there, the breathless elders were as qualified as anyone else to evaluate whether Galileo's telescope made distant enemy ships visible.[163] They didn't need to consult experts from outside Venice, nor did they want to do that because it was in

162. The Medicean Stars' market was literally global and their utility to the house of Medici was symbolic, not material (and therefore difficult to assess). Similarly, the risk the Medici took by accepting the Stars was more symbolic than financial (but serious nevertheless). Had the Stars proved artifactual, the laughter of other princes would have hurt much more than wasting money on Galileo's salary. Galileo's framing his discoveries in the Medici's dynastic mythologies did help the grand duke realize what he could gain from his Stars (*GC*, pp. 103–57), but that still left the assessment of their utility for the Medici a highly conjectural matter.

163. A description of this event, including the spotting of distant ships two hours ahead of their naked-eye sighting, are in an August 29 letter by Galileo (*GO*, vol. X, p. 253). Favaro, however, questions the authenticity of this letter. Another description of this event is in Antonio Priuli's diary (*GO*, vol. XIX, p. 587).

their interest to keep that invention as secret as possible. The senators' ability to assess the value of the instrument did not result from their expertise in telescopes—expertise that ranged from minimal to inexistent. All they needed to do was to decide whether, based on what they could see through Galileo's instrument, it was an invention worth rewarding for its military advantages. In any case, all they were risking was a limited amount of money—Galileo's salary raise at Padua.

By contrast, Galileo's success at showing Cosimo II the Medicean Stars in April 1610 was not sufficient to seal his court appointment.[164] Although the grand duke was sufficiently impressed by Galileo's demonstration to let him know that a position for him was in the making, he waited until July to formalize the offer.[165] As we have seen, Kepler's endorsement of Galileo's claims in the *Dissertatio* and the positive reception of the *Nuncius* played a crucial role in moving the Medici toward the final contract.

But we should not assume that the Medici would have been unable to trust their own eyes unless some external expert like Kepler told them to do so. By late April the grand duke had become familiar with the telescope, observing the Medicean Stars on a number of occasions. If he felt he could not trust what he had seen, he could have easily found professionally qualified local talent to assess Galileo's claims—something he did not do.[166] I believe that by late April the question was no longer whether the Medici believed in Galileo's discoveries but whether they *valued* them enough to

164. In a June 25 letter, Galileo remarks that the grand duke "col proprio senso ha più volte veduto" (*GO*, vol. X, p. 382). This claim was repeated in August. There Galileo added that Giuliano de' Medici and many others had witnessed the satellites as well (*GO*, vol. X, p. 422). In a May letter Galileo reminds Vinta of what he had told him during his visit to Pisa regarding the possibility of a position at the Florentine court (*GO*, vol. X, p. 350).

165. *GO*, vol. X, pp. 400–401. For a discussion of the grand duke's protracted hesitation, see *GC*, pp. 133–39.

166. Not only had the grand duke seen the satellites of Jupiter several times in April with his courtiers, but he might have observed the moon together with Galileo with one of his very early telescopes in the autumn of 1609 ("Che la luna sia un corpo similissimo alla terra, già me n'ero accertato, et in parte fatto vedere al Serenissimo Nostro Signore, ma però imperfettamente, non avendo ancora occhiale della eccellenza che ho adesso" (*GO*, vol. X, p. 280). Since then, the grand duke showed himself extremely interested in the telescope and its development (well before Galileo's discoveries) and even helped Galileo's work by sending him in Padua glass blanks made to his specifications by Medici artisans in Florence ("gli si mandano i cristalli conforme all'avviso suo"; *GO*, vol. X, p. 259). Galileo's visit to Florence in the fall of 1609 is not mentioned explicitly in any of his letters but hints can be found in *GO*, vol. X, pp. 262, 265, 268.

offer him the (expensive) position he sought from them. The issue was not the Stars' epistemological status *at that time* but what they could do for the Medici *in the future*. Because of the wide range of locations where the Medici expected their Stars and glory to be admired—all European courts and beyond—I believe that the question faced by the grand duke was not so much *who* had the necessary expertise to certify the Stars, but *where* that person needed to be, what "market niche" that person could be representative of.

The telescope was to be used locally in Venice and the senators had a clear idea of its value—the material advantages it could provide. But no matter how certain the Medici may have been of the truth of Galileo's claims, they could not have assessed the value of their Stars simply because they were not going to use them the way the Venetians used the telescope. Patrons can pay for monuments—including natural monuments like the Medicean Stars—but their value cannot be actualized in the absence of appreciative viewers. The Medici would not have got much of a return on Galileo's lifelong court stipend unless others praised the Stars as the exceptional natural monuments Galileo has construed them to be. The Medici were more like developers in search of investors than discoverers in search of corroborations.

The views of foreign experts mattered not necessarily because they were more technically competent than the Medici but because they were foreign—because they spoke from where the Stars were supposed to be appreciated. Kepler's testimonial was key not only because of his credibility as the Imperial Mathematician, but also because it indicated that the Medicean Stars were likely to be well appreciated at a key court like Prague. That Kepler's letter was not just a simple short statement like "I confirm that Jupiter has satellites," but went on and on about how great a discovery this was, and how wonderful Galileo's skills were, told the Medici that the Stars and their "astronomical artist" had great value, not just epistemological robustness or trustworthiness. That Kepler was so enthusiastic about the Stars despite the fact that he could not even observe them only reinforces the suggestion that his letter fit the genre of "product endorsement" better than that of "corroboration." [167]

After sending Kepler's endorsement to Florence, Galileo was told that

167. Even in the worst-case scenario that the Medicean Stars melted into thin air, Kepler's enthusiastic support would have maintained its relevance as it would have implicated the emperor in the Medici flop. We could say that it helped the Medici to spread the risk from their investment in the Stars.

his position at the Florentine court was almost a fait accompli.[168] It was finalized in July, and Galileo moved back to Florence over the summer. With the dream job secured, Galileo did not seem to worry about finding further witnesses to his discoveries. When, in August 1610, Kepler wrote asking him for names of people who had seen the satellites (as well as for the usual telescope) so that he could use them to silence the critics who were still active in Prague, Galileo replied:

> You, dearest Kepler, ask me for other witnesses. I will mention the grand duke of Tuscany, who, a few months ago, observed the Medicean Stars with me at Pisa, and generously rewarded me [. . .] I have been called back to my fatherland, with a stipend of one-thousand scudi a year, with the title of Philosopher and Mathematician of His Highness, with no duties but plenty of free time.[169]

One could argue that Galileo cast the grand duke as the only witness worth mentioning because, given his remarkably high social status, he was the most powerful witness Galileo or Kepler could use to convince the remaining skeptics. But Galileo might have invoked the grand duke for a very different reason: not because of the high epistemic value of his endorsement, but because he had given Galileo the position he wanted. Secure and well paid in Florence, Galileo did not feel compelled to rush to convince other people in other places. In this sense, Galileo may have invoked the grand duke not as a witness but precisely to make the point that he did not need witnesses anymore.[170]

At this point Galileo started acting as if the difficulties his "colleagues" were having while trying to replicate his discoveries were to be read as signs of his own authority, not of the possibly problematic status of his claims. If people still had difficulties with replication, it was their problem—a problem that confirmed the superiority of Galileo's telescopes. Now he could wait comfortably and work at producing more discoveries (as he did). The Jesuits' endorsement was still important, but at this point Galileo could wait for them to corroborate his findings and go to Rome (as he did in the spring of 1611) to be celebrated.

168. *GO*, vol. X, p. 350.

169. *GO*, vol. X, p. 422. Kepler's request is in *GO*, vol. X, p. 416.

170. Received views of the role of testimonials as resources for the epistemological stabilization of knowledge claims do not seem to capture the intricate and perhaps even counterintuitive roles of Kepler's report on the *Nuncius* or of Galileo's invocation of the grand duke as witness to the Medicean Stars.

It does not come as a surprise that a couple of years later, monopoly achieved, Galileo became more frank about the difficulties posed by satellite observation. He wrote to Mark Welser in December 1612 that it was indeed very difficult to see the satellites when they were close to Jupiter, due to the planet's brightness, and that one should not be surprised to see them emerge or disappear in and out of the blue because (he only now understood) the satellites may be eclipsed by Jupiter's own shadow. One still needed an "excellent instrument," but also the "sharpest eyesight." The measurement of angular distances necessary for the plotting of the satellites positions and the calculations of their periods that had been presented as so unproblematic in the *Nuncius* became, less than two years later, a "source of many possible errors."[171] But by that time it did not quite matter anymore.

CONCLUSION

One always discloses in order to gain credit. Today, those who publish a discovery or invent a device always give something away in the process of getting their claims recognized or their devices patented. Competitors may be able to use the publication of a patent to circumvent it or use it for free after it expires, and the publication of a scientific claim can allow other scientists to make further related discoveries and take credit for them. At the same time, discoverers or inventors receive something in return for disclosure. A scientist receives professional credit from his/her publications, and a patent holder is granted a temporary monopoly on his/her invention. Galileo worked in a different economy, one in which the checks and balances between credit and disclosure were drawn and managed quite differently.[172]

The analysis of Galileo's monopolistic tactics has provided a window onto the dynamics of the field in which he operated. I have tried to show that these tactics were part of economies that construed the objects they rewarded, not just the modalities of their crediting. Depending on the economies in which it circulated, Galileo's work could be put in different boxes (invention, discovery, artwork), each of them attached to different stan-

171. These remarks were attached to a long list of tabulated observations of the Medicean Stars ranging, with some gaps, from March to May 1612. It was sent to Welser as an appendix to Galileo's third and last letter on sunspots (*GO*, vol. V, pp. 247–49).

172. For a discussion of these two economies in the present context see Mario Biagioli, "Rights or Rewards?" in Mario Biagioli and Peter Galison (eds.), *Scientific Authorship: Credit and Intellectual Property in Science* (New York: Routledge, 2003), pp. 253–79.

dards of visibility, disclosure, and secrecy—practices that framed the conditions of acceptance and reward of that work. If Galileo's tactics changed drastically from 1609 to 1610 it was not because his work had neatly evolved from invention to discovery or from mathematics to natural philosophy. The telescope and the Medicean Stars were different kinds of objects not by virtue of some essentialist taxonomy, but because their features made them potentially suitable for different economies with different market sizes, notions of utility and value, and kinds of reward.

As a result of the later development of intellectual property law, the line between invention and discovery has come to epitomize the alleged dichotomy between two regimes of knowledge—interest-based technology and interest-free science. While technology is seen as linked to financial interests and thus as part of the economy, science tends to be seen as a noneconomy.[173] This chapter has questioned the traditional assumption about inventions following from discoveries by showing that the concept of discovery (in the sense exemplified by the Medicean Stars) emerged within a radical reconfiguration of the system that rewarded inventions. It has also shown that "discovery" is as much a term of art as is "invention." The two are not separated by the nature/economy divide, but by a line dividing two economies (one of which is cast as a noneconomy).

At a more specifically historical level, I have argued that the making and reception of Galileo's findings were both simpler and more complicated than previously thought, and that they do not readily fit received models about the closure of disputes about observations and experiments. Simpler because their observation did not depend on opaque perceptual dispositions or on tacit instrument-making skills. More complicated because the process required substantial investment of time and money and took place in a field divided by different social and disciplinary economies that construed evidence, utility, and reward in substantially different manners. This does not mean, however, that Galileo's discoveries were unproblematic facts that could be recognized as such once the "accidental" obstacles produced by his uncooperativeness could be removed, or after his critics had decided to drop their "obscurantist" stance and simply take the time to observe. His tactics (as well as those of his competitors and critics) were not unnecessary obstacles on the path to truth, but constitutive elements of the production of the objects he called "Medicean Stars."

173. Sharon Traweek, *Beamtimes and Lifetimes: The World of High Energy Physicists* (Cambridge: Harvard University Press, 1988), p. 162 makes a similar point about representations of science as a nonculture.

CHAPTER THREE

Between Risk and Credit

Picturing Objects in the Making

THIS chapter continues to analyze the nexus between discoveries, their pictorial representations, and credit. It does so by looking at a dispute that flared up in 1612 between Galileo and the Jesuit astronomer Christoph Scheiner around the discovery of sunspots.[1] Reflecting their different institutional affiliations, the two held different investments in the cosmological implications of this new discovery, as well as different views about the proper relationship between astronomy and natural philosophy. Galileo thought of natural philosophy as symbiotically connected to mixed mathematics and therefore alien to Aristotelian natural philosophy. Scheiner accepted the basic Thomistic framework that placed natural philosophy above mixed mathematics, but did so reluctantly, showing his eagerness to expand the domain of mathematics at the expense of philosophy.[2] But if Scheiner and Galileo were both philosophically heterodox their brands of heterodoxy were not compatible, or at least they became incompatible during this dispute. Both saw the new telescopic discoveries as prime material

1. A comprehensive analysis of the discovery of sunspots and a complete translation of Galileo and Scheiner's texts, is in Albert van Helden and Eileen Reeves, *Galileo and Scheiner on Sunspots* (Chicago: University of Chicago Press, forthcoming). All translations of Galileo and Scheiner cited in this chapter are by van Helden.

2. Scheiner's leanings toward cosmological subversion are cogently discussed in Rivka Feldhay, "Producing Sunspots on an Iron Pan," in Henry Krips, J. E. McGuire, and Trevor Melia (eds.), *Science, Reason, and Rhetoric* (Pittsburgh: University of Pittsburgh Press, 1998), pp. 119–43.

for the expansion of the mathematicians' domain, and were very keen to claim priority over the discovery of sunspots. But while Scheiner wished or needed to minimize their cosmological implications, Galileo opted for the opposite course of action, trying to make the sunspots as offensive as possible to Aristotelian cosmology. I use this tension as a key to understand their different textual and visual arguments about sunspots and, more generally, their different positions about the role of representation in the astronomer's knowledge.

FROM STILL LIVES TO SEQUENCES

Historians of early modern science and medicine have opposed two kinds of illustrations: schematic, diagrammatic, and normative pictures on one side, and realistic, mimetic, descriptive illustrations on the other. The former are said to abstract from the physical details of the object, while the latter identify an object as a specific, singular physical entity.[3] In the case of astronomical illustrations, this distinction has been recast as that between the mathematical and the physical: between the geometrical line diagrams used by astronomers to model planetary motions and the pictorial

3. Literature that proposes some version of the typifying/identifying dichotomy includes: James Ackerman, "Early Renaissance 'Naturalism' and Scientific Illustration," in Allan Ellenius (ed.), *The Natural Sciences and the Arts* (Stockholm: Almqvist, 1985), pp. 1–17; Samuel Edgerton, "Galileo, Florentine 'Disegno,' and the 'Strange Spottedness' of the Moon," *Art Journal* 44 (1984): 225–32; William Ashworth, "Natural History and the Emblematic Worldview," in David Lindberg and Robert Westman (eds.), *Reappraisals of the Scientific Revolution* (Cambridge: Cambridge University Press, 1990), pp. 303–32; William Ivins, "What about the 'Fabrica' of Vesalius?" in S. W. Lambert (ed.), *Three Vesalian Essays* (New York: McMillan, 1952), 45–99; Glenn Harcourt, "Andreas Vesalius and the Anatomy of Antique Sculpture," *Representations* 17 (1987): 28–61; Luca Zucchi, "Brunfels e Fuchs: L'illustrazione botanica quale ritratto della singola pianta o immagine della specie," *Nuncius* 18 (2003): 411–65; Martin Kemp, "Temples of the Body and Temples of the Cosmos," in Brian Baigrie (ed.), *Picturing Knowledge: Historical and Philosophical Problems Concerning the Use of Art in Science* (Toronto: University of Toronto Press, 1996), pp. 40–85. The overall argument of William Ivins, *Prints and Visual Communication* (Cambridge: MIT Press, 1953), also relies on the dichotomy between identifying and typifying images, which it links to the opposition between mechanically produced and hand-drawn pictures. The distinction between typifying and identifying images dovetails with another important theme in the recent historiography of early modern science: the transition from the Aristotelian notion of evidence (generic, everyday, nonspecialistic, etc.) to a more specialized one, often linked either to singular events or unusual objects or to experiments.

representations of celestial bodies as specific physical objects like Galileo's engravings of the Moon.[4]

One could offer a philosophical critique of such dichotomies and of the notion of representation-as-mimesis on which they rest, but that is not my main project here. My more immediate concern is that these dichotomies fail to support satisfactory interpretations of the pictorial genre utilized by both Scheiner and Galileo during the debate on sunspots: visual sequences. I believe that an understanding of the workings of visual sequences provides an empirical critique of assumptions about the opposition between diagrammatic and mimetic images, or about the feasibility of using codes of pictorial realism as paradigmatic of realism in general. Visual sequences provide a window on the specific epistemological issues raised by the discovery of sunspots, as well as on more general problems of visual representation.

Simply put, dichotomies between diagrammatic and mimetic representations (and other related oppositions) have been developed by giving paradigmatic status to the single, static image of an equally static object: anatomical tables, botanical illustrations, pictures of microscopic structures, instruments, rock formations, fossils, and so forth. The very question of whether an image presents a certain specimen as typical of a class or as a specific individual reflects the further assumption that these images operate within taxonomical projects, pedagogical texts, or field guides—genres where the still-life model reigns.[5] But if the epistemic stability of botanical or anatomical objects hardly hinges on our ability to represent their motions, movement and change are central to constituting other kinds of objects and processes.[6] Sequences of sunspots do not simply trace the successive appearances of a moving object, nor do they make visible temporal

4. Mary Winkler and Albert van Helden, "Representing the Heavens: Galileo and Visual Astronomy," *Isis* 83 (1992): 195–217. On mathematicians' use of diagrams, and their different meanings in statics and dynamics, see Michael Mahoney, "Diagrams and Dynamics: Mathematical Perspectives on Edgerton's Thesis," in John Shirley and David Hoeniger (eds.), *Science and the Arts in the Renaissance* (Washington: Folger Books, 1985), pp. 198–220.

5. See for instance the atlases analyzed in Lorraine Daston and Peter Galison, "The Image of Objectivity," *Representations* 40 (1992): 81–128, and the herbals discussed in Luca Zucchi, "Brunfels e Fuchs," *Nuncius* 18 (2003): 411–65.

6. If sometimes field guides picture either plants or animals at different stages of their development it is not because they want to trace their growth and change per se, but only to facilitate the identification of those plants and animals when the naturalist encounters them at a nonadult stage.

processes that, due to their high velocity, would escape human vision (like, say, Etienne-Jules Marey's photographic sequences of galloping horses, or other examples of the nineteenth-century "graphical method").[7] Galileo's visual narratives are instead deployed *to make a case for an object's existence based on its periodic, cyclical patterns of change.*

The objects of Galileo's sequences are remote and accessible to vision only—the Medicean Stars, the topographical features of the Moon, the sunspots. But similar sequences could be used also for objects and processes much closer to home. William Harvey's 1628 virtual representation of the circulation of the blood through a sequential depiction of the functioning of the valves in the veins provides such an example (fig. 7). The circulation of blood is not something whose existence Harvey could prove with scattered images of pulsating arteries and veins any better that Galileo could demonstrate the satellites of Jupiter, the irregularities of the lunar surface, or the sunspots without showing their regular, cyclical patterns of change. Related (but not identical) considerations apply to Fabricius of Aquapendente's attempt to depict the embryological *development* of the chick (not the chicken as a species) through a chronologically ordered, thirteen-day visual sequence in his 1621 *De formatione ovi et pulli* (fig. 8).[8] As we will see, Galileo's visual sequences of the constantly repeated life cycle of sunspots—their emergence, expansion, fragmentation, and disappearance—have much in common with Aquapendente's embryological narratives.

7. Among the many discussions of these material, see Joel Snyder, "Visualization and Visibility," in Caroline Jones and Peter Galison (eds.), *Picturing Science, Producing Art* (New York: Routledge, 1998), pp. 379–97; and Robert Brain and Norton Wise, "Muscles and Engines: Indicator Diagrams and Helmholtz's Graphical Methods," in Mario Biagioli (ed.), *The Science Studies Reader* (New York: Routledge, 1999), pp. 51–66.

8. Hieronymi Fabrici ab Aquapendente, *De ovi et pulli tractatus accuratissimus* (Padua: Benci, 1621), plate III. The difference I see between Aquapendente's table and Harvey's and Galileo's sequences is that Aquapendente was primarily trying to map the temporal dimension of a process whose existence was not contested while the other two were using phenomenological periodicities to argue for the existence of a previously unknown process (circulation of the blood) or previously unknown objects (satellites of Jupiter, topographical features on the Moon, sunspots, etc.). Aquapendente's text is translated in Howard Adelmann, *The Embryological Treatises of Hieronymus Fabricius ab Aquapendente*, 2 vols. (Ithaca: Cornell University Press, 1942), vol. I, pp. 231–33. The narrative flows left to right, top to bottom. Usually Aquapendente included two pictures per day (one in the first and second day, and four in the thirteenth day) depicting both the changing appearance of the opened egg and, when visible, of the fetus. I thank Claus Zittel for this reference.

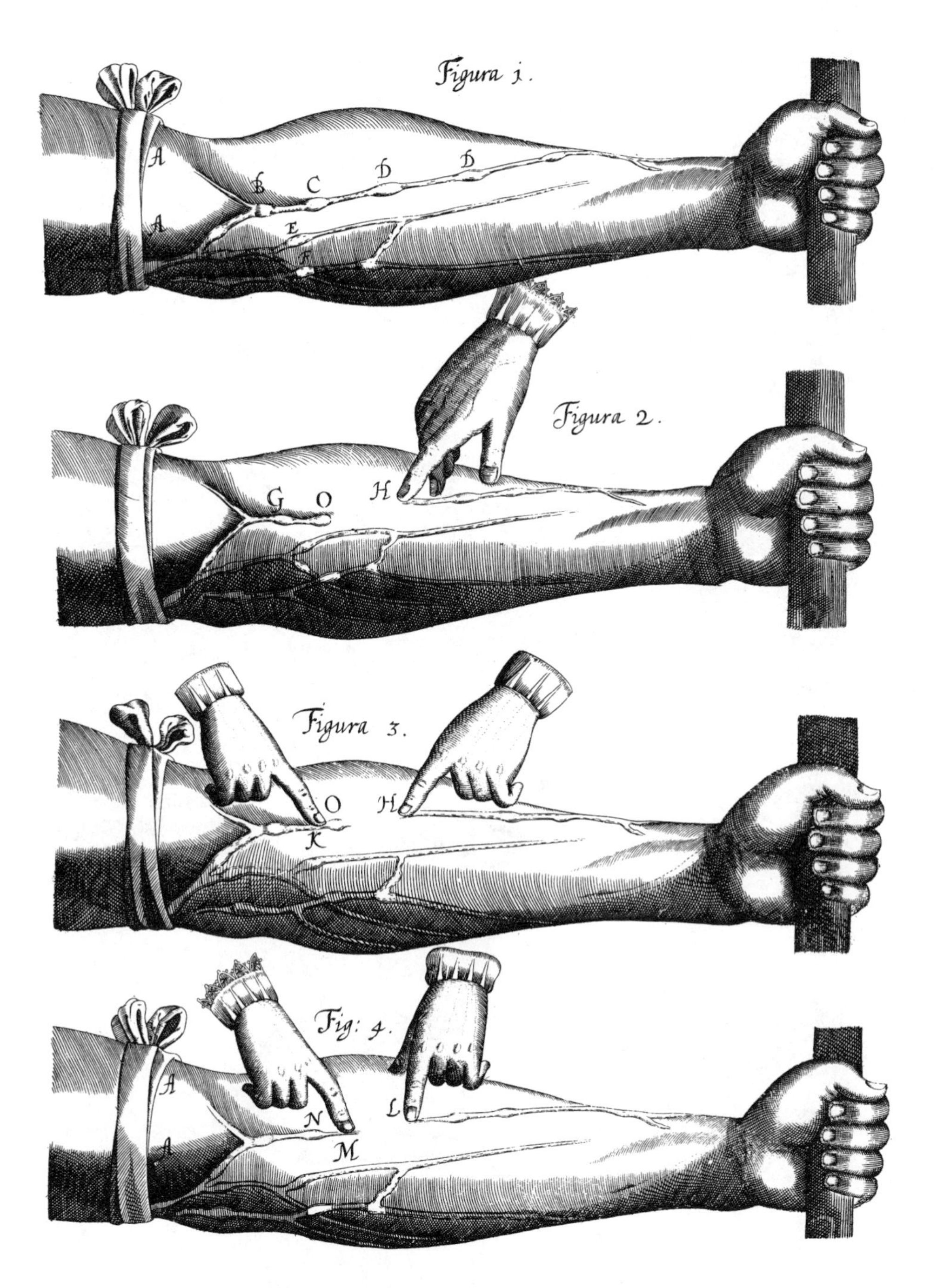

FIGURE 7. Harvey's illustration of the functioning of the valves in the veins in *Exercitatio anatomica de motu cordis et sanguinis in animalibus* (1628).

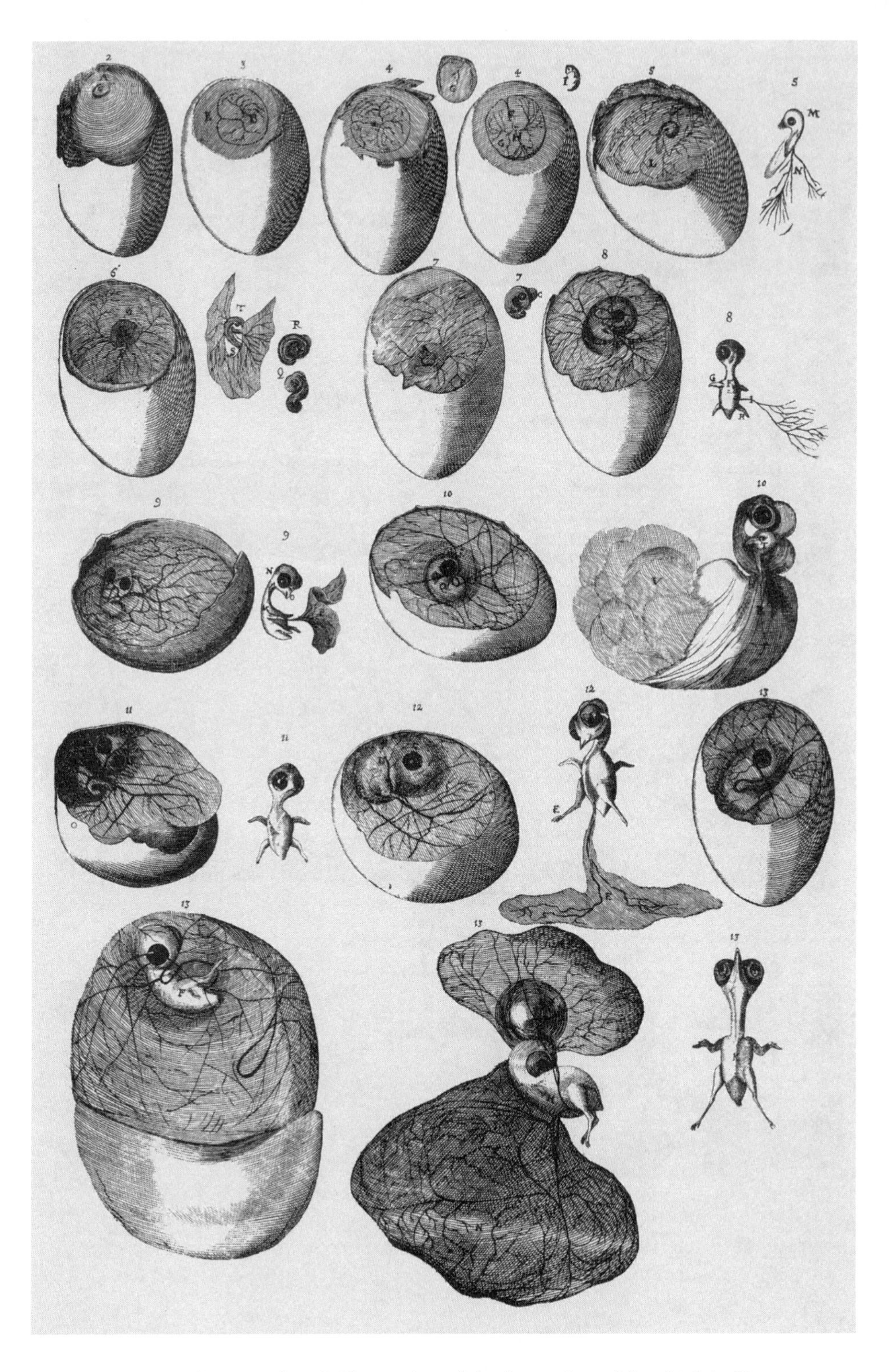

FIGURE 8. Aquapendente's illustration of the formation of the chick in *De formatione ovi et pulli* (1621).

Taken as a kind, these narratives do not necessarily operate within homogeneous temporal frameworks.[9] Galileo's sequences of the Medicean Stars or Aquapendente's sequences of developing embryos represent scenarios at one-day intervals, but other pictorial sequences are not structured by specific sampling rates. Harvey's pictures of the workings of the valves in the veins are arranged sequentially, but the length of time separating the scenarios represented in each image is irrelevant to an understanding of the process being depicted. What matters in Harvey's case is the sequence's temporal order, not the actual amount of time separating the configurations represented in the pictures. As a concept, the circulation of the blood does not depend on a specific heart rate any more than the concept of pendulum depends on a specific frequency of oscillation.

Harvey's sequence depicted simple experiments involving the arm of a living man, with a tourniquet loose enough to let arterial blood into the arm, but tight enough not to let venous blood back out. By showing the effects of pressing a finger on the engorged veins and sliding it back and forth over the valves, these images indicate that the valves functioned like on-off devices, thus making the blood flow in one direction only. Having done that, Harvey invited his readers to replay the same process over and over in their minds to produce a virtual visualization the *cyclical* flow of the blood from the periphery back to the heart and out again—a flow he could not actually picture.[10] Harvey's sequence makes a visual argument for the circulation of the blood without actually depicting it.[11]

9. Linkages between visual styles and patterns of discipline formation have been presented in Mary Winkler and Albert van Helden, "Johannes Hevelius and the Visual Language of Astronomy," in Judith Field and Frank James (eds.), *Renaissance and Revolution: Humanists, Scholars, Craftsmen, and Natural Philosophers in Early Modern Europe* (Cambridge: Cambridge University Press, 1993), pp. 97–116; and Martin Rudwick, "The Emergence of a Visual Language for Geological Science, 1760–1840," *History of Science* 14 (1976): 149–95. A more practice-oriented discussion is Michael Lynch, "Discipline and the Material Form of Images: An Analysis of Scientific Visibility," *Social Studies of Science* 15 (1985): 37–66.

10. His pictures functioned like the representation of a clock's escapement's positions: once you see its movements through a cycle you can visualize the rest without needing to map them out ad infinitum. William Harvey, *Exercitatio anatomica de motu cordis at sanguinis in animalibus* (Frankfurt: Fitzer, 1628), pp. 57–58. Nicholas King, "Narrative and the Effacement of the Visual in the *De motu cordis*" (unpublished manuscript, Department of History of Science, Harvard University, 1996), has given me key insights on the visualization challenges faced by Harvey.

11. Harvey modeled his images after an illustration from Aquapendente, *De venarum ostiolis* (Padua: Pasquati, 1603), Tabula ii, Figura i, but because Aquapendente wished

Pictorial sequences, in fact, function like arguments—arguments that are obviously framed by temporality but do not need to follow a homogeneous chronological narrative. The length, sampling, and image quality of such sequences is specifically tied to the features of the object or process at hand (accessibility being one of them), not to the requirements of a pictorial genre. These sequences do not amount to a discipline's "visual style" or "pictorial language," nor do they follow traditional codes of pictorial realism (like, say, Brunelleschian perspective). The images Galileo employed to track the motions of the satellites of Jupiter looked diagrammatic, while those of the lunar phases seemed realistic. The distinctive feature of these sequences is the mapping of certain periodicities or recurring patterns of differences in the object they depict, no matter what the style or medium of those depictions might be—oil painting, line or chiaroscuro drawing, woodcut, engraving, and so forth. In some cases the sampling frequency of a visual sequence may be more important than its pictorial aspects.[12]

Another key feature of sequences is that they do not function like illustrations; they do not provide an example (or an exemplar) of a known object or process. Sequences come with the fact, not after the fact. They are part and parcel of arguments about the existence of that which they claim to depict. Their specific epistemological role also frames their relationship to the text. While still life images tend to be linked to the text through a caption (typically the name of the species, variety, or object that image stands for), sequences are connected to a whole argument. Galileo's lunar sequences and Harvey's tabulation of the valves' functioning are so integrated into their textual arguments—and their arguments structured around those images—that neither text nor image would be effective without the other.

Due to such a high degree of intertextuality between image and text, the very notion of referent or signified (used here under erasure) changes as we

to demonstrate the presence of the valves throughout the body (not the circulation of the blood), he included a single picture (identical in scheme to Harvey's "Figura i" in our fig. 7) showing the valves' positions in the arm, not their functioning. Pursuing process rather than location, Harvey expanded that single representation into a four-image sequence.

12. The diagrams of the satellites of Jupiter in the *Nuncius*, for instance, could have been even more abstract without damaging their effectiveness. A lower sampling frequency, on the other hand, might have hampered the detection of their periodicities, thus destabilizing their epistemological status.

move from a still life to a sequence.[13] The referent of a sequence is not a stable and well-delineated object but may also be a network of arguments about what that object could or could not be, how it has been previously represented, and so forth. As the object is stabilized, the intertextuality of its referent/signified decreases until it reaches the point where the object may be represented through a single, static image. At that time, the picture becomes attached to a specific name and referent.

CINEMATIC PRESENTATIONS

News of the existence of sunspots began to circulate toward the end of 1611. Their discovery was an important coda to the first wave of astronomical findings that followed the introduction of the telescope in 1609. Not only were sunspots independently detected by a number of people involved in or mobilized by previous telescopic discoveries, but their interpretations and visual representations were closely related to those of the other new astronomical objects observed between 1609 and 1611.

As we have seen, the *Nuncius* was the first printed report of telescopic observations, as well as a trend-setting text concerning the use of visual representations of celestial bodies. Although philosophical in nature, most of Galileo's claims were constructed and presented through images. His anti-Aristotelian argument about the irregularities of the lunar surface was based on a sequence of wash drawings, each showing how rugged the Moon appeared as it went through its phases (fig. 9).[14] Galileo's use of contemporary chiaroscuro techniques in these wash drawings testifies to his training in the arts of *disegno* and his familiarity with codes of pictorial

13. I am aware of the terminal difficulties besieging the notion of referent. I have not found a comprehensive critique of the notion of reference and referent in scientific imagery with the partial exception of Ronald Giere, "Visual Models and Scientific Judgment," in Baigrie, *Picturing Knowledge,* pp. 269–302. I believe, however, that arguments developed by sociologists and philosophers against the tendency to anchor notions of truth on linguistic notions of reference may be easily expanded to cover images as well. Among these, see Michael Lynch, "Representation Is Overrated: Some Critical Remarks about the Use of the Concept of Representation in Science Studies," *Configurations* 2 (1994): 137–49.

14. Galileo's tactics, however, were not exclusively pictorial. He based other arguments for the existence of topographical features on the lunar surface and on the size and length of shadows cast by those irregularities to calculate their height (*SN*, pp. 51–52).

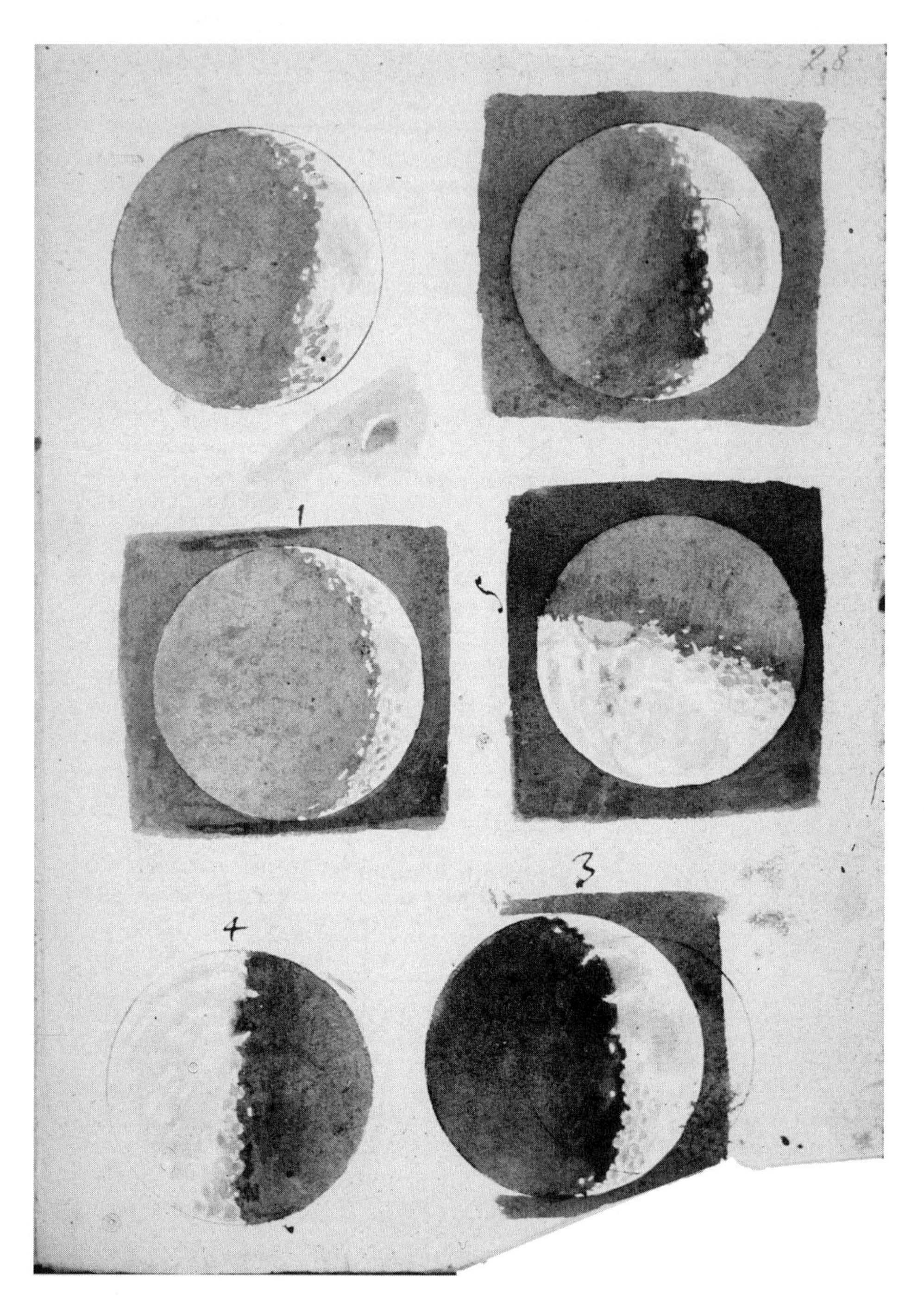

FIGURE 9. Galileo's 1609 wash drawings of the Moon. Ms. Gal 48, fol. 28r, courtesy of the Biblioteca Nazionale Centrale, Florence. (Reproduction courtesy of Owen Gingerich.)

realism.[15] It is puzzling, then, that the *Nuncius*' pictures of the Moon are both so realistic looking and remarkably inaccurate by modern standards.[16]

As Galileo's original wash drawings were turned into engravings, a very large crater emerged almost out of nowhere in the Moon's southern hemisphere (fig. 10). This addition may have been the result of the engraver's poor skills or of the rush Galileo imposed on him, but it may have also reflected Galileo's desire to stress as much as possible the Earth-like appearance of the Moon. Be that as it may, the difference between the wash drawings and the engravings is so noticeable that, were Galileo a modern astronomer, he would have probably been charged with scientific misconduct.[17] None of his contemporaries, however, seemed to notice or commented on Galileo's apparent exaggerations (even when they were further amplified in the unauthorized reprint of the *Nuncius* in the Fall of 1610).[18] Despite their realistic look, these pictures conveyed a philosophical point about the physical nature of the Moon not by representing it "the way it was" (that is, in its specificity), but by exaggerating the irregularities of the lunar surface, thus making them more generic. They conveyed something about the physical nature of the Moon without being mimetic.

In contrast to Galileo's pictures of the Moon, his illustrations of the movements of the satellites of Jupiter look diagrammatic (fig. 11). They

15. Horst Bredekamp, "Gazing Hands and Blind Spots: Galileo as Draftsman," in Jürgen Renn (ed.), *Galileo in Context* (Cambridge: Cambridge University Press, 2001), pp. 153–92, provides the most comprehensive discussion of Galileo's background in *disegno*.

16. Galileo himself complained about them and hoped to produce a new edition of the *Sidereus nuncius* with more extensive and better illustrations of the Moon (*GO*, vol. X, pp. 299–300). Such an edition, however, never appeared.

17. Guglielmo Righini, "New Light on Galileo's Lunar Observations," in Maria Luisa Righini Bonelli and William Shea (eds.), *Reason, Experiment, and Mysticism* (New York: Science History Publications, 1975), pp. 59–76; Owen Gingerich, "Dissertatio cum Profesor Righini and Sidereo Nuncio," ibid., 77–88; Samuel Edgerton, "Galileo, Florentine 'Disegno,' and the 'Strange Spottedness' of the Moon," *Art Journal* 44 (1984): 225–32; Ewen Whitaker, "Galileo's Lunar Observations and the Dating of the Composition of the *Sidereus nuncius*," *Journal for the History of Astronomy* 9 (1978): 155–69; Winkler and van Helden, "Representing the Heavens: Galileo and Visual Astronomy"; Feyerabend, *Against Method*, pp. 99–143. See also chapter 2, this volume.

18. Galileo Galilei, *Sidereus nuncius* (Frankfurt: Paltheniano, 1610). Not only was the quality of the pictures of the Moon much worse in this edition, but some images were out of order and one was printed upside down. One of Galileo's interlocutors in the debate over the irregularities of the lunar surface, Johann Brengger, used the Frankfurt reprint but did not comment on the bad quality of the pictures (*GO*, vol. X, p. 461).

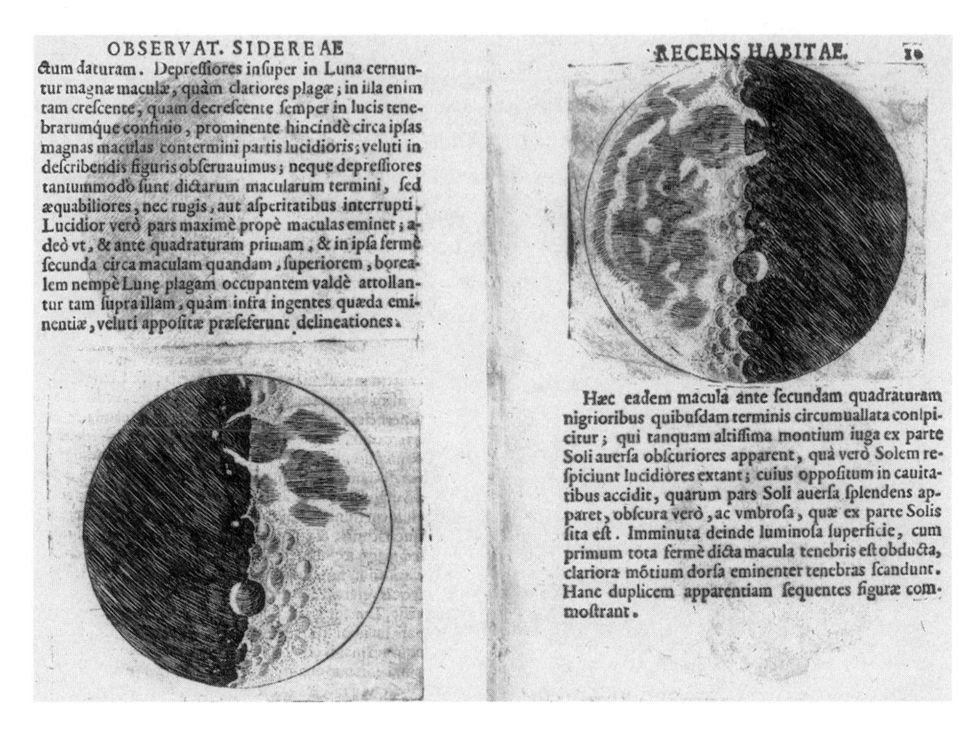

OBSERVAT. SIDEREAE

ctum daturam. Depreſſiores inſuper in Luna cernuntur magnæ maculæ, quàm clariores plagæ; in illa enim tam creſcente, quam decreſcente ſemper in lucis tenebrarumque confinio, prominente hincindè circa ipſas magnas maculas contermini partis lucidioris; veluti in deſcribendis figuris obſeruauimus; neque depreſſiores tantummodò ſunt dictarum macularum termini, ſed æquabiliores, nec rugis, aut aſperitatibus interrupti. Lucidior verò pars maximè propè maculas eminet; adeò vt, & ante quadraturam primam, & in ipſa fermè ſecunda circa maculam quandam, ſuperiorem, borealem nempè Lunę plagam occupantem valdè attollantur tam ſupra illam, quàm infra ingentes quæda eminentiæ, veluti appoſitæ præſeferunt delineationes.

RECENS HABITAE. 10

Hæc eadem macula ante ſecundam quadraturam nigrioribus quibuſdam terminis circumuallata conſpicitur; qui tanquam altiſſima montium iuga ex parte Soli auerſa obſcuriores apparent, quà verò Solem reſpiciunt lucidiores extant; cuius oppoſitum in cauitatibus accidit, quarum pars Soli auerſa ſplendens apparet, obſcura verò, ac vmbroſa, quæ ex parte Solis ſita eſt. Imminuta deinde luminoſa ſuperficie, cum primum tota fermè dicta macula tenebris eſt obducta, clariora mōtium dorſa eminenter tenebras ſcandunt. Hanc duplicem apparentiam ſequentes figuræ commoſtrant.

FIGURE 10. Galileo's engravings of the Moon in *Sidereus nuncius* (1610). The crater is visible in the lower half of the two illustrations near the terminator. (Reproduction courtesy of Houghton Library, Harvard University.)

were printed from woodcuts, not engravings. Jupiter is represented by a capital "O" and its satellites are designated by simple asterisks of four different sizes depending on the satellites' luminosity. But despite their schematic look they construe the Medicean Stars as material and physical, not abstract and mathematical. They achieve that by mapping the movements of the satellites of Jupiter over forty-four days, thus indicating that they were neither optical artifacts nor fixed stars.

If we now compare Galileo's "realistic" pictures of the lunar surface and the "diagrammatic" ones of the Medicean Stars we see that both these pictures make the *same kind of claim* about the existence of specific new objects, but at the same time they seem to belong to *different pictorial genres*. This suggests that the distinction between diagrammatic and identifying images is not conceptually useful. Furthermore, that physical claims could be conveyed either through diagrammatic-looking pictures or through

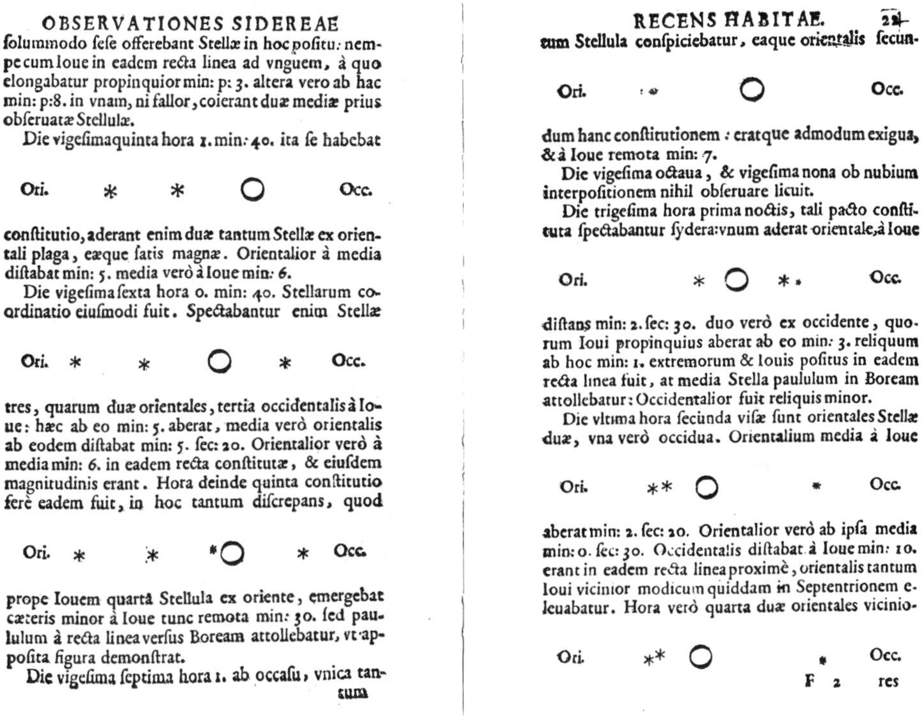

OBSERVATIONES SIDEREAE

ſolummodo ſeſe offerebant Stellæ in hoc poſitu: nempe cum Ioue in eadem recta linea ad vnguem, à quo elongabatur propinquior min: p: 3. altera vero ab hac min: p:8. in vnam, ni fallor, coierant duæ mediæ prius obſeruatæ Stellulæ.

Die vigeſimaquinta hora 1. min: 40. ita ſe habebat

Ori. * * O Occ.

conſtitutio, aderant enim duæ tantum Stellæ ex orientali plaga, eæque ſatis magnæ. Orientalior à media diſtabat min: 5. media verò à Ioue min: 6.

Die vigeſima ſexta hora o. min: 40. Stellarum coordinatio eiuſmodi fuit. Spectabantur enim Stellæ

Ori. * * O * Occ.

tres, quarum duæ orientales, tertia occidentalis à Ioue: hæc ab eo min: 5. aberat, media verò orientalis ab eodem diſtabat min: 5. ſec: 20. Orientalior verò à media min: 6. in eadem recta conſtitutæ, & eiuſdem magnitudinis erant. Hora deinde quinta conſtitutio ferè eadem fuit, in hoc tantum diſcrepans, quod

Ori. * * *O * Occ.

prope Iouem quartà Stellula ex oriente, emergebat cæteris minor à Ioue tunc remota min: 30. ſed paululum à recta linea verſus Boream attollebatur, vt appoſita figura demonſtrat.

Die vigeſima ſeptima hora 1. ab occaſu, vnica tantum

RECENS HABITAE. 21

tum Stellula conſpiciebatur, eaque orientalis ſecun-

Ori. ·* O Occ.

dum hanc conſtitutionem: eratque admodum exigua, & à Ioue remota min: 7.

Die vigeſima octaua, & vigeſima nona ob nubium interpoſitionem nihil obſeruare licuit.

Die trigeſima hora prima noctis, tali pacto conſtituta ſpectabantur ſydera: vnum aderat orientale, à Ioue

Ori. * O *· Occ.

diſtans min: 2. ſec: 30. duo verò ex occidente, quorum Ioui propinquius aberat ab eo min: 3. reliquum ab hoc min: 1. extremorum & Iouis poſitus in eadem recta linea fuit, at media Stella paululum in Boream attollebatur: Occidentalior fuit reliquis minor.

Die vltima hora ſecunda viſæ ſunt orientales Stellæ duæ, vna verò occidua. Orientalium media à Ioue

Ori. ** O * Occ.

aberat min: 2. ſec: 20. Orientalior verò ab ipſa media min: o. ſec: 30. Occidentalis diſtabat à Ioue min: 10. erant in eadem recta linea proximè, orientalis tantum Ioui vicinior modicum quiddam in Septentrionem eleuabatur. Hora verò quarta duæ orientales vicinio-

Ori. ** O * Occ.

F 2 res

FIGURE 11. Example of Galileo's maps of Jupiter and its satellites in *Sidereus nuncius* (1610).

realistic-looking illustrations suggests that the notion of referent or signified may need some serious amendment.[19] The referent of Galileo's illustrations is not limited to the object they purport to represent. Because an emerging object cannot, by virtue of being emerging, be reducible to an accepted kind, its representations cannot just refer to a kind the way a botanical illustration refers to a known plant species. The "referent function," then, must be supplemented by something else.

Although no one had seen the satellites of Jupiter before, they were not, conceptually speaking, a completely new kind of object. They could be put, as Galileo did, in a traditional category—"wandering stars"—that until

19. My statement about Galileo's images being "substantially incorrect" does not rest on any general notion of accuracy, but only on the comparison between his wash drawings and the engravings derived from them.

then had included only planets.[20] Relying on previous beliefs about wandering stars, accepted protocols to identify them, and the fact that there was no explicit authoritative objection to the possibility of finding wandering stars orbiting other planets, Galileo could construct a legitimate argument about the physical existence of the satellites of Jupiter by tabulating their periodical motions (no matter how schematically Jupiter or the satellites were presented in print).[21]

Galileo was trying to extend the category of wandering star rather than introducing a completely new one.[22] As a result, the referent of these pictures was not the satellites in and of themselves (an object that was not established yet), but the features that made them classifiable within a category of traditionally accepted kind: the wandering stars. The features traditionally attributed to wandering stars informed the parameters of accuracy according to which Galileo's pictures were to be evaluated. His pictures did not need to be as detailed as, say, photographs. And if that kind of detail was not necessary in the illustration of the Medicean Stars, it was not necessary in their observations either. For instance, in September 1610, Kepler and his associates were able to corroborate Galileo's observations with a telescope that showed the satellites as square dots.[23] It was sufficient for Galileo's pictures to indicate that the satellites behaved neither ran-

20. The term "satellite" was introduced by Kepler, not Galileo.

21. The discovery of satellites orbiting planets posed philosophical problems to Aristotelian philosophers who wished to reconcile the Aristotelian belief in crystalline spheres with the existence of these new bodies that, presumably, were carried around by their own spheres. It was not clear what kind of arrangements could make the two spheres coexist. This, however, did not appear to be a cosmological nightmare (James Lattis, *Between Copernicus and Galileo* [Chicago: University of Chicago Press, 1994], p. 201). While the existence of satellites was not philosophically unproblematic, it was not a claim that had been ruled out by Aristotelian philosophy (like the existence of the vacuum, action at a distance, etc.).

22. In the *Nuncius* Galileo compared the Medicean Stars to Venus and Mercury in the sense that they were "wandering stars" that did not orbit the Earth (*SN*, p. 36). Although, according to Ptolemy and Aristotle, Mercury and Venus orbited the Earth, there was a plausible minority position among astronomers (and a nonheretical one) according to which Mercury and Venus went around the Sun. This explained the fact that their elongation was always limited. Tycho's model incorporated that position. Galileo, in sum, tried to "normalize" the Medicean Stars by arguing that they were like Venus and Mercury. One could argue that this was also a way to point to the fact that more wandering stars were found that did not orbit the Earth—an oblique pro-Copernican claim.

23. Kepler, *Narratio,* in *KGW,* vol. IV, p. 319.

domly nor like fixed stars, and that they displayed periodical patterns that could not be associated with any known optical artifact. The referent of these sequences included what those bright dots could *not* be.

Assumptions about the possibility or impossibility of finding new objects played a direct role in Galileo's representation of the irregularity of the lunar surface too. In this case, however, the result was a very different-looking kind of picture. One could say (as Galileo did) that the valleys and mountains the telescope detected on the Moon belonged to the same category as the topographical features of the Earth (the same way the satellites of Jupiter could be said to be wandering stars).[24] But there was a problem. While there was no explicit theological or philosophical veto against the possibility of satellites orbiting Jupiter, in the case of the lunar surface Galileo had to confront the authoritative Aristotelian assertion of the incorruptibility of the heavens that ruled out the possibility of topographical irregularities outside the Earth. This constraint, however, came with a few interesting loopholes.

It may be puzzling that Galileo used pictorially realistic but apparently inaccurate pictures to make the particularly controversial claim about the irregularities of the lunar surface. The inaccuracy of Galileo's pictures, however, was of a kind that did not matter much in this case. They worked as a short sequence showing that the changing patterns of bright and dark areas cast by the topographical features of the Moon through different phases displayed a distinct periodicity. Galileo did not include an image of the full moon (a standard feature of later selenographies) because such an image would have been useless to him.[25] He was not mapping the Moon per se but rather the movement and changing appearance of the line that divided the dark and light areas of the Moon. Like the illustrations of the satellites of Jupiter that, although "schematic," were effective at casting the satellites as nonartifactual by being able to represent their periodic motions, the pictures of the lunar surface needed to be accurate only in the sense that they

24. *SN*, p. 47. In fact Galileo applied optical methods for the computation of terrestrial mountains' height to the features of the lunar surface, the same way he had applied the method for sorting errant from fixed stars to the satellites of Jupiter (*SN*, pp. 51–52). On the measurement of the height of the lunar mountains see Florian Cajori, "History of Determinations of the Heights of Mountains," *Isis* 12 (1929): 482–514; C. W. Adams, "A Note on Galileo's Determination of the Height of Lunar Mountains," *Isis* 17 (1932): 427–29.

25. Ewen Whitaker, *Mapping and Naming the Moon* (Cambridge: Cambridge University Press, 1999), p. 21.

could sustain a visual narrative about the periodicity of the changing patterns of lights and shadows on the Moon. They did not need to be mimetic maps of the Moon to support the claim that its surface was irregular.[26]

Because the Aristotelians denied the possibility of the Moon having an irregular surface, all Galileo had to do was to show that, contrary to Aristotelian wisdom, it did display such irregularities. The strength of that claim did not hinge on how accurately Galileo depicted a specific crater (like the one whose dimensions were exaggerated in the engraving). He was arguing for the existence of a new kind, not for a particular instance of such a kind. The referent of Galileo's pictures of the Moon, therefore, was not only the irregularity of the lunar surface (as a kind), but the erroneousness of Aristotle's authoritative assumptions about incorruptibility as well. Because Aristotle had been so categorical in ruling out celestial corruptibility, one just needed to represent some degree of "corruption" to make the point that Aristotle was wrong. Paradoxically, the Aristotelians' denial of corruptibility eased the constraints on the pictorial representation of corruption.

The argument applies, in a somewhat weaker form, also to Galileo's remarkably coarse woodcuts of the Belt and Sword of Orion, the constellation of the Pleiades, the nebula in the head of Orion, and the nebula of Praesepe which he represented as a congeries of asterisk-like marks in three or four different sizes (figs. 12 and 13). The illustration was able to effectively convey a physical claim about various nebulae and constellations not by providing a mimetic representation but simply by highlighting that they were *not* the undifferentiated bodies they were assumed to be (in the case of the nebulae) or that they contained thousands more stars than previously believed (in the case of the constellations). The artificiality of the representation was explicitly acknowledged by Galileo: "For the sake of distinction, we have depicted the known or ancient ones larger and outlined by double lines. And the other inconspicuous ones [visible with the telescope] smaller and outlined by single lines."[27]

Such a contrast between pictorial representations and received beliefs

26. See chapter 2, this volume, pp. 107–11.

27. *SN*, p. 61. Galileo arranged his pictorial codes to highlight the difference between the known stars (whose position he tries to depict as accurately as possible so as to function as a frame of reference for the others) and the new ones observed with the telescope. His representational scheme, therefore, is geared toward highlighting the existence of new stars rather than mapping them out as extensively and accurately as possible.

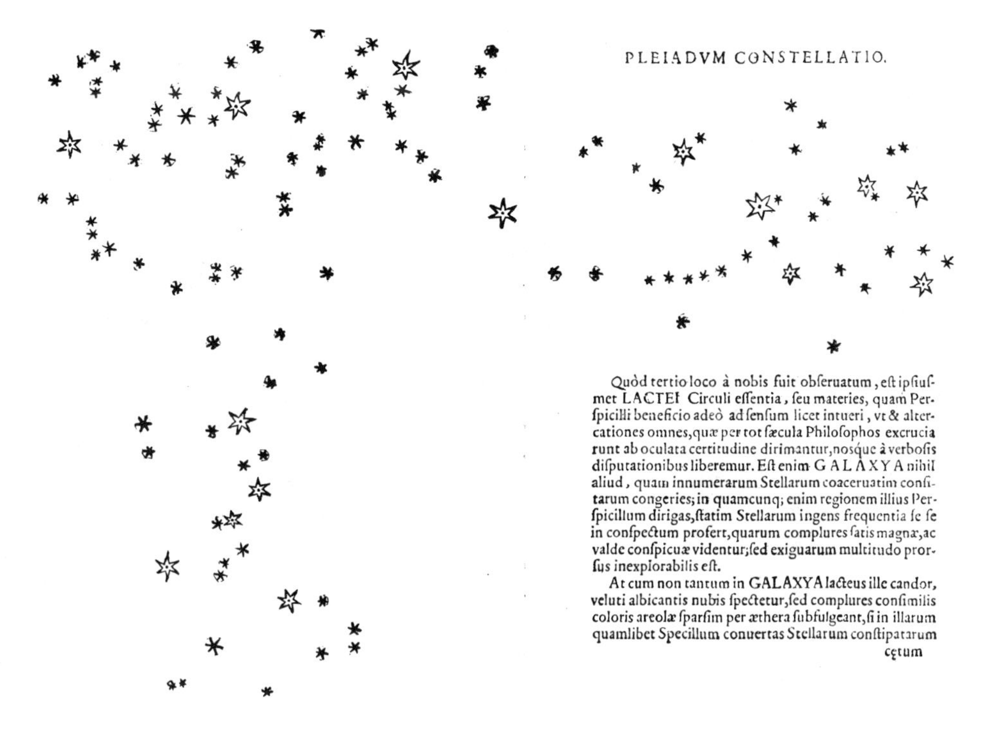

PLEIADVM CONSTELLATIO.

Quòd tertio loco à nobis fuit obſeruatum, eſt ipſiuſmet LACTEI Circuli eſſentia, ſeu materies, quam Perſpicilli beneficio adeò ad ſenſum licet intueri, vt & altercationes omnes, quæ per tot ſæcula Philoſophos excruciarunt ab oculata certitudine dirimantur, nosq́ue à verboſis diſputationibus liberemur. Eſt enim GALAXYA nihil aliud, quam innumerarum Stellarum coaceruatim conſitarum congeries; in quamcunq; enim regionem illius Perſpicillum dirigas, ſtatim Stellarum ingens frequentia ſe ſe in conſpectum profert, quarum complures ſatis magnæ, ac valde conſpicuæ videntur; ſed exiguarum multitudo prorſus inexplorabilis eſt.

At cum non tantum in GALAXYA lacteus ille candor, veluti albicantis nubis ſpectetur, ſed complures conſimilis coloris areolæ ſparſim per æthera ſubfulgeant, ſi in illarum quamlibet Specillum conuertas Stellarum conſtipatarum

cętum

FIGURE 12. Belt and Sword of Orion (left) and the constellation of the Pleiades (right) in *Sidereus nuncius* (1610). (Reproduction courtesy of Houghton Library, Harvard University.)

about their object may also explain why Galileo's pictures of the Moon were very effective in 1610 but became the target of criticism only a few decades later. That criticism did not come from an Aristotelian, but from a fellow astronomer. Van Helden has discussed the fact that in his monumental 1647 *Selenographia*—the most detailed and extensive seventeenth-century description of the lunar surface—Hevelius commented on the inaccuracy of Galileo's pictures and attributed it to the poor quality either of Galileo's telescope or of his drawing skills.[28] Because, as shown by Horst Bredekamp, Galileo had substantial pictorial skills and because telescopes improved only incrementally between 1610 and 1647, Hevelius' statement reflects not so much the evolution of the visual language of astronomy as the transition from a phase in which new kinds of objects were being in-

28. Winkler and van Helden, "Johannes Hevelius and the Visual Language of Astronomy."

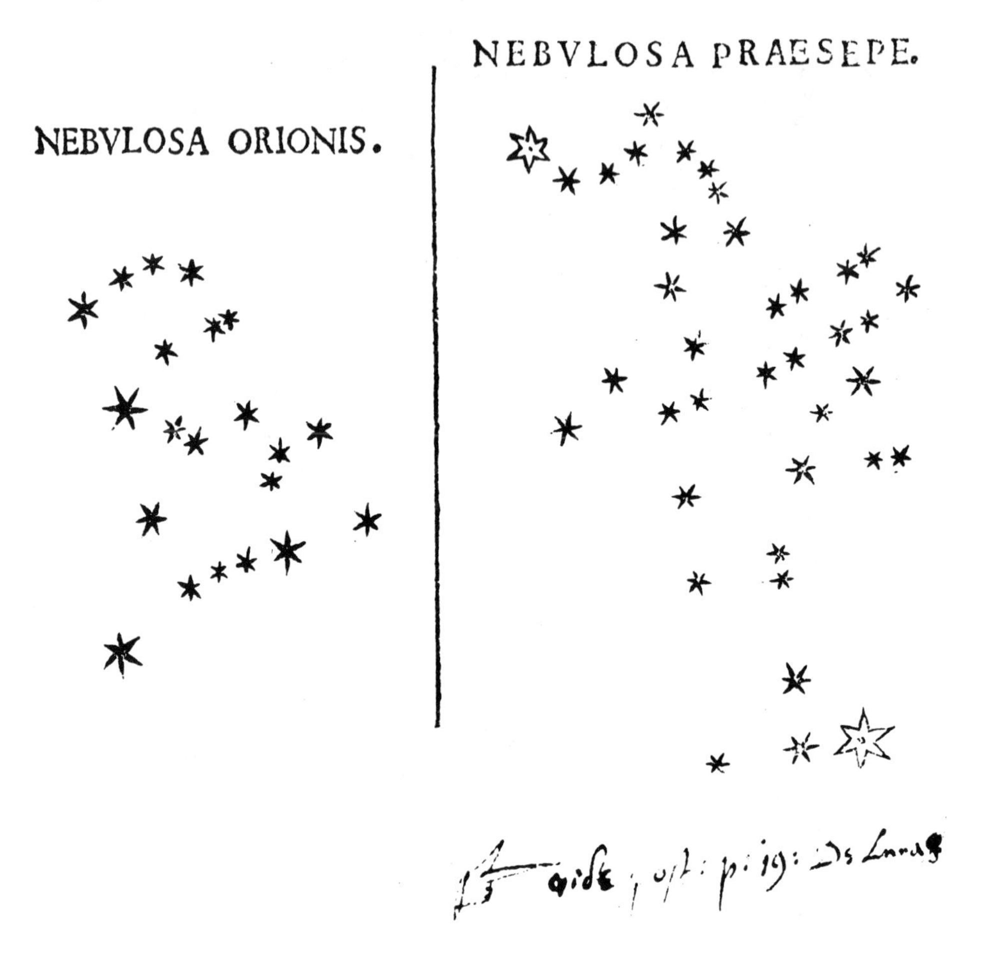

FIGURE 13. Nebula of Orion (left) and Nebula of Praesepe (right) in *Sidereus nuncius* (1610). (Reproduction courtesy of Owen Gingerich.)

troduced against opposing cosmological beliefs to one in which, having been accepted, those objects could be simply mapped and catalogued.[29] Galileo's *Nuncius* belongs to the former phase, while Hevelius' *Selenographia* exemplifies the latter (fig. 14). This shift is reflected in the early history of lunar nomenclature. Galileo was concerned with proving the existence of lunar topographical features in general, not with naming specific ones. But

29. Bredekamp, "Gazing Hands and Blind Spots," pp. 153–92.

FIGURE 14. Map of the full moon from Hevelius' *Selenographia* (1647). (Reproduction courtesy of Houghton Library.) The double circle marks the extent of the librations during Hevelius' observations.

later selenographers who did not need to fight Galileo's battle again simply started to give names to specific lunar features.[30]

We cannot say, therefore, that Galileo's pictures of the Moon were less realistic or less detailed than Hevelius'. What we find between 1610 and 1647 is not a simple trajectory toward better astronomical illustrations, but rather a shift in the definition of what counted as relevant detail—the

30. Whitaker, *Mapping and Naming the Moon*, pp. 17–18, 25–35.

difference that made a difference. Once you accept that the Moon is not a perfect body, you can then use your newly constituted physical assumptions about the existence of lunar craters and mountains to argue that your pictures, telescopes, and drawing methods are better than those of somebody else because they can resolve more craters and mountains. Similarly, once you have accepted that there might be satellites orbiting planets (that is, after you have accepted the category "satellite"), you can then engage in the search for and naming of new satellites, and the telescope that spots the most may then be represented as the best. Observations, of course, could still be contested, but the contestations would most likely be about the presence of a specific object (e.g., a lunar crater in a certain spot), not about the category that object belonged to (e.g., "lunar crater").

PURSUING NOVELTY, CONTROLLING CHANGE

If the *Sidereus nuncius* marked astronomy's shift toward questions of a more physical nature, it also epitomized its transformation into a discovery-intensive discipline.[31] New astronomical objects that were almost unthinkable in 1609 had become commonplace by 1611. As people awaited more discoveries, high-power telescopes were built and used throughout Europe in a hunt for astronomical novelties. This atmosphere of competition informed the debate on sunspots between Galileo and Scheiner.

The pursuit of new discoveries was not necessarily connected to a support of the new astronomy. While not everyone shared Galileo's strategy of dedicating discoveries to patrons, discoveries did become standard material for patronage and were keenly pursued by astronomers of different cosmological commitments, Jesuits included.[32] New discoveries could provide

31. The discovery of novae in 1572, 1600, and 1604 and of the superlunar trajectory of the comets of 1577 are the exception to the meager list of discoveries one can attribute to ancient and early modern astronomy before the introduction of the telescope in 1609.

32. For Jesuit astronomers, the production or dedication of scientific discoveries was not necessarily a means for developing personal patronage, but a way to open the door for other forms of institutional patronage on behalf of the Society. For instance, Scheiner's work on sunspots and his relationship with Welser was the first step in a long patronage-intensive life that culminated in his position of confessor of Archduke Maximilian (brother of Emperor Rudolph II) and then of Archduke Karl (brother of Emperor Ferdinand). He developed patronage connections also with Archduke Leopold. Through these appointments, Scheiner managed to gain financial support for the building of churches (Innsbruck) and the establishment of new colleges (Neisse). Michael John Gorman, "A Matter

evidence against the dominant Aristotelian cosmology by uncovering new objects or phenomena that may not be accommodated within that framework, but it was sometimes possible to control their disruptive potential, as the Jesuits did with the phases of Venus.[33] Furthermore, not all of the new findings were directly threatening to traditional philosophy, and some that challenged Aristotelian cosmology did not necessarily challenge Ptolemaic astronomy. The topographical irregularities of the lunar surface, for instance, were of no relevance to Ptolemaic astronomy but contradicted the Aristotelian assumptions about the nature of nonterrestrial bodies. But some skillful philosophical footwork could control that anomaly too. A few Aristotelians and Jesuits argued that what one saw from the Earth was not the Moon's actual surface but an inner surface—a messy layer encased within a perfectly smooth, transparent (and thus invisible) outer sphere. One could still acknowledge the rugged look of the Moon while denying that it reflected actual topographical features.[34] In sum, while telescopic discoveries had the potential to challenge the cosmological status quo, the actualization of that threat was by no means automatic.

If it was often possible to include some buffer between a discovery and its cosmological implications, it was also possible, to some extent, to separate the discoverer's credit from the philosophical and theological liability his findings might accrue down the line.[35] Most of the audience shared such an attitude about discoveries as it allowed them to be curious and feel pious at the same time. The main difference between Copernicans and non-

of Faith? Christoph Scheiner, Jesuit Censorship, and the Trial of Galileo," *Perspectives on Science* 4 (1996): 307.

33. The Jesuits did not accept the Tychonic system until 1620, but their public acceptance of Galileo's discovery of the phases of Venus in the spring of 1611 suggests that they were aware of the possibility of placing those discoveries in a Tychonic framework to defuse their antigeocentric potential. It would have been most unlikely that Clavius and his students would endorse a discovery that could effectively refute the philosophical framework in which the Church operated. On the Jesuits' slow path toward Tycho see Baldini, *Legem impone subactis* (Rome: Bulzoni, 1992), pp. 217–50.

34. Such an interpretation seems to still have had supporters twenty years later as Galileo described and criticized it in the *Dialogo* (*GO,* vol. VII, pp. 94–95).

35. Sometimes one could receive credit for a discovery whose philosophical significance was still debatable. For instance, in the case of the changing appearance of Saturn (announced by Galileo in 1610 after the publication of the *Nuncius*), Galileo received credit for the discovery despite the fact that for a long time it was not clear what the object behind those observations may have been.

Copernicans was not in their interest in new astronomical objects—an interest they usually shared—but in the way they packaged their discoveries.

All of them wanted to be first, but while Galileo hoped to find pro-Copernican evidence in forthcoming discoveries, some of the Jesuits and other non-Copernican astronomers hoped to be able to interpret their discoveries in ways that would allow them to gain priority credit without requiring more than minor readjustments to the established cosmological framework. The way a discovery was labeled, the way it was severed from or connected to cosmological claims was a subtle exercise in philosophical marketing that could have a great impact on how much credit or trouble a discoverer would incur. Early modern cosmology was such a philosophical and theological minefield that all discoverers, no matter what their party affiliation may have been, had to walk a fine line between credit and risk. If Copernican advocacy put Galileo in a difficult position vis-à-vis the Church, the Jesuits' "philosophical laundering" of discoveries may have put them in an even more complex and slippery predicament.[36]

FROM REALISM VS. NOMINALISM TO CREDIT VS. RISK

As the discovery of artifacts carried no credit, both Galileo and non-Copernican astronomers like the Jesuits were eager to claim that what they discovered was something quite real (while occasionally arguing that it was not real if others had claimed priority over it).[37] Their registers of the

36. Much has been made of the tensions between Galileo and the Jesuits—interpretations that cast Galileo as the progressive and the Jesuits as the conservative element. But it may be worth considering that in some cases they may have played an unplanned symbiotic "good cop, bad cop" routine. For instance, it is very unlikely that the Jesuits would have been able to publish a text like the *Sidereus nuncius* had they wished to do so. At the same time, the discoveries listed in the *Nuncius* and in Galileo's later letters allowed the Jesuits to endorse claims they may have not been able to make on their own. The enthusiasm with which they endorsed Galileo in 1611 indicates that they were not just "corroborating" his claims, but using them as resources for their internal debates with or against the philosophers of their order. Similarly, while the Jesuits were probably unhappy when Galileo went public with the discovery of the phases of Venus just before they did, they may have also been relieved that they did not have to be the ones to refute Ptolemaic astronomy.

37. In the dispute on comets of 1619–23, Galileo argued that the comets observed by the Roman Jesuits may have been not real physical objects but optical artifacts (*GC*, pp. 273–80). Galileo's claim emerged in a context in which the Jesuits had been first to publish observations of the comets while Galileo had been sick and unable to produce a comparable body of observations.

"real," however, diverged substantially as they reflected differences in their positions about the philosophical status of mixed mathematics and the relationship between astronomy, philosophy, and theology. In that regard, Galileo has been cast in a realist position, while the Jesuits have been placed on the more nominalist side of the methodological spectrum.[38] But the positions of Galileo and the Jesuits appear inverted when we look at how they conceptualized the sunspots. While Galileo claimed to know how sunspots moved but not what they were, Scheiner argued that they were solar satellites.

The Jesuits' apparently realist stance about discoveries was closely tied to their practice of controlling the disruptive philosophical implications of new findings by ad hoc rearrangements and extensions of the orthodox cosmology. They acknowledged the phases of Venus (knowing that their antigeocentric implications could be controlled by framing them in the Tychonic system) as well as the satellites of Jupiter (which posed no direct threat to geocentrism). Both discoveries were later used by Scheiner to model more controversial discoveries (like the sunspots) in a cosmologically safe manner. In turn, one of Scheiner's students went so far as to claim that comets (whose apparently ephemeral nature cast them as serious challenges to the doctrine of celestial incorruptibility) were made up of small, incorruptible bodies like those his teacher had employed to explain away the sunspots.[39] In the 1660s, the satellites of Jupiter were invoked again as

38. Different views on the Jesuit mathematicians' (especially Clavius') complex attempt to give epistemological status to the mixed mathematics while skirting mathematical realism—a stance that would have brought them into conflict with the philosophers and theologians of their order—appear in: Peter Dear, "Jesuit Mathematical Science and the Reconstitution of Experience in the Early Seventeenth Century," *Studies in History and Philosophy of Science* 18 (1987): 133–75; Nicholas Jardine, "The Forging of Modern Realism: Clavius and Kepler against the Sceptics," *Studies in History and Philosophy of Science* 10 (1979): 141–73; Baldini, *Legem impone subactis,* pp. 36–56; Giuseppe Cosentino, "Le matematiche nella Ratio Studiorum della Compagnia di Gesù," *Miscellanea storica ligure,* n.s., 2 (1970): 171–213; Alistair Crombie, "Mathematics and Platonism in the Sixteenth-Century Italian Universities and in Jesuit Educational Policy," in Y. Maeyama and W. G. Saltzer (eds.), *Prismata* (Wiesbaden: Franz Steiner Verlag, 1974), pp. 63–94; Lattis, *Between Copernicus and Galileo,* pp. 30–38.

39. William Donahue, *The Dissolution of the Celestial Spheres: 1595–1650* (New York: Arno Press, 1981), pp. 110–12. By the time Cysat used configurations of individual incorruptible stars to explain the physical nature of comets, there was little analogy left between these bodies and the objects after which they were ultimately patterned: the satellites of Jupiter. For instance, it was not even clear whether the bodies invoked by Cysat orbited anything. What's interesting is precisely how the Jesuits' stretching the boundaries

a model to account for the rings of Saturn—another cosmologically unsettling discovery.[40]

The Jesuits' bricolages of celestial objects could be seen as a cosmological version of the astronomers' "save the phenomena" tactics.[41] Traditionally, the astronomers' task had been to break down complex planetary motions into combinations of eccentrics, epicycles, and equants. Although all planetary motions appeared to be philosophically anomalous (as they did not display the uniformity and circularity expected of proper celestial motions), the anomaly was then claimed to be solved by showing that these motions could be simulated by a combination of circular and uniform mo-

---

of safe objects to cover new discoveries led to a progressive modification of the original objects so as to maximize the range of discoveries they could help normalize.

40. Another satellite-based model of the sunspots was published in *Austriaca Sidera heliocyclica . . .* in 1633. Its author, the Belgian Jesuit Charles Malapert, had completed the text in 1628, and died in 1630. He had named the spots after the Hapsburg family. Jean Tarde (not a Jesuit) published his *Borbonia sidera . . .* in 1620 in an attempt to show (with more technical details than Scheiner had provided) how the phenomenon of the sunspots could be saved by an arrangement of satellites. On Tarde and, to a lesser extent, Malapert see Frederic Baumgartner, "Sunspots or Sun's Planets: Jean Tarde and the Sunspots Controversy of the Early Seventeenth Century," *Journal for the History of Astronomy* 18 (1987): 44–54. Concerning the peculiar appearance of Saturn, Scheiner himself hinted at its possible explanation through a configuration of satellites in his early texts on sunspots, but had provided no details about his hypothesis. When Christiaan Huygens put forward the claim that Saturn was surrounded by a ring in his *Systema Saturnium* (The Hague, 1659), the Jesuit Honoré Fabri responded with a satellite-based model of the ring, published under Eustachio Divini's name as *Brevis annotatio in Systema Saturnium Christiani Hugenii* (Rome, 1660). On that debate and on Fabri's model see Albert van Helden, "The Accademia del Cimento and Saturn's Ring," *Physis* 15 (1973): 237–59. The Jesuits did not limit their bricolages to celestial objects, but included mathematical models too. After the trial of Galileo, they lifted Kepler's elliptical planetary orbits out of their original heliocentric framework and plugged them into the more pious geocentric Tychonic system, boosting its performance at no theological cost.

41. Pierre Duhem, *To Save the Phenomena* (Chicago: University of Chicago Press, 1969). Subsequent literature on the epistemological boundaries between astronomy and cosmology or natural philosophy and mixed mathematics includes Robert Westman, "The Astronomer's Role in the Sixteenth Century," *History of Science* 18 (1980): 105–47; Peter Dear, *Discipline and Experience* (Chicago: University of Chicago Press, 1995); Nicholas Jardine, *The Birth of the History and Philosophy of Science* (Cambridge: Cambridge University Press, 1984); Mario Biagioli, "The Anthropology of Incommensurability," *Studies in History and Philosophy of Science* 21 (1990): 183–209; Peter Barker and Bernard Goldstein, "Realism and Instrumentalism in Sixteenth-Century Astronomy: A Reappraisal," *Perspectives on Science* 6 (1998): 232–58, and the references listed in note 38 above.

tions made possible by eccentrics, epicycles, and equants. In a similar fashion, the Jesuits tended to reduce new discoveries to a combination of unproblematic astronomical objects and gain credit for the discoveries (or for their cosmological normalization) without destabilizing the philosophico-theological status quo.[42]

To play this normalization game, the Jesuits had to define what object category a new phenomenon was being reduced to (e.g., that sunspots were satellites). Galileo, on the other hand, tended to limit himself to showing that his discoveries could not be dismissed as artifacts. While in some cases he attached physical labels to these new phenomena (the wandering stars orbiting Jupiter or the mountains of the Moon), in some others (the sunspots or the peculiar appearance of Saturn) he simply argued that what he had observed was "real" but that he did not quite know, physically speaking, what it was or what caused it. He stressed quite emphatically that the linguistic labels he attached to his discoveries were not signifiers of physical essences.[43]

This was not philosophical humility, though. By convincing his readers that, due to their periodic features, his discoveries were not artifacts, he could secure credit for them without having to make pronouncements about their physical nature. His phenomenological stance could be seen as a long-term philosophical investment. Processes that were understood neither by him nor by the Aristotelians at the time of their discovery were not likely to be explained in Aristotelian terms at a later time. An unexplained new phenomenon was not likely to be a philosophical anomaly for someone who, like Galileo, did not cast himself as a system builder. The Aristo-

42. Ironically, much of their initial toolbox was constituted by those discoveries of Galileo's that had been publicly endorsed by the mathematicians of the Collegio Romano in the spring of 1611: the satellites of Jupiter, the phases of Venus, the existence of many more fixed stars, the rough appearance of the Moon, and the noncircular shape of Saturn. As their endorsement of Galileo's controversial findings was not challenged by the Church establishment, the Jesuit mathematicians probably assumed that these objects could be used in pious cosmological repair work. A former poison (what Clavius had called "monsters") had been turned into an antidote (Clavius to Welser, January 29, 1611, Archivio Pontificia Università Gregoriana 530, cc. 183r–184v).

43. In the first letter to Welser, Galileo stated that "names and attributes must accommodate themselves to the essence of the things, and not the essence to the names because the things come first and the names afterwards" (*GO*, vol. V, p. 97). He also commented on the arbitrariness of the term "star" in an unpublished fragment related to the *Istoria* in *GO*, vol. V, pp. 257–58.

telians were in a much different position. Because of the comprehensiveness, authority, and interconnectedness of their system, they may have perceived new unexplained phenomena as philosophical time bombs.[44]

The ways the Jesuits and Galileo framed discoveries seem to contradict the former's generally nominalist and the latter's generally realist stance about the epistemological status of astronomy and mixed mathematics. The contradiction disappears if we treat nominalism and realism (or related terms like instrumentalism or conventionalism) as traces of philosophical tactics. The Jesuits, for instance, tended to reduce new discoveries to known physical objects the way they would reduce apparently anomalous celestial motions to fictional geometrical devices. While satellites were physical objects, the Jesuits deployed them the way they would have deployed epicycles, eccentrics, or equants, that is, as tools for "saving the appearances." Depending on the context, the defense of cosmological orthodoxy could be achieved through an apparently realist stance (the reduction of potentially anomalous discoveries to known objects) or through an apparently nominalist position (the accounting of anomalous motions through fictional devices).

This also helps to resolve the apparent contradiction between Galileo's aggressive stance about the epistemological status of mixed mathematics and his apparently humbler stance about the physical nature of astronomical discoveries.[45] While conservative tactics join the Jesuits' apparent realism about discoveries and nominalism about astronomical models, Galileo tried to avoid the nominalism/realism dichotomy altogether. In one of the sunspots letters Galileo harshly criticized those who discussed geometrical devices as if they were physical, material objects.[46] This apparently nominalist view did not lead him to conclude that these devices were fictions useful only for astronomical calculations. He argued instead that epicycles

44. Galileo made a closely related point in a June 16, 1612, letter to Paolo Gualdo: "These sunspots and my other discoveries are not things that will go away not to return anytime soon, like the new stars of 1572 and 1604, or the comets. [New stars and comets] eventually go away thus giving—through their disappearance—a chance to rest to those who, when these [phenomena] were present, experienced some anxiety. But these [discoveries of mine] will torment them forever because they will always be visible" (*GO*, vol. XI, pp. 326–27). I thank Claus Zittel for this reference.

45. See the 1613 "Letter to Castelli" (*GA*, 47–54) and the 1615 "Letter to the Grand Duchess Christina" (*GA*, 87–118).

46. Galileo Galilei, *Istoria e dimostrazioni intorno alle macchie solari e loro accidenti . . .* (Rome: Mascardi, 1613), in *GO*, vol. V, pp. 102–3.

were not real the way mechanical gears were, but they were nevertheless very real in the sense that their steady, periodical effects (the orbits of planets) could be clearly observed. How could one familiar with those periodical phenomena "deny that eccentrics and epicycles can really exist in nature?"[47] His stances on the astronomers' geometrical devices and his telescopic discoveries were identical: they were both real not as essences but in the sense that their regular periods showed them to be nonartifactual.

It was only by rejecting the essentialist notion of the real that informed the dichotomy between real and fictional used by philosophers and theologians to marginalize astronomy that Galileo could sustain a noninstrumentalist view of the astronomer's knowledge. This, I believe, gives us a key to understanding the epistemological status he attributed to pictures. For the very same reason that he treated both astronomical devices and his discoveries as not unreal, he did not feel compelled to produce mimetic pictures of the objects he had discovered. He did not need to picture their "essence" but only map their periodic motions or patterns of change to show that they existed. From astronomical devices, to discoveries, to pictorial evidence, to naming, Galileo's positions were structured around a nonessentialist working definition of the real as the nonartifactual.[48]

CHANGE IN THE SUN?

The tension between a desire to gain credit for discoveries and the need to present them as cosmologically unthreatening framed much of Scheiner's conceptual choices during the debate on sunspots. Scheiner had just become teacher of mathematics and Hebrew at the Jesuit college at Ingolstadt when, in the spring of 1611, he briefly observed with a student "some rather blackish spots like dark specks" on the Sun.[49] When he returned to

47. Galileo, "Considerations on the Copernican Opinion (1615)," in *GA*, p. 77 (but see also pp. 73, 78). Identical views are in Galileo to Dini, March 23, 1615 (*GA*, pp. 61–62).

48. It also allowed him to avoid unnecessary risks. Venturing into definitions could have brought him into conflict with philosophers and possibly even with theologians. For instance, he did not know what epicycles or sunspots were, but he also had nothing to gain (and much to lose) by saying what they were. Galileo's nonessentialism was also tactically astute. It allowed him to stay clear of the traps of the realist/nominalist dichotomy, and to introduce new astronomical objects without having to define them—something that, given the state of his knowledge, he could not do.

49. Christoph Scheiner, *Tres epistolae de maculis solaribus scriptae ad Marcum Welserum* (Augsburg: Ad insigne pinus, 1612), in *GO*, vol. V, p. 25.

observe the Sun in October, he began to draw pictures of the spots to map their movements and changing appearances as they slowly moved across the solar disk. He was probably unaware that by that time the sunspots had already been observed with telescopes by several people in various parts of Europe.[50]

Thomas Harriot was probably the first to view them in December 1610.[51] Galileo showed them to a few friends in April 1611 (including some Roman Jesuits), though he later claimed to have observed them before August 1610.[52] Kepler in Prague, gentlemen in Padua, and painters in Rome were observing the spots in the fall of 1611.[53] While the discovery of sunspots has been traditionally identified with the priority dispute between Scheiner and Galileo, neither of them was first to publish that finding, nor

50. A survey of pretelescopic observations of dark spots in the Sun is in Albert van Helden, "Galileo and Scheiner on Sunspots: A Case Study in the Visual Language of Astronomy," *Proceedings of the American Philosophical Society* 140 (1995): 368–69. Van Helden argues that while spots were frequently observed in the west, they were usually read not as sunspots but as transits of Mercury or Venus across the solar disk. See also K. C. C. Yau and F. R. Stephenson, "A Revised Catalogue of Far Eastern Observations of Sunspots," *Quarterly Journal of the Royal Astronomical Society* 29 (1988): 175–97; Hosie Alexander, "The First Observations of Sun-Spots," *Nature* 20 (1879): 131–32; Justin Schove, "Sunspots and Aurorae," *Journal of the British Astronomical Association* 58 (1948): 178–90; Bernard Goldstein, "Some Medieval Reports of Venus and Mercury Transits," *Centaurus* 14 (1969): 49–59; George Sarton, "Early Observations of the Sunspots?" *Isis* 37 (1947): 69–71.

51. John North, "Thomas Harriot and the First Telescopic Observations of Sunspots," in John Shirley (ed.), *Thomas Harriot: Renaissance Scientist* (Oxford: Clarendon Press, 1974), pp. 129–57.

52. It is documented that Galileo showed the spots to people in Rome during his visit in the spring of 1611. If and when he observed them prior to his trip to Rome is a more open question. It is also puzzling that he never published the discovery or communicated it to friends through letters as he did the phases of Venus or the three-bodied Saturn. In the *Istoria* he claimed that he had been observing sunspots since November 1610 (*GO*, vol. V, p. 95). In a letter to Barberini, however, he put that date at December 1610 (*GO*, vol. XI, p. 305). Later on, in the *Dialogue on the Two Chief World Systems,* he claimed to have observed the spots when he was still at Padua, that is, before September 1610 (*GO*, vol. VII, p. 372). Favaro discusses these various statements in "Sulla priorità della scoperta e della osservazione delle macchie solari," *Memorie del Reale Istituto Veneto di Scienze, Lettere, ed Arti* 13 (1887): 729–90.

53. Maria Luisa Righini Bonelli, "Le posizioni relative di Galileo e dello Scheiner nelle scoperte delle macchie solari nelle pubblicazioni edite entro il 1612," *Physis* 12 (1970): 405–10. On Kepler's early observations of sunspots see Massimo Bucciantini, *Galileo e Keplero* (Turin: Einaudi, 2003), pp. 214–15.

could either prove he was the first to observe them.[54] The first to go to print about the discovery was the German Johannes Fabricius. His specific observations were not dated, but Fabricius claimed to have made his first observation on March 9 and to have completed his book on June 13, 1611.[55]

Based on the observations he conducted in the fall of 1611, Scheiner wrote three letters to Mark Welser, a politically influential patrician of Augsburg and a patron of the Society of Jesus.[56] In January 1612, Welser printed Scheiner's letters as the *Tres epistolae de maculis solaribus*. The two had corresponded before about other telescopic discoveries, and Scheiner was eager to pursue the connection. Described as ambitious (and being once reproached by his superiors for that reason), Scheiner had already perfected his patronage skills with Duke William V of Bavaria and knew how much curious instruments, objects, and discoveries were appreciated in aristocratic circles.[57] Since hearing about the new telescopic discoveries,

54. Most likely this was a case of multiple independent discovery. Telescopes had become quickly available since late 1610 or early 1611, and there had been so many discoveries about so many different planets (Moon, Jupiter, Venus, Saturn) that it would have occurred to many people to aim a telescope to the Sun. As Welser himself put it in March 1612, "One should not think it a novelty that in natural philosophy one can find various inventors, each of them ignorant about the other" (*GO*, vol. XI, p. 282). Slow means of communication and book distribution and especially the lack of a standardized concept of what counted as an acceptable priority claim fueled the practitioners' sense of priority.

55. Johannes Fabricius, *De maculis in sole observatis et apparente earum cum Sole conversione . . .* (Wittemberg: Typis Laurentii Seuberlichii, 1611). Substantial portions of Fabricius' booklet are reproduced in Antonio Favaro, "Sulla priorità della scoperta e della osservazione delle macchie solari," pp. 767–76. Favaro argues that Fabricius' text was written in the middle of June 1611 (p. 777). However, it does not seem to have been distributed until the Frankfurt book fair in the fall.

56. On Welser see R. J. W. Evans, "Rantzau and Welser: Aspects of Later German Humanism," *History of European Ideas* 5 (1984): 257–72; Antonio Favaro, "Sulla morte di Marco Velsero e sopra alcuni particolari della vita di Galileo," *Bullettino di bibliografia e storia delle scienze matematiche e fisiche* 17 (1884): 252–70; Giuseppe Gabrieli, "Marco Welser Linceo augustano," *Rendiconti della Reale Accademia Nazionale dei Lincei, Classe di Scienze Morali, Storiche e Filologiche*, 6th ser., 14 (1938): 74–99.

57. The standard but limited biography of Scheiner is Anton von Braunmuehl, *Christoph Scheiner als Mathematiker, Physiker, und Astronom* (Bamberg: Buchnersche Verlagsbuchhandlung, 1891), but see also Antonio Favaro, *Oppositori di Galileo, III: Cristoforo Scheiner* (Venice: Ferrari, 1919), and, for the later period of his life, Franz Daxecker, *Briefe des Naturwissenschaftlers Christoph Scheiner SJ an Erzherzog Leopold V von Österreich Tirol 1620–1632* (Innsbruck: Publikationsstelle der Universität Innsbruch, 1995). The reproach from the general came later in his life, when Scheiner was perceived

Scheiner had been working at building telescopes to test the claims of Galileo and others.

The brevity and fragmentary nature of Scheiner's three letters on sunspots testifies to his concern about establishing his priority over the discovery. One of the letters had been written in less than an hour, and the entire book filled a meager twelve printed pages.[58] It came off the press ten days after the completion of the third and last letter.[59] Scheiner seemed so concerned with priority that he sent a fourth letter to Welser on January 16, 1612 (only eleven days after the publication of the *Tres epistolae*), asking him to print this new letter as soon as he could:

> I have sent this letter, which has matured for a long time [*sic*], to you especially as a matter of priority, so that [. . .] you will preserve undiminished this glory of our Germany and your Augsburg, which I trust can be done if the publication is in no ways delayed [. . .] Hence I fear that, unless you will anticipate them, they will almost be forced from our hands.[60]

Scheiner did not write the *Tres epistolae* as a book but, quite literally, as three letters. These letters were not installments of a book-length argument, but time-sensitive periodical reports of his work and findings. Each was dated so as to register the time of the claim or discovery and was addressed to an internationally known figure who could testify to having received those letters at that time. Welser was Scheiner's publisher, but he also functioned as the register of his discoveries (a role not unlike that assumed by the Royal Society later in the century). Between January and July 1612, Scheiner wrote three more letters on sunspots that Welser then published as the *De maculis solaribus et stellis circa Iovem errantibus accura-*

---

as overextending his duties as archducal confessor to include those of political advisor (Gorman, "A Matter of Faith?" p. 308).

58. Scheiner's text occupies eight pages (illustrations excluded) in the version reissued together with Galileo's *Istoria e dimostrazioni* (Rome: Mascardi, 1613). I have not been able to access the original Augsburg edition of his work. Scheiner's mention of having written one of the three letters in less than one hour is at the very beginning of the *Accuratior* (*GO*, vol. V, p. 39).

59. Scheiner's last letter to Welser is dated December 26, 1611. The book came off the press on January 5, 1612.

60. Christoph Scheiner, *De maculis solaribus et stellis circa Iovem errantibus, accuratior disquisitio ad Marcum Welserum . . .* (Augsburg: Ad insigne pinus, 1612), in *GO*, vol. V, pp. 53–54.

tior disquisitio in September 1612. As the pace of Scheiner's output exceeded the speed of scholarly communication, the comments on the *Tres epistolae* that Welser had gathered from various European practitioners (including Galileo's first letter on sunspots) reached Scheiner only as he was writing the third and last letter of his second book.

The fragmentarity of Scheiner's arguments and the somewhat hurried nature of his sunspots illustrations was, therefore, anything but accidental. His simultaneous concern for gaining priority credit while maintaining cosmological orthodoxy was inscribed in his argument. First, he sought to establish the sunspots as real objects, not artifacts produced by the telescope, the eyes, or meteorological conditions. He reported observing the spots with eight different telescopes to ensure they were not optical illusions. He tried "turning and moving the tubes back and forth," but even these drastic interventions "never moved the spots along with the tubes, which ought to happen if the tube produced this phenomenon."[61] He also had several witnesses confirm his observations.[62]

Fearing perhaps that less pious minds could use his discovery as evidence for the existence of change and corruption in the Sun—a claim that was opposed by the Jesuit theologians and philosophers—he stated that the spots were not on the solar surface. Scheiner had already taken a stance against the possibility of corruptibility in the heavens in 1610, when, with other Jesuits, he disputed Galileo's claims about the irregularities of the lunar surface.[63] It seemed impossible to him to admit that the Sun could dis-

61. *GO*, vol. V, p. 26.

62. *GO*, vol. V, p. 25.

63. When the mathematicians of the Collegio Romano were asked by Cardinal Bellarmine to assess Galileo's discoveries, they disagreed over how to interpret the irregularity of the lunar surface—the finding that was most directly related to the issue of celestial change. All of them agreed that the Moon looked rough, but while some of the younger mathematicians thought that its rough appearance reflected actual topographical features, Clavius did not draw that conclusion: "non si può negare la grande inequalità della Luna, ma pare al P. Clavio che più probabile che non sia la superficie inequale, ma più presto che il corpo lunare non sia denso uniformemente, et che habbia parti più dense, et più rare, come sono le macchie ordinarie, che si vedono con la vista naturale. Altri pensano essere veramente inequale la superficie: ma infin hora noi non habbiamo intorno a questo tanta certezza, che lo possiamo affermare indubitatamente" (Christopher Clavius et al. to Cardinal Bellarmine, April 24, 1611, in *GO*, vol. XI, pp. 92–93). The hypothesis about the transparent sphere was proposed by the Florentine Aristotelian Ludovico Delle Colombe to Clavius in a May 27, 1611, letter (*GO*, vol. XI, p. 118). We know that Scheiner shared

play even darker spots or, even worse, to provide ammunition for those who may have wanted to say that change in the Sun could justify the presence of corruption on the lunar surface:

> It has always seemed to me unfitting and, in fact, unlikely, that on the most lucid body of the Sun there are spots and that these are far darker than any ever observed on the Moon [. . .] Moreover, if they were on the Sun, the Sun would necessarily rotate on its axis and cause them to move, and those seen first would at length return in the same arrangement and in the same place with respect to each other on the Sun. But so far they have never returned, yet successive new ones have run their course across the solar hemisphere visible to us. This proves that they are not on the Sun. Indeed, I would judge that they are not true spots but rather bodies partially eclipsing the Sun from us and are therefore stars.[64]

According to Scheiner, the spots were not dark stains on the solar surface but the shadows of opaque bodies close to the Sun's surface. What people observed as black spots were, in fact, partial eclipses of the Sun. Scheiner had to pull the spots out of the Sun to defend the incorruptibility of the heavens, but taking that step significantly narrowed the range of acceptable conceptual boxes in which he could place the sunspots. Identifying the spots as stars, therefore, was almost an obligatory move.[65]

The second letter dealt with yet a new discovery, this one about the or-

---

Delle Colombe's opinion about the Moon. On January 7, 1611 (before the sunspots debate), Welser wrote Galileo that an unnamed friend of his (Scheiner) did not believe that the lunar surface was physically irregular (*GO*, vol. XI, pp. 13–14). Other debates concerning the Moon's surface involved Jesuits, like Giovanni Biancani at Mantua (*GO*, vol. XI, pp. 126–27, 130–31; vol. III, pt. 1, pp. 301–7). On the broader philosophical and religious implications of claims about the irregularities of the Moon, see Eileen Reeves, *Painting the Heavens: Art and Science in the Age of Galileo* (Princeton: Princeton University Press, 1997), pp. 138–83.

64. *GO*, vol. V, p. 26.

65. Another implication Scheiner wanted to control by denying the existence of spots on the solar surface was the rotation of the Sun on its own axis. If the spots were found to be on the Sun (not just very close to it) then their motion across the solar disk would be strong evidence of the Sun's own rotation—another sign of celestial change. Scheiner's anxieties would have substantially increased had he realized that the Sun's velocity of rotation itself changed in time and it was going through a substantial acceleration in the years Scheiner was observing it (John Eddy, Peter Gilman, and Dorothy Trotter, "Anomalous Solar Rotation in the Early 17th Century," *Science* 198 [1977]: 824–29; Richard Herr, "Solar Rotation Determined from Thomas Harriot's Sunspots Observations of 1611 to 1613," *Science* 202 [1978]: 1079–81).

bit of Venus. Scheiner seemed strangely unaware of both Galileo's and the Roman Jesuits' observations of the phases of Venus at the end of 1610—a discovery that was commonly taken to imply that Venus orbited the Sun. Scheiner came to that conclusion too, though he claimed to have arrived at it not by observing the phases of Venus but by trying (and failing) to observe that planet's transit across the solar disk. He read his failure to observe Venus' transit as unequivocal evidence that at that time Venus must have been behind the Sun—a position it could have occupied only if its orbit were centered on the Sun, not the Earth.[66] It is virtually impossible, however, that Scheiner had not heard of the discovery of the phases of Venus either from Welser (who was in continuous correspondence with Clavius) or from his former mathematics teacher, Johann Lanz, who had been receiving letters from the Collegio Romano about the progress in telescopic astronomy since early 1611.[67] Furthermore, Kepler's 1611 *Dioptrice* reported Galileo's post-*Nuncius* discoveries, even reproducing the letters in which Galileo communicated his findings to Kepler.[68]

I believe Scheiner remained silent about his knowledge of the phases of Venus in order to defend or constitute a small priority claim. He was trying to create the impression that he had independently discovered the true orbit of Venus, and that he had reached that conclusion through evidence different from that used by Galileo and the Roman Jesuits.[69] He could, in

66. The orbit of Venus came up as a topic because Scheiner claimed that if Venus went around the Earth, it should have shown itself against the solar disk when at upper conjunction. But when Scheiner tried to observe what he expected to be a partial eclipse, he saw nothing. That led him to believe that Venus at that time was on the other side of the Sun.

67. A student of Clavius in Rome, Paul Guldin, sent a long letter to Lanz at the Jesuit college in Munich on February 13, 1611 detailing the corroboration of Galileo's discoveries by his group. The letter is in August Ziggelaar, "Jesuit Astronomy North of the Alps: Four Unpublished Jesuit Letters, 1611–1620," in Baldini (ed.), *Christoph Clavius e l'attività scientifica dei gesuiti nell'età di Galileo* (Rome: Bulzoni, 1995), pp. 117–21. Lanz reported on Guldin's letter to Tanner at the Jesuit college at Ingolstadt, adding "I would like to have these things also communicated to Father Scheiner and others who are interested in these things" (Lanz to Tanner, March 1, 1611, Graz, Universitätsbibliothek, MS 159, no.17, p. 2 [trans. Albert van Helden]). Scheiner's silence about the phases of Venus in the *Tres epistolae* is all the more puzzling given that, in a January 16, 1612, letter to Welser (written only three weeks after completing the *Tres epistolae*) he did mention both Galileo's and the Roman Jesuits' observations of the phases of Venus (*GO*, vol. V, p. 46).

68. Johannes Kepler, *Dioptrice* (Augsburg: Franci, 1611), in *KGW*, vol. IV, pp. 344–54.

69. This reading is confirmed by Scheiner's own discussion of the discovery of the orbit of Venus at the end of the *Accuratior*. There he lists his "discovery" of the heliocentric orbit

a sense, claim priority over the method and evidence, if not over the finding itself.[70] Unfortunately, the alleged discovery turned into an embarrassment when Scheiner realized that he had misread the time of the conjunction of Venus and the Sun in the tables he was using.[71]

Scheiner went back to sunspots in his third letter, reporting that "at most, they spend no more than fifteen days on the Sun." That, however, did not mean that the same spot came full circle in about thirty days: "as is apparent from a course of observations of about two months, no spot has returned to the same place and arrangement." Surprisingly, he took this to confirm that "it is impossible that any spot is on the Sun."[72] Scheiner's con-

---

of Venus first although it is clear from his own narrative that he did not mean to claim to have been the first to have discovered it: "For if Venus goes around the Sun, as was made known in the first painting by Apelles, and gradually established from its daily transformations, and as Tycho Brahe taught some time ago, and as the Roman mathematicians and Galileo observed at about the same time . . . " (*GO,* vol. V, p. 69).

70. Scheiner's competitiveness seemed directed to his own senior colleagues in Rome as much as toward lay astronomers like Galileo. His ambition to be treated on a par with the Society's mathematical elite in Rome (and his displeasure with their criticism) comes through in his correspondence. He tried to walk a thin line. As a Jesuit, he was not supposed to let his private ambition override his complete allegiance to the order. Therefore, his discoveries were expected to support the Society's interests, not his personal fame as an astronomer. But there was always a slippage between the two ethoses. For instance, in a letter to Paul Guldin in March 1613, Scheiner distinguished priority credit within and without the Society. He cast his discoveries as his own within the boundaries of the order, but as belonging to the order when they were presented to people outside of the Society: "It never came to my mind to be afraid that you would appropriate mine for yourselves. But what I feared has happened, and therefore I was eager to guard against others rushing into the harvest—not mine (for it is mine inside the Society) but ours." The distinction between intra- and extra-Society credit reemerges later on in the same letter: "I write this, Paul, not out of ambition—I have none—but so that we will disclose our [discoveries] to others as *ours,* for in this way ten times more esteem and honor will accrue to the Society in the eyes of others, I beg that on this score you do not fail me, and that if you deem it worthy of light you rescue it from obscurity" (Scheiner to Guldin, March 31, 1613, Graz, Universitätsbibliothek, MS 159, fasc. 1, no. 3 [trans. Albert van Helden]). The context of the letter, however, shows that Scheiner was trying to enlist the Roman Jesuits to help him establish his priority claims. Scheiner's emphasis on "our discoveries," therefore, may have been something of a carrot to mobilize Clavius' students' support so that he could establish those discoveries as his own.

71. He tried to control the damage from this (rush-induced?) mistake by acknowledging that the second letter was "incomplete and not perfect" in a postscript that, however, could not be included in the original printing (*GO,* vol. V, p. 32). Welser had this short addendum printed and then mailed to the people to whom he had previously sent the *Tres epistolae.*

72. *GO,* vol. V, p. 29.

clusion could be easily undermined by considering that the Sun might not be a rigid body, or that the spots could change so much in fifteen days so as to be unrecognizable when they came around again.[73] That he ignored those possibilities suggests that he could conceptualize the presence of spots in the Sun only as fixed features on a rigid planet, like the topographical features of the Moon—a phenomenon he much opposed.[74]

Confident of having demonstrated that the spots could not be on the solar surface, in the sublunary sphere, or in the orbs of the Moon, Mercury, or Venus, Scheiner concluded that "what remains is that these shadows must revolve in the heaven of the Sun."[75] Having then ruled out that the spots could be either clouds—"who would assume clouds there?"—or comets, Scheiner stated that they must be "solar stars."[76] He then addressed one aspect of the sunspots' behavior that could have been easily turned against his claim that they were above the solar surface. According to Scheiner's hypothesis, when the solar stars reached the limb of the Sun, they should have been observable as distinct bodies located right next to the Sun, not on it. Instead, observations showed them very flat (not round) and extremely close to the solar surface (if not actually on it). That is, they looked exactly like spots would have looked, due to foreshortening, had they been on the Sun.

To control this serious anomaly, Scheiner drew an analogy between the appearance of his solar stars and that of Venus as it went through its phases.[77] He argued that the intense sunlight reflected by one side of the spots made it so bright that it became virtually indistinguishable from the nearby Sun. Under those circumstances all we can observe is, at best,

73. Galileo remarked on the circularity of Scheiner's argument in his first letter to Welser (*GO*, vol. V, p. 101).

74. He made this assumption explicit in the later *Accuratior:* "Sole invariabili et duro posito, sive rotetur interim sive non" (*GO*, vol. V, p. 49; see also p. 64). This suggests that Scheiner conceptualized new objects in terms of old objects not only when he was trying to figure out what they were (sunspots as stars or satellites) but also when he was thinking about what they *could not be* (sunspots like spots on the Moon). As shown by his correspondence with Welser in 1610 (prior to the sunspots debate), Scheiner did not believe there were mountains and valleys on the Moon. At the same time, he seemed to assume that the only way someone could think of the spots on the Sun would be by analogy to the irregularities of the lunar surface.

75. *GO*, vol. V, p. 29.

76. *GO*, vol. V, p. 30.

77. Scheiner, however, did not acknowledge the analogy in this first text, perhaps to pretend he did not know about the phases of Venus.

the darker half of the solar stars.[78] The perceived flattening of the spots toward the limb was, in fact, an optical effect of the solar satellites going through phases. Scheiner was using two of Galileo's previous discoveries to support his attempt to explain the sunspots in a cosmologically safe manner: the satellites of Jupiter (at least as a model of starlike bodies going around other planets), and the phases of Venus (as a way to explain the effacement of the solar stars). By doing so, he was also constructing the sunspots as physically real—objects he should receive credit for discovering.

Scheiner's use of analogies and models kept oscillating between a defensive function (the normalization of discoveries) and a proactive one (the maximization of his credit as a discoverer or, as we will see, even as a predictor of future discoveries). He started by modeling the sunspots after the satellites of Jupiter in order to domesticate them, but he then redirected the analogy in other directions, expanding and turning it into a distinctly proactive tool.[79] Having argued that the sunspots were solar satellites that became visible when they crossed over the solar disk (also like the transit of Venus he had tried, and failed, to observe), it seemed only natural to him to see them as exemplars of a much larger category of yet undiscovered objects orbiting the Sun:

> It occurs to me to think that from the Sun all the way to Mercury and Venus, at proper distances and proportions, very many wandering stars turn, of which to us only those have become known whose motions fall in with the Sun.[80]

This may explain Scheiner's quizzical statement at the end of the *Tres epistolae* that "the Sun will also give signs; who would hear the Sun speak a falsehood?"[81] Far from being corrupted, the Sun functioned as a "detector" of new astronomical bodies that it made visible by projecting their shadows toward the observer. The Sun's apparent stains were actually projections of true knowledge. Like the Sun turned from a stained corrupted body into a knowledge projector, Scheiner's defensive analogies turned into proactive ones.

78. *GO*, vol. V, pp. 30–31.

79. "It is also consistent that the companions of Jupiter are by no means of an unlike nature as far as motion and place is concerned" (*GO*, vol. V, p. 31).

80. *GO*, vol. V, p. 31.

81. *GO*, vol. V, p. 32.

This was not the end of Scheiner's discovery forecast. Inverting the analogy between the satellites of Jupiter and the sunspots, Scheiner assumed that what held true for the sunspots could apply to the satellites as well. If the sunspots' orbital planes around the Sun were not limited to the solar equator, most likely the satellites of Jupiter would do the same. People had been looking for them around Jupiter's equator, but they may be in for a surprise because "it [is] almost certain that of these [satellites of Jupiter] there are not just four but many, and not carried just in one circle but many."[82] (Indeed, he proceeded to discover a fifth satellite that did not disappear quickly enough to prevent him from dedicating it to Welser's family).

A last example of Scheiner's turning defensive analogies into proactive ones concerns the phases of Venus. After using them as a model for the changing appearance of the solar stars as they approached the solar limb, he presented the "thinning out" of the solar satellites as a way to reduce other well-known puzzles like the changing appearances of Saturn to optical illusions: "I am not altogether afraid to believe something similar about Saturn, namely that it appears at one time of an oblong shape and at other times accompanied by two lateral stars touching it."[83] Nothing cosmologically strange was happening around Saturn.

Scheiner's analogies did not work like a conceptual scheme that directs its believers toward other discoveries (the way a Copernican would tend to look for the phases of Venus). The discoveries foretold by Scheiner could bring credit but were not embedded in any specific cosmological paradigm. More satellites of Jupiter, more bodies between the Sun and Venus, an explanation of the appearance of Saturn as an optical effect were by no means banal claims, but neither would they have provided crucial evidence to either cosmological camp. That's why, I believe, Scheiner was so eager to predict them.

The problem with Scheiner's analogy-based thought style was that it had a tendency to multiply patterns of similarity while expanding their scope. His initial analogical impulse was a conservative one, but the play of analogies (coupled with his desire for recognition) could quickly lead him to make increasingly sweeping claims that, while apparently safe, had risky implications. Scheiner's tendency to think fast and publish even faster further complicated things.

82. *GO*, vol. V, p. 31.

83. *GO*, vol. V, p. 31.

THE RISKS OF RUSHING

Like Galileo in 1610, Scheiner needed to publish as quickly as possible to claim priority. As a Jesuit, however, he was required to submit his book manuscript to an internal review committee composed of mathematicians, philosophers, and theologians.[84] It could take months.[85] In addition, the Society was going through a period of doctrinal retrenchment, making it unlikely that the theologians and philosophers among the reviewers would have approved Scheiner's manuscript.[86] Nor is it obvious that the hurried and fragmentary character of his text would have gained him much support among the mathematicians of the Collegio Romano either.[87]

It was in this context that his connection with Welser, and Welser's willingness to publish his letters, made a big difference. Following the orders of his superiors at Ingolstadt who worried about the controversial nature of the topic, Scheiner published pseudonymously as "Apelles latens post tabulam."[88] By having Scheiner publish pseudonymously and without mentioning his institutional affiliation, his superiors, willing to please a powerful patron like Welser, allowed the *Tres epistolae* to be printed with-

84. Baldini, *Legem impone subactis,* pp. 75–119.

85. The central censorship board was in Rome, with satellite boards in the various provinces. Had it gone through a normal review, Scheiner's manuscript could have been sent to the censors of the German Province. But given its controversial topic, it could also have been sent to Rome.

86. On the Society's general renewed call to doctrinal orthodoxy in these years, see Richard Blackwell, *Galileo, Bellarmine, and the Bible* (Notre Dame: University of Notre Dame Press, 1991), pp. 135–40. Additionally, I doubt that, given the preliminary and rushed nature of Scheiner's text, the mathematicians of the censoring board would have gone to war with the philosophers and theologians in the committee to get the book published. Baldini's study of the *censurae librorum* shows that the mathematicians on the board were careful at picking their battles (Baldini, *Legem impone subactis,* pp. 217–50).

87. Although much less controversial, Scheiner's next book, the 1615 *Sol ellipticus,* was indeed reviewed in manuscript by Grienberger and Guldin in Rome. They requested a number of changes. A 1614 letter from Scheiner and Guldin suggests a certain strain between the Roman mathematicians and their colleague in Ingolstadt (Ziggelaar, "Jesuit Astronomy North of the Alps," pp. 104–5, 122–27).

88. This is a pun on the fact that, like Apelles who hid behind his paintings to hear what people had to say about his work, Scheiner hoped to elicit more candid comments from his readers by not revealing his identity. Scheiner discussed the reasons behind his use of pseudonymity in his later *Rosa Ursina* (Bracciano: Apud Andream Phaeum, 1630), bk. 1, chap. 2, pp. 6–7.

out undergoing the internal review and censoring process.[89] The months Scheiner saved by bypassing the review process strengthened his priority claim over the discovery of sunspots, but they also deprived him of feedback from the mathematicians at the Collegio Romano. For instance, no astronomer (Jesuit or otherwise) or publication concerning the first wave of telescopic discoveries was mentioned in the *Tres epistolae,* giving an impression of either ignorance or ungenerosity.[90] Feedback from senior Jesuit mathematicians could have fixed that problem as well as the blunder concerning the transit of Venus. It could have also given him some pointers on how to align his cosmological claims with the positions being entertained in Rome at that time—a mismatch that was to play a key role in the development of Scheiner's claims.

This was an intense period for Clavius and his group. The discovery of the phases of Venus just a few months earlier had undermined Clavius' defense of Ptolemaic astronomy, while the observations of astronomical "monsters" (as he termed new phenomena like the irregular surface of the Moon, the nonspherical appearance of Saturn, the satellites of Jupiter, and most recently the sunspots) were challenging traditional Aristotelian cosmology.[91] These developments reconfigured not only the Jesuit mathematicians' own positions, but also their delicate relation to the philosophers and theologians within the order.[92]

89. In a January 17, 1612, letter to Guldin in Rome, Scheiner relayed that "Mr. Welser has prevailed on the Father Provincial that he might publish it without the Society or the name of any of us mentioned [. . .] You in Rome may not reveal Apelles hiding behind the painting, for it would not please the superiors, but neither does Apelles himself desire it" (Scheiner to Guldin, Graz Universitätsbibliothek, MS 159, 1, 1, cited in van Helden, "Scheiner").

90. The exception is Giovanni Antonio Magini, who was mentioned only as the author of the tables (incorrectly) used by Scheiner to find the time of conjunction of Venus and the Sun.

91. "And thus I believe that gradually other *mostrosità* about the planets will be discovered" (Clavius to Welser, January 29, 1611, Archivio Pontificia Università Gregoriana 530, cc. 183r–184v). The topos of "monster"—*fabulosa monstrorum prodigia*—was also used in a public lecture at the Collegio in February 1612 to refer to the continuing emergence of new astronomical objects (*GO*, vol. XI, p. 274).

92. In the *Nuntius sidereus Collegii Romani,* the oration given in Rome in honor of Galileo during his visit in the spring of 1611, Oto Maelcote (Clavius' student) subscribed to Galileo's conclusion that the phases of Venus showed that it went around the Sun. There is no evidence that Maelcote was deviating from the consensus position within Clavius' group. An observer, however, reported that the philosophers were less pleased, saying that

Scheiner's near-explicit endorsement of Tycho's planetary model shows that he seriously underestimated the amount of work and negotiations still required to get the Society to accept those positions.[93] And his cheerful prediction of the discovery of more new astronomical objects did not jibe with the Society's increasing conservatism on philosophical matters. While he seemed aware of the scope and stakes of his claims, he did not seem to understand how much the hurried nature of his text (not to mention his reluctance to cite anyone who was anyone in telescopic astronomy) could have turned him into an easy target for theologians and mathematicians alike.[94]

As soon as the *Tres epistolae* came off the press, Welser sent copies to various European astronomers and savants asking for their comments.[95] Galileo was just one of them. Welser's request initiated an exchange that produced three letters from Galileo and a second publication from Scheiner. Galileo's letters were eventually collected and printed in 1613 in Rome by the Accademia dei Lincei as *Istoria e dimostrazioni intorno alle macchie solari.*[96] Half of the edition included a reprint of Scheiner's two previous publications. As the small Augsburg editions of the *Tres epistolae* and the *Accuratior* went immediately out of print, most people became familiar with Scheiner's arguments only by reading them as an appendix to Galileo's

---

Clavius' students had demonstrated "to the scandal of the philosophers, that Venus circles about the Sun" (Lattis, *Between Copernicus and Galileo,* pp. 193–94, 197). Clavius' terminal illness—he died in February 1612 right as Scheiner's *Tres epistolae* were arriving in Rome—made things more complicated by adding a generational shift to an already delicate scenario.

93. At the end of the *Accuratior,* Scheiner said, "And therefore Christopher Clavius, the choragus of mathematicians of his age, should rightly and deservedly be heeded. In the final edition of his works he warns astronomers that, on account of such new and hitherto invisible phenomena (although it is a very old problem), they must unhesitatingly provide themselves with another system of the world" (*GO,* vol. V, p. 69). But in the 1611 edition of his textbook on Sacrobosco's *Sphaera,* after referring to Galileo's discoveries, Clavius made the much more modest and conservative claim that "since things are thus, astronomers ought to consider how the celestial orbs may be arranged to save these phenomena" (Clavius, *Sphaera* [1611], p. 75, cited in Lattis, *Between Copernicus and Galileo,* p. 198).

94. It is difficult to gauge the reaction of the mathematicians at the Collegio Romano to the *Tres epistolae,* except that they were probably critical in private and supportive in public. See note 153 below.

95. *GO,* vol. V, p. 93.

96. Galileo Galilei, *Istoria e dimostrazioni intorno alle macchie solari e loro accidenti . . .* (Rome: Mascardi, 1613), reprinted in *GO,* vol. V, pp. 72–249.

book. This bit of publication history greatly contributed to constituting Galileo and Scheiner's texts as a dispute—an effect that has since been canonized by the historiography of science.

Welser's letter made an impression on Galileo. Galileo did not rush to answer Welser's letter but included a few lines about *his* discovery of the sunspots (but no mention of Scheiner's *Tres epistolae*) in the manuscript of his *Discourse on Floating Bodies,* which was licensed on April 5, 1612.[97] When he eventually wrote back to Welser on May 4 Galileo apologized for the delay, saying he could say little about the spots, and certainly nothing definitive, without conducting more systematic observations.[98] He endorsed Scheiner's claim that the spots were neither meteorological phenomena nor optical illusions produced by the telescope, and agreed with his observation of how complicated the motions of the spots appeared to be.[99] Galileo had no problem with all these claims because, as he told

97. The *Discorso intorno alle cose che stanno in su l'acqua, o che in quella si muovono* (Florence: Giunti, 1612) came off the press toward the end of May. Its licensing process lasted from March 5 through April 5 (*GO,* vol. IV, p. 141). Welser's letter to Galileo was dated January 6, and by early February Galileo had heard of Scheiner's discoveries from other friends as well. So he would have had the time to add a mention of the sunspots in the manuscript before he applied for a license. In the book Galileo claimed that "aggiungo a queste cose l'osservazione d'alcune macchiette oscure, che si scorgono nel corpo solare, le quali mutando positura in quello, porgono grand'argomento, o che il Sole si rivolga in se' stesso, o che forse altre stelle, nella guisa di Venere e di Mercurio se gli volgano intorno, invisibli in altri tempi, per le piccole digressioni e minori di quella di Mercurio, e solo visibili, quando s'interpongono tra il Sole e l'occhio nostro, o pur danno segno, che sia vero e questo e quello; la certezza delle quali cose non debbe disprezzarsi, o trascurarsi" (*GO,* vol. IV, p. 64). This shows that although Galileo had already observed the sunspots since 1611 or perhaps 1610, he had no clear position about them. It also indicates that Galileo had read Scheiner's *Tres epistolae* and acknowledged, as a hypothesis, the possibility of the sunspots being solar satellites. He did not, however, mention Scheiner's publication in order, I believe, not to support his priority claim. In the second edition of the book (published toward the end of 1612) he added a paragraph to the effect that, having conducted further observations, he had concluded that spots were contiguous to the solar surface and thus not stars (*GO,* vol. IV, p. 64).

98. Galileo started to observe the spots on a quasi-regular basis only after he got Welser's letter. The first observation recorded is from February 12, 1612. If he did observe the spots regularly before then, the records are lost. The first set of observations stops on May 3. Very likely, this is when Galileo switched to Castelli's observational apparatus, which I describe later in this chapter. Between February 12 and May 3, Galileo recorded only twenty-three observations (*GO,* vol. V, pp. 253–54).

99. *GO,* vol. V, p. 95. Although Galileo always refers to "Apelles" in his letters to Welser, I have decided to drop Scheiner's pseudonym throughout this chapter to avoid confusion.

Welser, he had been observing the spots and showing them around for the last eighteen months.[100] Nor did he have any qualms about Scheiner's claims about the orbit of Venus because, as he reminded Welser, he had reached the same conclusion "almost two years ago." But he was surprised that "it has not come to his ears, or if it has that he has not relied on the most exquisite and judicious means that can often be used, discovered by me about two years ago and communicated to so many that by now it has become well known; and this is that Venus changes its shapes in the same way as the Moon."[101] He must have been equally surprised not to see his name or work ever mentioned in the *Tres epistolae* despite the fact that it discussed some of his other discoveries.[102]

Concerns about priority and credit colored the interaction between Scheiner and Galileo since its inception. While Galileo argued that observation and verbal communication were sufficient to determine priority, Scheiner continued to stress (not unreasonably) a link between priority and publication. Galileo grew particularly vocal about his priority claims in the 1623 *Assayer* and in the 1632 *Dialogue on the Two Chief World Systems,* but already the *Istoria* of 1613 included a long discussion of Galileo's observations in Rome in April 1611 and the names of high-ranking witnesses.[103] Scheiner did not back down in the least and confirmed all his priority claims (and more) in a lengthy introduction to his 1630 *Rosa Ursina*—an introduction that was followed by a 800-page (folio) text on sunspots. Obviously both Galileo and Scheiner had problems letting go of the

100. *GO,* vol. V, p. 95.

101. *GO,* vol. V, p. 98.

102. Galileo's surprise is recorded in a letter to Cesi in *GO,* vol. XI, p. 426.

103. Galileo claimed his priority in the *Dialogue,* giving 1610 as the date of his first observations. He also claimed that Scheiner was eventually convinced by Galileo's interpretation of the phenomena (*GO,* vol. VII, pp. 372–73). For a synopsis of the priority dispute between Galileo and Scheiner, see Antonio Favaro, "Sulla priorità della scoperta e della osservazione delle macchie solari," *Memorie del Reale Istituto Veneto di Scienze, Lettere, ed Arti* 13 (1887), pp. 729–90. Favaro's text, however, is biased in Galileo's favor. Maria Righini Bonelli, "Le posizioni relative di Galileo e dello Scheiner nelle scoperte delle macchie solari," pp. 405–10, reviews what Galileo knew about other people's observations of sunspots as he was writing his *Istoria.* In the preface to the *Istoria,* Angelo de Filiis (Galileo's fellow member of the Accademia dei Lincei) stated that Galileo had shown the spots in various places in Rome, including the Quirinale, to Cardinal Bandini and other prelates and gentlemen. He then continued with a lengthy discussion of priority issues (*GO,* vol. V, pp. 81–88). This introduction was promoted and reviewed by Galileo.

matter. Galileo kept calling the Jesuit a "beast," a "pig," and a "malicious ass" as late as 1636.[104]

PICTURING SUNSPOTS

Scheiner and Galileo agreed that sunspots were not telescopic artifacts, but disagreed about their physical nature.[105] Claiming that the question of the spots' nature could exceed human comprehension, Galileo chose to limit himself to an apparently phenomenological discussion of their location, appearances, and motions.[106] He consequently saw Scheiner's confident claims about the spots being solar stars as implying an a priori commitment to the incorruptibility of the heavens.[107] Parenthetically, Galileo assumed a com-

104. "Ma a che metter mano a registrar le fantoccerie di questo animalaccio, se elle sono senza numero? Il porco e maligno asinone" (*GO*, vol. XVI, p. 391). For a critical reassessment of the claims about Scheiner's involvement in Galileo's trial see Gorman, "A Matter of Faith?" pp. 283–320.

105. On Galileo's and Scheiner's different uses of visual representations of sunspots see Dear, *Discipline and Experience*, pp. 100–107, and especially Albert van Helden, "Galileo and Scheiner on Sunspots: A Case Study in the Visual Language of Astronomy," *Proceedings of the American Philosophical Society* 140 (1995): 357–95.

106. "It remains to consider that which Apelles decides about the essences and substances of these spots, which is that they are neither clouds nor comets, but rather stars that revolve about the Sun. About such a decision, I confess to Your Most Illustrious Lordship that I do not yet have enough certainty to dare to establish and affirm any conclusion as certain, for I am very certain that the substance of the spots can be a thousand things unknown and unimaginable to us, and the accidents that we observe in them, that is, the shape, the opacity, and the motion, which are very common, can provide us with no, or very little, or too general information. Therefore, I do not believe that the philosopher who confesses that he does not know—and cannot know—the material nature of sunspots deserves any reproach" (*GO*, vol. V, pp. 105–6).

107. "But that they cannot be on the solar body does not appear to me to have been demonstrated with entire necessity. For it is not conclusive to say, as he does in the first letter, that because the solar body is very bright it is not credible that there are dark spots on it, because as long as no cloud or impurity whatsoever has been seen on it we have to give it the title of most pure and most bright, but when it reveals itself to be partly impure and spotted why should we not have to call it spotted and impure? Names and attributes must accommodate themselves to the essence of the things, and not the essence to the names because the things come first and the names afterwards" (*GO*, vol. V, p. 97). Scheiner had remarked that "it has always seemed to me unfitting and in fact unlikely that on the Sun, a very bright body, there are spots, and that these spots are far darker than any ever seen on the Moon" (*GO*, vol. V, p. 26).

parable stance a few years later during the dispute on comets when, as in this case, he was not the first to publish nor to put forward an interpretation of that phenomenon.[108]

Galileo's argument was structured as a critique of Scheiner and therefore of Aristotle. By tracking the spots' motions and location, Galileo tried to show that although he did not know what the spots were about, he could tell that they were not the solar stars Scheiner had claimed they were. If the spots were not solar satellites, it meant that they were on the Sun or contiguous to it. In turn, this meant that, no matter what the nature of the spots might have been, Aristotle was wrong in his fundamental assumptions about the property of the elements.[109] As he sent off his letter to Welser, Galileo wrote Cesi that

> I believe this discovery will be the funeral or perhaps the last rites of the pseudo-philosophy [. . .] I hope the mountuosity of the Moon is about to turn into a joke or a minor irritation compared to the scourge of the clouds, smoke, and vapors that are produced, moved, and dissolved continuously on the very face of the Sun.[110]

108. A discussion of Galileo's phenomenological stance during the dispute on comets is in *GC*, pp. 267–311. Perhaps in these two cases a phenomenological stance was a way to cast doubt on (and reduce the opponent's credit for) his claims without having to counter them with an alternative physical explanation. It also gave Galileo more time to come up with an alternative explanation, if any.

109. The protracted negotiations between Cesi, the censors, and Galileo leading to the publication of the *Istoria* shows that the issue of the corruptibility of the heavens was a sensitive one. As he was writing his text, Galileo asked a friendly theologian whether the corruptibility of the heavens went against scriptural teachings. When he was told by Cardinal Conti that not only did the corruptibility of the heavens not contradict the Scripture, but that it was actually closer to biblical teachings than the Aristotelian veto on it (*GO*, vol. XI, pp. 354–55, 376), Galileo went on the attack and tried to deploy the Bible against the Aristotelians. This move, however, was blocked by the censors, who required the elimination of those passages (*GO*, vol. XI, pp. 428, 437–39, 446, 453, 460, 465). One of the contested passages is reproduced in the notes in *GO*, vol. V, pp. 138–39. It seems, therefore, that the issue of corruptibility was a sensitive one for the philosophers, but the theologians objected mostly to the use of the Scripture in support of an attack on Thomistic philosophy. Welser also expressed the possibility of "various oppositions" against the publication of Galileo's text in Augsburg, were he to decide to publish there (*GO*, vol. XI, p. 361). It is not clear, however, whether Welser thought those "oppositions" were going to come from censors.

110. Galileo to Cesi, May 12, 1612, *GO*, vol. XI, p. 296. An almost identical remark is in a letter to Cardinal Barberini, *GO*, vol. V, p. 311.

Scheiner's stakes in ruling out the corruptibility of the Sun were as high as those Galileo had in proving it, though Galileo's investment in the spots grew even higher as he articulated an increasingly pro-Copernican (not just anti-Aristotelian) interpretation of these phenomena.[111]

Galileo saw the sunspots much more like nonhomogeneous, blurry, fickle, and fast-changing clouds than like satellites:

> The sunspots appear and vanish in variably short periods of time. Some shrink and draw apart greatly from one day to the next. Their shapes change, and most of them are very irregular and display varying degrees of darkness. Because they are on the solar body or very close to it, they must be very large. Because of their varying opacity, they impede the transmission of sunlight to different degrees. Sometimes many appear, other times only a few, and then again none. Now, very large and immense masses that appear and disappear in brief times, that sometimes last longer and at other times shorter, that rarify and condense, that easily change their shapes, and that display varying degrees of density and opacity cannot be found near us except for clouds.[112]

Galileo's emphasis on the exceptional mutability of the spots was reflected in the tool he was to employ so extensively in the second letter: the use of pictorial evidence.[113] He approached the study of the sunspots through the same sequential mapping of their movements he had used in the *Nuncius* to map the periods of the satellites of Jupiter and of the irregularities of the lunar surface.

Scheiner too had used pictures in the *Tres epistolae* to map out the movement and appearance of the spots (fig. 15). These illustrations were

111. Initially, the movement of the spots provided Galileo with evidence of the Sun's rotation on its own axis. He later hypothesized that such a rotation could be related to the movement of the other planets around the Sun. In the 1632 *Dialogue,* he developed a more directly pro-Copernican reading of the phases of Venus, though stating that it developed from a series of observations of an unusually large spot conducted at Salviati's villa well after 1613 (*GO*, vol. VII, pp. 373–83). The strength of Galileo's argument has been debated in Mark Smith, "Galileo's Proof of the Earth's Motion from the Movement of Sunspots," *Isis* 76 (1985): 543–51; Keith Hutchinson, "Sunspots, Galileo, and the Orbit of the Earth," *Isis* 81 (1990): 68–74; and David Topper, "Galileo, Sunspots, and the Motions of the Earth," *Isis* 90 (1999): 757–67.

112. *GO*, vol. V, p. 106.

113. This was not an afterthought because he already informed Welser in the first letter that he was going to use images of the spots' motions.

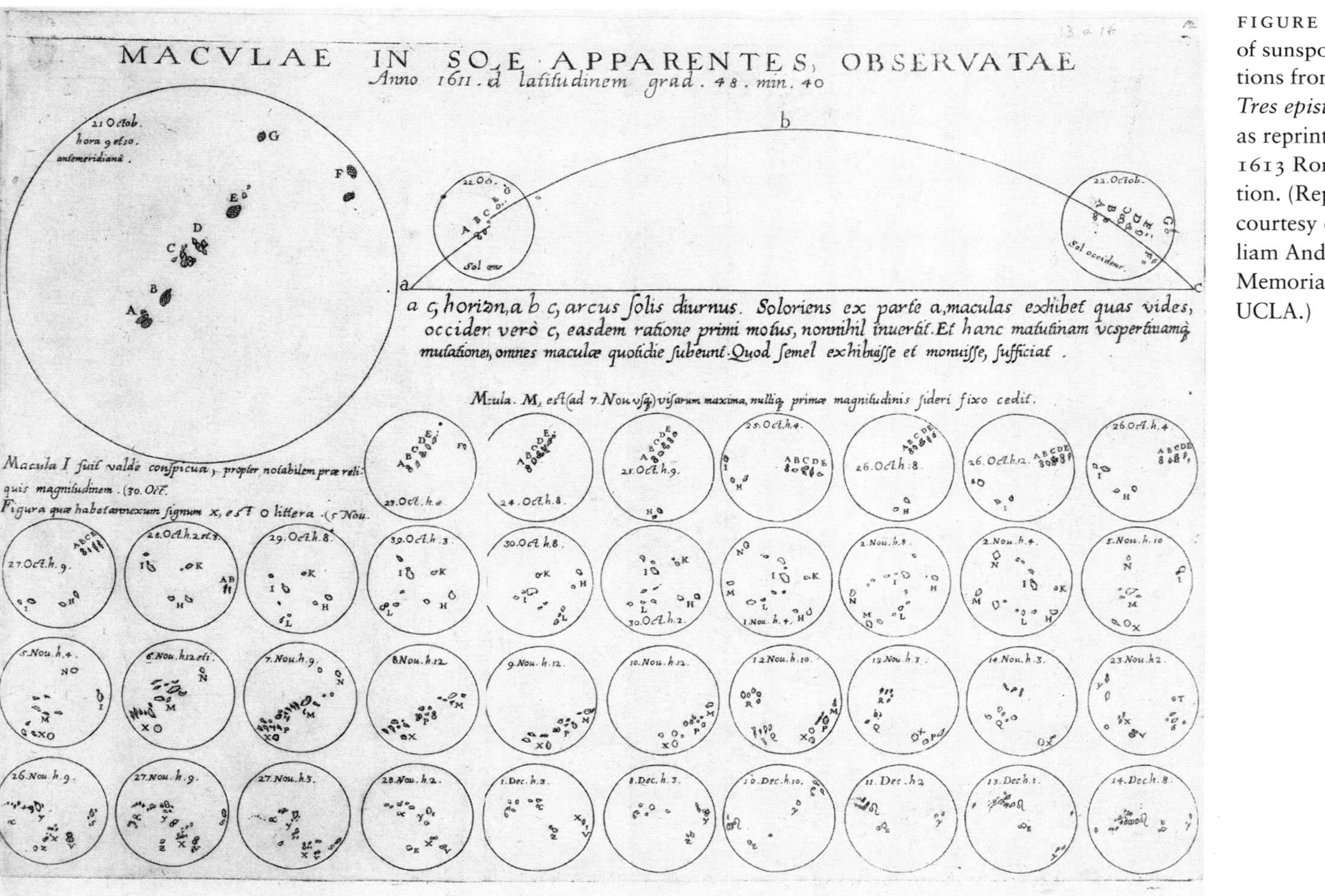

FIGURE 15. Foldout of sunspot illustrations from Scheiner, *Tres epistolae* (1612), as reprinted in the 1613 Roman edition. (Reproduction courtesy of the William Andrews Clark Memorial Library, UCLA.)

small (about one inch in diameter), crammed into one foldout page and, by Scheiner's own admission, not very accurate:

> About the observations shown, I have the following admonitions. They are not terribly exact, but rather hand drawn on paper as they appeared to the eye without certain and precise measurement, which could not be done sometimes due to the inclement weather, sometimes due to lack of time, and at other times due to other impediments.[114]

Accuracy was only part of the problem. Scheiner's one-page map of the sunspots' motions was hard to read both because of its size and because of the various conventions he had adopted while drawing and assembling the pictures:

> The more notable spots that appeared unchanged are marked by the same letters [. . .] If I added spots without letters they were either not seen constantly because of the turbulence of the air or, if they appeared constant, they did not need to be observed in comparison to the others because of their smallness. But this is to be noted as well: the proportion of the spots to the Sun should not be taken from the drawing, for I made them larger than they ought to be so that they would be more conspicuous [. . .] Frequently many small ones were conflated into one large one [. . .] The spots that always retain the same letters next to them are always the same, although they are depicted as they appeared at the time they were drawn. When some spots and their letters are no longer drawn, these had ceased at that time to appear on the Sun. But when different spots are designated by different letters, these are different newly appearing spots. When, however, spots not designated by any letters are at times represented and at times not, these either have entirely set [disappeared from the edge of the Sun] and so are not drawn, or (which happens often) they have not appeared due to thick air, since these kinds of spots only offer themselves to view when the Sun is very bright and the air very pure.[115]

Given the small size as well as the multiple codes inscribed on the spots (the lettering, scale relative to the Sun, graphical rendition in relation to their size and relative permanence, etc.), Scheiner's pictures were far from self-evident.[116] One had to study that page quite carefully to figure out how

114. *GO*, vol. V, pp. 26–7. Only one of Scheiner's illustrations was larger (86 mm diameter).

115. *GO*, vol. V, p. 27.

116. We do not know, however, if this was the original size of the pictures in Scheiner's manuscript. The foldout engraving included in the printed text was commissioned by

to read it.[117] This, however, was not the result of Scheiner's poor design skills or of his engraver's ability (which was excellent), but rather of the constraints of his observation system and, perhaps, of German winter weather.[118]

Scheiner pointed his telescope directly at the Sun and, not to be blinded, added heavily colored blue or green glass filters (but not sooted glass) between the eye and the eyepiece.[119] These filters, I believe, worsened the telescope's resolution and created more distortions.[120] These problems may have encouraged Scheiner to use filterless telescopes, trusting meteorological or atmospheric conditions to abate the Sun's luminosity.[121] For instance, he reported observing through thin clouds or at dusk or dawn.[122] In the latter case, however, the observational window shrank down to only a quar-

---

Welser, not Scheiner. Quite probably, the hand drawings and the engravings were the same size, but we cannot know for sure.

117. Van Helden, "Galileo and Scheiner on Sunspots," pp. 370–72.

118. On the effect of weather conditions, see van Helden, "Galileo and Scheiner on Sunspots," p. 378. Alexander Mayr, the engraver selected by Welser, was a well-known and accomplished artist (ibid., p. 372).

119. "The Sun can be observed everywhere through a tube equipped with a convex and concave lens and also a dark blue or green glass, plane on both sides and of the appropriate thickness, at the end that is applied to the eye. [A tube so equipped] will protect the eyes from injury even [when the Sun is] on the meridian" (*GO,* vol. V, p. 27).

120. In *A Description of Helioscopes and Some other Instruments* (London: Martyn, 1676), p. 3, Hooke strongly advised against the use of colored glasses: "The generality of the Observers have hitherto made use of either some very opacous thick Glasses next the Eye, whether of red, green, blew, or purple glass [. . .] As to the coloured Glasses, I cannot at all approve of them, because they tinge the Rayes into the same colour, and consequently take off the truth of the appearance as to Colour; besides, it superinduces a haziness and dimness upon the Figure, so that it doth not appear sharp and distinct." Given that colored glass was not part of standard astronomical equipment and that Scheiner does not mention having access to glassmaking facilities, he probably relied on glass made for ornamental purposes. It is therefore likely that his filters were quite thick, as commercial colored glass is not nearly as opaque as the smoked glass used to observe solar eclipses. Scheiner's concerns with optical glass are discussed in *GO,* vol. V, pp. 58–62.

121. Scheiner acknowledges that his colored filters could use some additional help from cloudy conditions: "even better if to this blue or green glass that is not sufficiently tempered [rendered opaque] a thin air vapor or mist is added, the Sun being wrapped as it were in shadow" (*GO,* vol. V, p. 27).

122. Under similar conditions, sunspots can be observed without telescopes, like when sunlight is abated by the smoke produced by forest fires or by dust storms (Justin Schove, "Sunspots, Aurorae, and Blood Rain," *Isis* 42 [1951]: 134).

ter of an hour.[123] It may have been difficult to draw detailed pictures of all the sunspots in such a short time.

Taken together, these constraints produced frequent gaps in Scheiner's observational record—gaps that further disrupted the flow of his visual narrative. Then, being produced at different times of the day, Scheiner's pictures reflected the observer's changing positions due to the Earth's daily rotation. As a result, the spots appeared to be aligned differently in different pictures—changes that were particularly confusing when comparing observations conducted at dusk with those made at dawn (see, for instance, the smaller images in the bottom half of fig. 15). The juxtaposition of pictures with different orientations added more confusion to an already complicated display.[124] The limitations of his system of observation imposed more constraints on the kind of visual narratives he could tell, which in turn hampered the intelligibility of those narratives.[125] The compounded effect was that he could not provide a viewer-friendly "movie" of the spots' motions, emergence, and disappearance over several consecutive days.

It is unclear whether Scheiner, driven by the assumption that the spots could not be anything but satellites, was not eager to produce the kind of images and visual narratives that could have provided evidence against his cosmological beliefs. What is clear is that Scheiner's small and viewer-unfriendly pictures did not weaken his argument.[126] While Galileo relied on visual representations of the spots that could highlight their complicated motions and metamorphoses, Scheiner benefited from simplifying the visual complexity of the sunspots by making them look like dots or patterns that appeared to be reducible to planets. On the contrary, Galileo needed to make the spots look as mutable as possible so that no conceivable arrangement of satellites could seem able to simulate their appearance. Schei-

123. "Near the horizon, the morning and evening Sun can be observed for the fourth part of an hour without any danger whatsoever with a simple [filterless] tube (but a good one) when [the sky] is cloudless and clear" (*GO*, vol. V, p. 27). That he tried to avoid filters is also confirmed by the fact that most of his recorded observations were taken at dusk or dawn.

124. Scheiner's introduction, in his 1630 *Rosa Ursina,* of the "heliotropic telescope" (the first equatorially mounted telescope) was a response to this problem.

125. There were substantial gaps in Scheiner's visual record of the sunspots. The sequence included thirty-one days (sometimes with more than one illustration per day) from October 21 to December 14, with twenty-four days left uncovered.

126. Only one illustration (that of October 21, 1611) was published in a size sufficiently large to detect with clarity the irregularities of the spots.

ner's "realist" claims were better served by "schematic" pictures while Galileo's nonessentialist stance about their nature were better supported by "realistic" representations.

The most conspicuous conceptual and visual element of Galileo's second letter was a long, day-by-day sequence of 35 large-format illustrations of the sunspots' changing positions and appearances from June 2 until July 8 (fig. 16).[127] There were only two one-day gaps in the monthlong sequence (compared to twenty-four gaps in Scheiner's illustrations between October 21 and December 14, 1611).[128] Each picture (five times larger than Scheiner's in diameter) occupied one full page in the book.[129] The whole set occupied almost forty pages. The size of the pictures, the virtual absence of gaps, the detail with which the spots' peculiarities were represented, coupled with the viewer-friendliness of the illustrations (which, unlike Scheiner's, required little or no decoding) turned that section of Galileo's letter into a virtual movie any viewer (not only an astronomer) could watch.[130]

Galileo's pictures of the sunspots functioned like the other sequences in the *Nuncius,* but the look of the images was substantially different. In the *Nuncius* Galileo could use plain diagrams of the satellites or exaggerated depictions of the Moon as evidence of his discoveries, but now he needed to show the details of the spots' irregular and ever-changing contours.[131] His earlier sequences cast the satellites and lunar mountains and valleys as physically real because of the periodical patterns they displayed, but now

127. Galileo added also a set of three drawings (August 19 to August 21) at the end of the sequence. He did so to include spots that, due to their unusually large size, could have been observed without the telescope.

128. Weather conditions certainly helped the tightness of Galileo's visual record. It could not hurt that he observed in the spring and summer in Italy while Scheiner had observed in the fall and winter in Germany (van Helden, "Galileo and Scheiner on Sunspots," p. 378).

129. Scheiner's pictures were 25 mm in diameter, while Galileo's were about 124 mm.

130. The claim that Galileo tried to develop an alternative audience for his discoveries and natural philosophy rather than convince the traditional philosophers was already introduced by Feyerabend, *Against Method,* pp. 141–43.

131. This new difficulty came with new resources. In 1610 Galileo could attach physical meaning to pictures of the Moon that (by later standards) were quite inaccurate both because he did not need anything better and because pictures that (by later standards) would have been better may not have been read as such at that time. Instead, by this time "telescopic accuracy" had become a recognizable category, as Scheiner's own claims were predicated on the acceptance of telescopic evidence. Consequently, Galileo could rely on these new established parameters and on the fact that, since 1610, he had managed to represent himself as the producer of the best telescopes.

his sequences argued that the spots were *not* satellites by showing that they did *not* display the periodical patterns of satellites. (As we will see in a moment, Galileo's sunspots images highlighted other, more complex patterns of change.)

The pictures Galileo produced against Scheiner were more detailed than those published in the *Nuncius* in the sense that they made visible more differences—differences that mattered in the specific context of the sunspots dispute. However, rather than being more accurate in any general sense of the term, Galileo's images of sunspots were remarkably ad hominem. His pictorial tactics were framed by the nature of Scheiner's arguments and pictures, and by the fact that they had been published before Galileo's. Scheiner's pictures and arguments were part of the referent of Galileo's pictures of sunspots.

While in 1610 Galileo crafted his illustrations of the Moon to refute the Aristotelians' sweeping veto against celestial change, his 1612 pictures of sunspots were produced to refute Scheiner's positive claim about the existence and nature of sunspots. If Galileo's pictures of sunspots seemed more accurate than those of the Moon, it is because they were deployed against an adversary whose claims about sunspots were more specific and nuanced than the Aristotelians' veto against the corruptibility of the Moon that had confronted the *Nuncius*. Had Welser not published Scheiner's *Tres epistolae* when he did, it is quite conceivable that Galileo could have made his point about the corruptibility of the Sun through images as "inaccurate" as those he had used to prove the irregularities of the lunar surface. Instead, coming after Scheiner, Galileo needed pictures that could refute both Aristotle's veto on corruptibility and Scheiner's claim that the spots were satellites.

Another key factor in determining the look of Galileo's pictures of sunspots was the observed behavior of the sunspots themselves. New spots emerged, fell apart, came together, and disappeared as they drifted eastward over the solar disk. If some lasted long enough to reemerge from the backside of the Sun, they were so changed that they could not be conclusively identified as an old spot. They did not appear to behave like the satellites of Jupiter or the mountains of the Moon—virtually unchanging bodies whose cyclical appearances one could observe ad infinitum. Galileo, therefore, tried to make two interrelated claims: (1) that the spots' irregular behavior showed that they could not be the satellites Scheiner had claimed they were, but (2) also that they were real despite their apparently chaotic appearances. This required a visual sequence able to show that the spots displayed certain kinds of periodical regularities (so as to show that

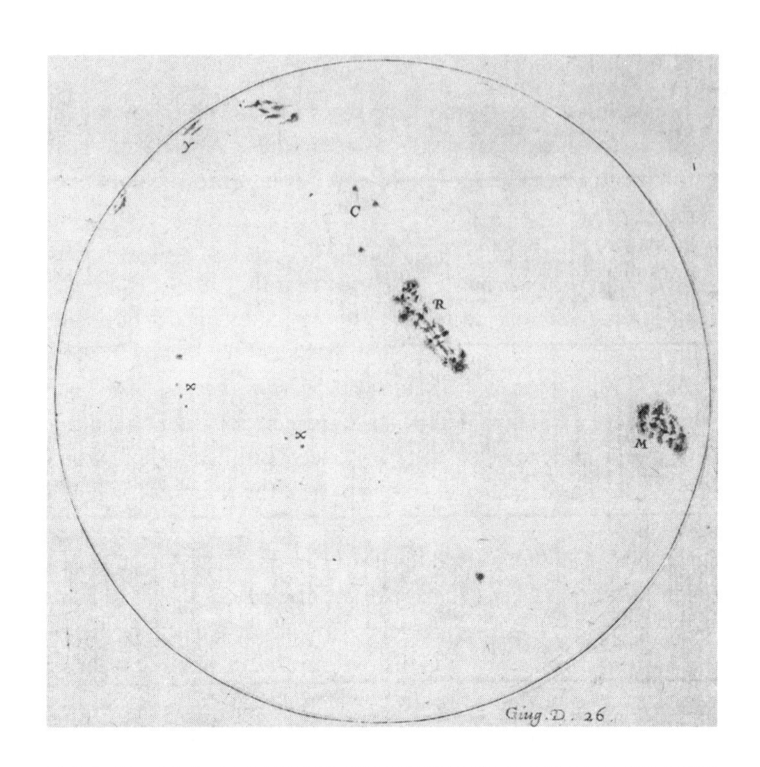
Y
C
R
x
x
M
Giug. D. 26

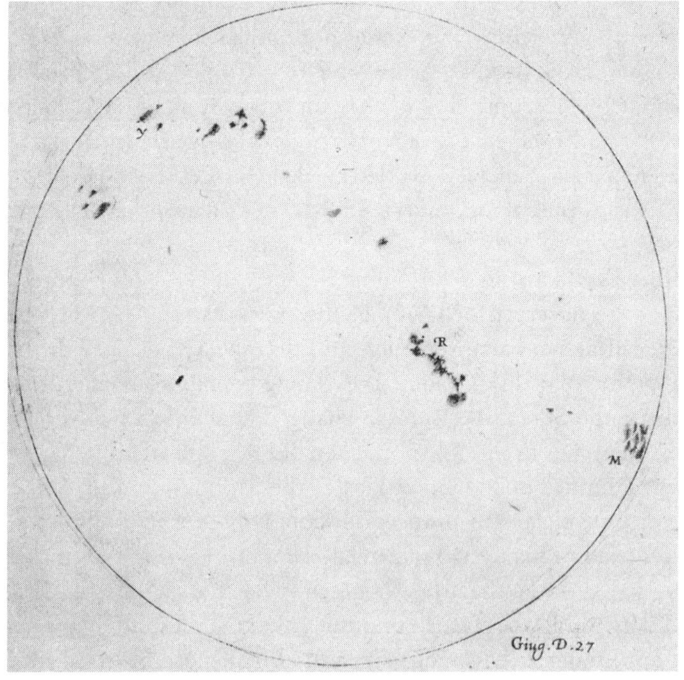
Y
R
M
Giug. D. 27

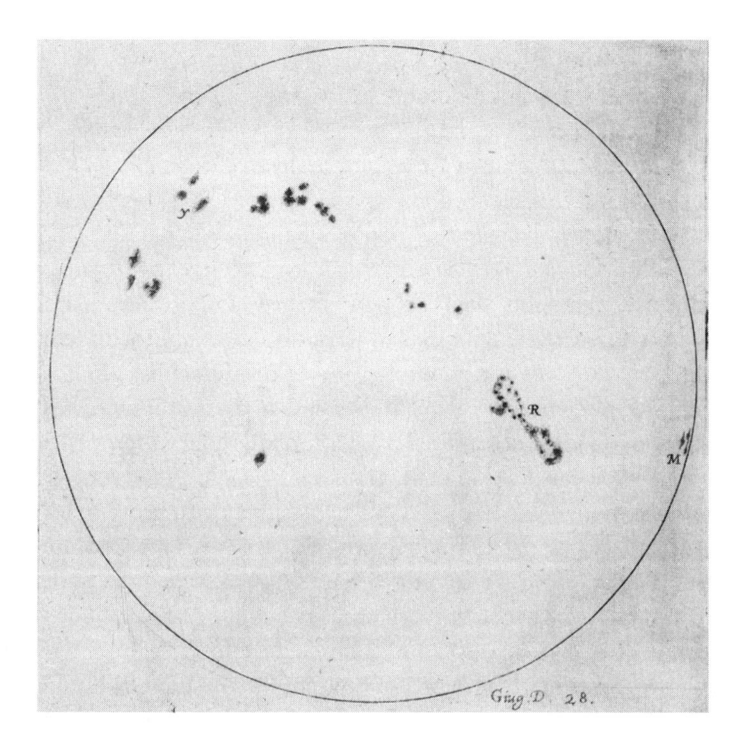

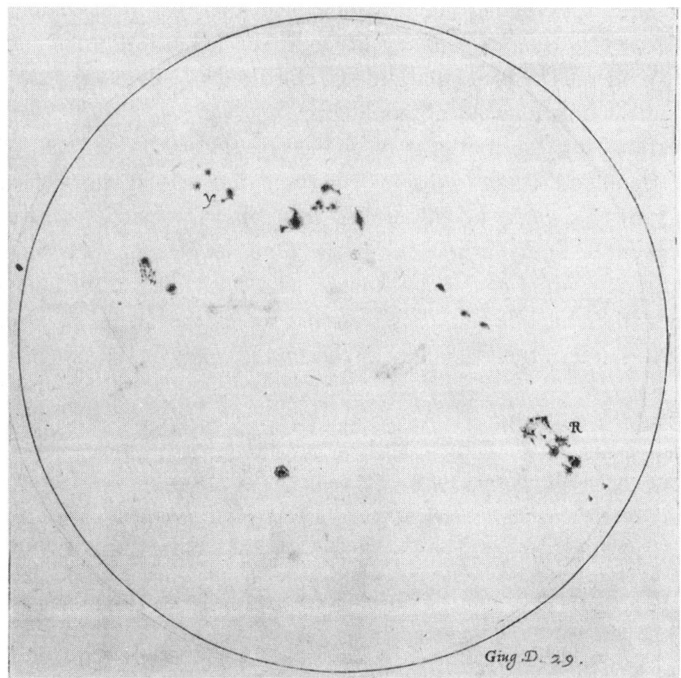

FIGURE 16. Galileo's illustrations of sunspots for June 26–29 from his *Istoria e dimostrazioni* (1613). (Reproduction courtesy of the William Andrews Clark Memorial Library, UCLA.)

they were real objects) but did not display either the kind of periodical regularities or the appearance of satellites (so as to refute Scheiner). Luckily for Galileo, these two goals were not mutually exclusive.

That the spots emerged and fell apart indicated that they could not be satellites—at least in the traditional, pre-Scheiner sense of the term. But because of the specificity of their cycle of emergence, growth, and eventual disappearance, the spots' lack of permanence actually testified to their nonartifactual nature. The periodical pattern that established them as objects was not just their fifteen-day trip over the solar disk, but their life cycle—their emergence, growth, and eventual disappearance.[132] (In this sense, Galileo's sunspots pictures are closer to Aquapendente's sequence of the embryological development of the chick than to Galileo's own images in the *Nuncius*).[133] Although Galileo was unable to prove that some of the bigger spots came back after surviving the trip over the backside of the Sun, he could offer a strong argument (based on the cycle of the spots' birth and decay) about why that was the case. He could not track one specific spot during its many trips over the solar surface in the same way he had been able to track the revolutions of the satellites of Jupiter, but he could determine the spots' half period (about fifteen days) and argue that the fact that they did not come around did not mean that they were artifacts, but simply objects with a short life cycle—a life cycle he could map and share with his readers. The spots that did not come back around were a problem for Scheiner's position, but not Galileo's.

The spots' intricate patterns of change needed to be represented with pictures detailed enough to display the stage of development or decay of a certain spot on a certain day, but also through a fast-paced narrative that could cover the spots' whole life cycle. One detailed picture taken, say, every fifth day would have not allowed a reader to observe the life cycle of the spots and conclude that they were not artifacts. "Detail" was in both the pictures and in the *frequency of the sampling*.[134] At the same time, we

132. Furthermore, the spots were not observed just anywhere on the Sun, but tended to concentrate within a band along the solar equator.

133. It would be interesting to know if Aquapendente was a direct source of inspiration for Galileo's sequence. Although I have found no evidence to support that, it is well documented that Galileo and Aquapendente were close friends, that Aquapendente treated Galileo and gave him medical recipes, and that Galileo lobbied the Medici to offer his friend a court position.

134. *GO*, vol. V, pp. 132–33.

have no context-independent yardstick to say that the pictures in the *Istoria* were more accurate than those in the *Nuncius.*[135] All we can say is that each set of pictorial sequences functioned as an argument that managed to convince most of its readers of the periodicities of its respective phenomena. That some of those pictures may appear "schematic" while others "realistic" is quite irrelevant.

Galileo admitted to Welser that not all issues of the sunspots debate could be answered by pictures.[136] And yet the impact of their narrative, of the myriad details and changes they put in motion, created a reality effect that undermined the plausibility of Scheiner's claims:

> Their different densities and blackness, the changes in shape, and the mingling and separation are in themselves manifest to vision without the need of further discussion, and therefore a few simple comparisons will suffice of those accidents in the drawings that I am sending you.[137]

As in Scheiner's case, the appearance, display, and kind of visual narratives produced by Galileo were closely connected to his observational system. His system, however, was radically different from Scheiner's:

135. Galileo's sunspots pictures appear remarkably more detailed than those of the satellites of Jupiter or of the Moon because they were asked to map differences that were more complex than those needed to establish the satellites of Jupiter or the mountains of the Moon as physical objects.

136. "A few simple comparisons will suffice of those accidents in the drawings that I am sending you [. . .], but that they are contiguous to the Sun and that they are carried around by its revolution, this must be deduced and concluded by reasoning from certain particular accidents provided to us by sensory observations" (*GO,* vol. V, p. 117). A discussion of those nonvisual arguments is in William Shea, *Galileo's Intellectual Revolution* (New York: Science History Publications, 1977), pp. 54-58.

137. *GO,* vol. V, p. 117. Also: "Convinced that it is a falsity to introduce such a sphere between the Sun and us [the sphere where Scheiner located the solar satellites] that alone could satisfy most of the phenomena, [. . .] it is not necessary to lose time in reexamining every other conceivable position, for each of them by itself will immediately encounter manifest impossibilities and contradictions, even if it were quite capable [of accommodating] all the phenomena I have recounted above and which continuously are truly observed in these spots. And so that Your Lordship [Welser] may have examples of all the particulars, I send you the drawings of thirty-five days [. . .] In these Your Lordship will first of all have examples of these spots appearing shorter and thinner in the parts very near the circumference of the solar disk, comparing the spots marked A of the second and third days" (*GO,* vol. V, p. 130). And: "And from all these and other accidents that Your Lordship will be able to observe in the same drawings, it can be seen to what irregular changes these spots are subject" (*GO,* vol. V, p. 132).

> Direct the telescope upon the Sun as if you were going to observe that body. Having focused and steadied it, expose a flat white sheet of paper about a foot from the concave lens [the eyepiece]; upon this will fall a circular image of the Sun's disk, with all the spots that are on it arranged and disposed with exactly the same symmetry as in the Sun. The more the paper is moved away from the tube, the larger this image will become, and the better the spots will be depicted. Thus will they all be seen without damage to the eye, even the smallest of them—which, when observed through the telescope, can scarcely be perceived, and only with fatigue and injury to the eyes. In order to picture them accurately, I first describe on the paper a circle of the size that best suits me, and then by moving the paper towards or away from the tube I find the exact place where the image of the Sun is enlarged to the measure of the circle I have drawn. This also serves me as a norm and rule for getting the plane of the paper right, so that it will not be tilted to the luminous cone of sunlight that emerges from the telescope. For if the paper is oblique, the section will be oval and not circular, and therefore will not perfectly fit the circumference drawn on the paper. By tilting the paper the proper position is easily found, and then with a pen one may mark out the spots in their right sizes, shapes, and positions. But one must work dexterously, following the movement of the Sun and frequently adjusting the position the telescope, which must be kept directly on the Sun.[138]

The apparatus adopted by Galileo (but developed by his pupil, Benedetto Castelli) was as close as one could get to a mechanically produced image. Galileo actually referred to the image of sunspots not as projected but as printed (*stampata*) by sunlight.[139] He did not provide an image of his apparatus, but its verbal description matches an image later published by Scheiner in his 1630 *Rosa Ursina* (fig. 17).[140] The advantages were many: no filters, a much higher level of detail, no need to limit observations to certain times of the day, no need to go back and forth between observing the Sun (in near-blinding conditions) and drawing from memory on a piece of paper (under very different lighting conditions), no problems with measuring the size and position of the spots or with maintaining the scale of the images constant, better visibility of weak sunspots, and minimization of the impact of personal drawing skills. It greatly routinized observation too.

Such de-skilling of sunspot observation facilitated their spread among

138. *GO*, vol. V, pp. 136–37.

139. *GO*, vol. V, p. 137.

140. An equivalent, but coarser image of this apparatus is found in Scheiner's earlier *Refractiones coelestes, sive solis elliptici . . .* (Ingolstadt: Eder, 1617), p. 91.

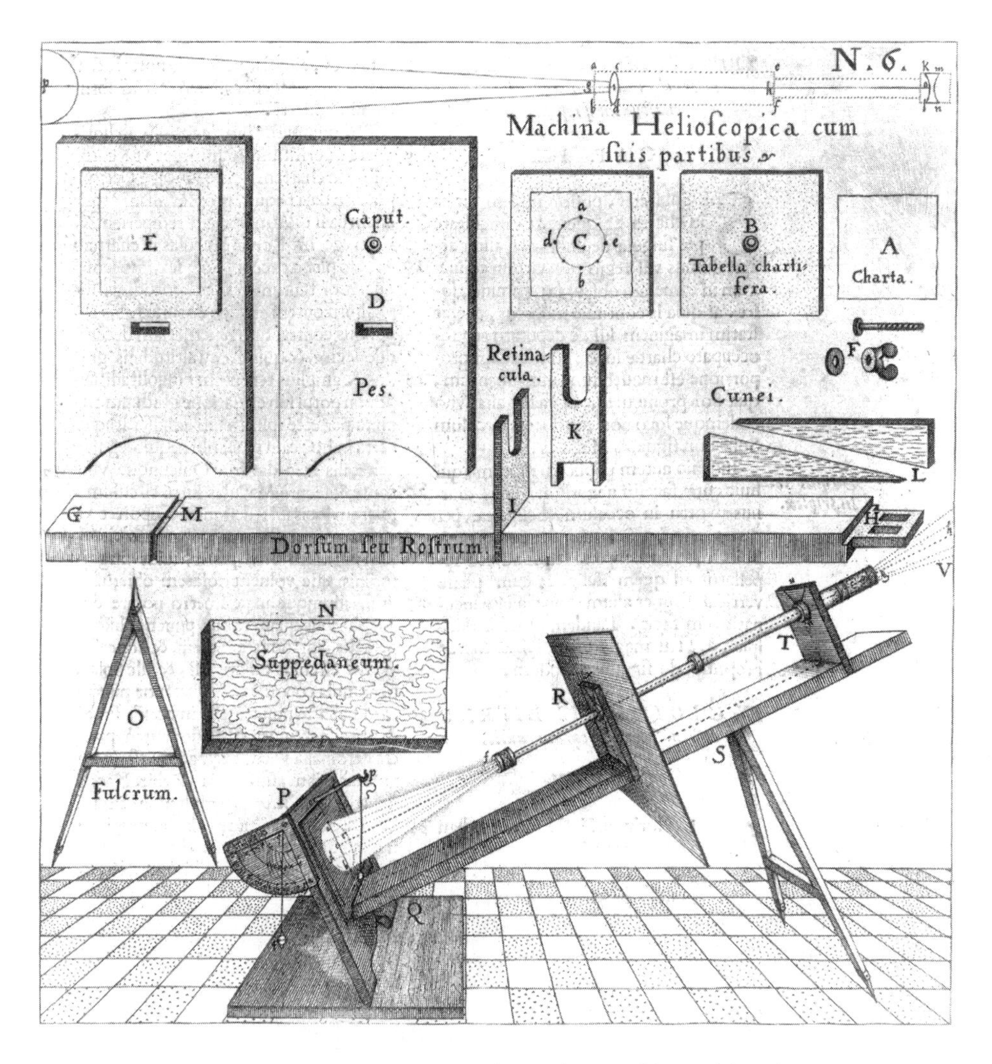

FIGURE 17. Projection apparatus similar to that used by Galileo, from Scheiner's *Rosa Ursina* (1630), 77. (Reproduction courtesy of Houghton Library, Harvard University.)

nonspecialists and helped solidify Galileo's claim that the spots were not satellites. He could encourage other people to observe and publicize the vagaries of sunspots, but he could also compare their drawings—produced in other parts of Europe—with his own:[141]

141. A few of Galileo's friends, especially Domenico Cresti da Passignano (a painter formerly at the Medici court in Florence, commonly known as Domenico Passignani), had been observing the spots since the fall of 1611 without Castelli's projection system (*GO*, vol. XI, pp. 208–9, 212). Galileo gave Passignani advice and asked for drawings of his observations to compare them with his own (*GO*, vol. XI, pp. 214, 229). Passignani sent the drawings on December 30, adding that he had shown them to Grienberger and Maelcote, who had asked about how his eyes could stand observing the Sun. Passignani replied that he used a blue filter attached to the eyepiece (*GO*, vol. XI, p. 253). At the beginning of February, Ludovico Cardi Cigoli wrote to Galileo that Passignani, probably after having heard of Scheiner's *Tres epistolae*, felt that he should be given priority credit for the discovery (*GO*, vol. XI, p. 268). On February 17, 1612, Passignani wrote Galileo that he had been observing the spots since mid-September and that he disagreed with Scheiner's claim that they were not in the Sun (*GO*, vol. XI, pp. 276–77). It is not clear whether Galileo ever told Passignani about Castelli's apparatus. While Passignani's observations appear to have started without Galileo's prodding, Daniello Antonini became interested in sunspots after receiving a letter from Galileo: "Cominciai, subito doppo hauta la lettera di V.S., a dipingere il sole" (*GO*, vol. XI, p. 363). After that, Antonini wrote him about his observations in Brussels in July 1612, enclosing a number of drawings. Antonini requested Galileo's own drawings, which he received in October and found a close match to his own (*GO*, vol. XI, p. 406). Antonini's early drawings were mentioned by Galileo in his *Istoria* (*GO*, vol. V, p. 140). At first, Cigoli acted as a trait d'union between Galileo and Passignani, but soon started his own observation, under Galileo's direct prodding. On March 23, 1612, he sent Galileo a number of drawings of the spots that he had produced without Castelli's system (which was unknown to Galileo himself at that time) but with a telescope equipped with thick green filters (*GO*, vol. XI, pp. 287–88). He was not too pleased with his illustrations because he could not frame the whole Sun in the telescope's field of vision; he needed to move it to observe and depict all the spots. He sent more drawings (which Galileo had requested) on June 30, saying that, due to time constraints, he had observed little, and that he had passed much of the observing and drawing task to Cosimino—most likely an assistant. A July 14 letter confirms that Cigoli was using the projection system (which he had heard of from another painter, Sigismondo Coccapani) and that he was still using Cosimino, who was "being trained" (*GO*, vol. XI, pp. 361–62). He added that his pictures were now drawn to the diameter specified by Galileo. More projective drawings were sent on July 28 (*GO*, vol. XI, p. 369). Cigoli sent more pictures to Galileo on August 31, still made to his specification and through the projections system. He also asked if Galileo needed more (*GO*, vol. XI, pp. 386–87). Galileo, who by this time had concluded the second letter to Welser, must have ended his requests for drawings, as Cigoli stopped mentioning them in their letters. That Galileo told him to stop (and that, therefore, he was using him as part of an "extended observatory" made possible by the drawing technology) is confirmed by Cigoli on May 3: "I told him [Virginio Cesarini] that you had made me observe them, and that Your Lordship then told me

> We must recognize the divine kindness because the means needed for such understanding [of the spots' changing nature] are very easily and quickly learned. And he who is not capable of more [mathematical and philosophical arguments] may arrange to have drawings made in far-flung regions and compare them with the ones made by himself on the same days because he will find them absolutely to agree with his own. And I have just received some made in Brussels by Mr. Daniello Antonini [. . .] that fit exactly with mine, and with others sent to me from Rome.[142]

A few months later, Antonini (who in the meantime had received some drawings from Galileo) confirmed the evidentiary power of these pictures: "[Your] images of the Sun [. . .] match to the dot those I made on the same days in Brussels, so that I do not need your [mathematical] demonstration to be sure that the spots that appear on the surface of the Sun are contiguous to it."[143] With another correspondent, Galileo requested that the drawings be made the same size as his own, probably to allow for easy comparison through superimposition.[144] On top of gaining external confirmations for his observations, Galileo was probably trying to make sure that, between his own drawing and those of his collaborators, he could secure an uninterrupted visual narrative spanning many weeks.[145]

---

not to observe them any longer, and how you had told me to observe them in that specific size, and told me how to do that" (*GO*, vol. XI, p. 501). Castelli, the inventor of Galileo's observational apparatus, also sent him drawings of sunspot observations on May 8, 1612 (*GO*, vol. XI, pp. 294–95). Other people send drawings to Castelli, possibly for delivery to Galileo (*GO*, vol. XI, pp. 412–13).

142. *GO*, vol. V, p. 140. The role of astronomical illustrations not only as a means of communicating a discovery, but also of calibration of remote observers and instruments, is discussed in Simon Schaffer, "The Leviathan of Parsontown: Literary Technology and Scientific Representation," in Timothy Lenoir (ed.), *Inscribing Science* (Stanford: Stanford University Press, 1998), pp. 182–222.

143. *GO*, vol. XI, p. 406.

144. Letters from Cigoli in Rome to Galileo in Florence refer to Galileo's instructions about the size of the drawings (*GO*, vol. XI, pp. 362, 502).

145. Scheiner too used the drawings of other people's observations to support his own. In the *Accuratior*, for instance, he included small reproductions of Guldin's observations (done in Rome) arguing that they matched his own. The practice of using drawings to use or check other people's observations had quickly become standard among telescopists. Galileo's correspondence documents exchanges of drawings of positions of the satellites of Jupiter. In the *Narratio*, Kepler discusses the use of drawings to check other observers' perceptions. What is specific about Galileo's apparatus for the production of such drawings is that, because of the kind of claims he was making about sunspots, he could make use of other observers' drawings only if they were detailed and skillfully drawn. This re-

Not only did Castelli's apparatus improve observation, but it also greatly improved the *dissemination* of images of sunspots, both as printed images and as drawings. The projection system was the first step toward producing convincing visual sequences like the one included in Galileo's second letter and then printed in his 1613 *Istoria.* But those pictures could also be easily copied by putting another piece of paper over them, placing the sandwich against the light, and tracing the contours of the original image. This system allowed him to make limited editions of these drawings and circulate them before they were printed, like the sets he sent to Prince Cesi and Cardinal Barberini in May and June 1612—sets that could be further copied and showed around without much quality loss.[146] Thanks to this picture-making system, Galileo could send two "originals" of his second letter and its pictorial appendix to Welser in Augsburg and to Cesi in Rome at the same time. While Welser shared one set with his northern European correspondents, Cesi readied the other for publication in Rome. The drawings attached to the second letter to Welser were copied once more during the letter's stopover in Venice, thus producing a third set for circulation in northern Italy.[147]

More important, Galileo's image-making system allowed for high-quality engravings, not just drawings. It was common practice among engravers to use "carbon paper" methods to transfer drawings onto plates. Typically, the artist would rub the verso of a drawing with some pigment and then attach it to a copperplate coated with a thin layer of wax. S/he would then go over the outlines of the drawing, pressing them down to transfer the pattern onto the plate, which would then be engraved accordingly.[148] We do not have the original drawings for the engravings that

---

quired a nontrivial amount of talent. His apparatus dumbed-down the skill required of observers, thus allowing him to use more of their drawings more effectively.

146. *GO,* vol. XI, pp. 297, 307–11. The set sent to Cesi was then copied (and perhaps further distributed) by Cigoli (*GO,* vol. XI, p. 302). I believe that Galileo's choice of Barberini and Cesi as recipients of these pictures reflects his perception that, given the large networks Barberini and Cesi were centers of, many other people would be able, so to speak, to "see" Galileo's argument before reading it.

147. In a September 22, 1612, letter to Galileo, Giovanfrancesco Sagredo reported that "I had the letter for Augsburg copied, illustrations included" (*GO,* vol. XI, p. 398).

148. Michael Bury, *The Print in Italy, 1550–1620* (London: British Museum Press, 2001), pp. 14–15. I thank Lisa Pon for this reference. Unlike woodcuts, engraving plates worked by capturing ink in the cavities cut into them and then releasing it to the paper that was pressed against it. Therefore, making a "carbon copy" of a drawing of sunspots on a cop-

Galileo included in his *Istoria,* but a contemporary set of drawings that Galileo sent to Cardinal Barberini have been preserved.[149] Because the Barberini drawings concerned observations not included in the *Istoria,* a comparison of their diameter (about 127 mm) with that of the printed images (about 124 mm) cannot tell us whether Galileo's drawings and prints were exactly the same size.[150] However, the match is close enough to support the hypothesis that the engravings in the *Istoria* were produced by attaching Galileo's drawings (or perhaps a direct copy of them) onto the copperplate. If so, the quality of the printed illustrations was not the result only of a detailed, quasi-mechanically produced drawing, but also of an equally quasi-mechanical transfer from drawing to plate.[151]

Judging from the reception of the *Istoria*—and from Scheiner's later work on sunspots—Galileo's strategy paid off.[152] Few people outside of the Society of Jesus seemed to believe that the complex patterns presented by his images could be accounted for by Scheiner's acrobatic satellites.[153] It

---

perplate and engraving through the dark parts of the drawing would produce a plate that would yield positive prints of the original drawing.

149. The originals of the drawings sent to Cesi are lost, but those to Barberini are at Biblioteca Apostolica Vaticana, MSS Barberini Latini 7479.

150. The dimensions of the printed images vary slightly from page to page. I thank Ill.mo Signor Christoph Luthy for checking the size of the drawing at the Vaticana for me.

151. The production of the copperplates for Galileo's *Istoria* was a careful and expensive business. Prince Cesi recognized the importance of these illustrations and spared little money, time, or talent to produce and revise the plates according to Galileo's desires (*GO,* vol. XI, pp. 404, 409, 416, 418, 422, 424, 472, 475).

152. Scheiner used much larger images in the 1630 *Rosa Ursina* (210 mm in diameter), that is, almost twice the diameter of Galileo's (124 mm) and almost nine times that of his original pictures (25 mm).

153. This does not mean that everyone gave Galileo credit for the discovery of sunspots (nor that they should have). In the decade after the debate, the views on the priority claims tended to be distributed along party lines, with Jesuit authors tending to cite Scheiner as the discoverer of and authority on sunspots. There is no question that Scheiner's later opus, the *Rosa Ursina* (Bracciano: Apud Andream Phaeum, 1630, although the printing began in 1626) was the most detailed and comprehensive seventeenth-century text on sunspots. Similarly, in the period between the publication of Scheiner's *Tres epistolae* and Galileo's *Istoria,* the Jesuits of the Collegio Romano publicly sided with Scheiner although some expressed serious qualms about Scheiner's positions in private. On February 17, 1612, Passignani told Galileo that Grienberger "is of the same opinion of the writer [Scheiner], that is, that the spots one sees are stars like those seen around Jupiter" (*GO,* vol. XI, p. 276). A week earlier, in a letter to Galileo, Grienberger had taken a more am-

may not have helped Scheiner's case that most people became familiar with his argument and illustrations through the reprint of the *Tres epistolae* and *Accuratior* appended to Galileo's 1613 *Istoria.* While Prince Cesi—the sponsor of Galileo's publication—invested much time and money to produce the best plates for the *Istoria*'s illustrations, he did not apply the same standards to the reproduction of some of Scheiner's images.[154]

PUBLIC MOVIES AND PRIVATE DARK ROOMS

Scheiner had not seen either Galileo's images of the sunspots or the description of his apparatus by the time he wrote the *Accuratior.* All he had heard was Galileo's announcement, included in the very last paragraph of his first letter to Welser, that

> in a few days I will send him [Scheiner] some observations and drawings of the solar spots of absolute precision, indeed the shapes of these spots and of the places that change from day to day, without an error of the smallest hair, made by a most exquisite method discovered by one of my students.[155]

---

biguous stance saying that Scheiner's claims, which he had just read, were "not improbable" and that he had managed to keep the stars off the Sun. But, Grienberger continued, he could not, at that point, either certify or refute Scheiner's claims (*GO*, vol. XI, p. 273). In September 14, during a public disputation at the Collegio Romano, a Dominican friar defended the claim that the Sun was at the center of the cosmos and invoked the observation of sunspots to buttress his argument. The Jesuits replied that the sunspots were very minute stars that were visible only when grouped together but became invisible when isolated (*GO*, vol. XI, p. 395). The Jesuits had assumed a pro-Scheiner position (though a more muted one) at a similar event in late January 1612 (*GO*, vol. XI, p. 274). On October 19, 1612, Cigoli wrote Galileo that Grienberger still defended the position that the sunspots were stars (*GO*, vol. XI, p. 418). However, on November 23, 1612, Johannes Faber wrote Galileo that, about a week earlier, Grienberger had visited him at home and that "he agrees more with you than with Apelles, as he finds very convincing the arguments Your Lordship uses to refute the assumption that they are not [sic] stars. However, as a child of holy obedience, he dare not say it" (*GO*, vol. XI, p. 434).

154. Cesi decided to reprint Scheiner's two texts together with Galileo's because Scheiner's original editions were already rare in 1612. Without a reprint people would read Galileo's text without being familiar with its counterpart. Welser did not send the original plates to Rome, so Cesi had all the images reengraved. More precisely, he used engravings for Scheiner's pictures of sunspots (and kept to the same size—25 mm diameter), but used woodcuts for all other images (and made those images significantly smaller) (*GO*, vol. V, pp. 404, 472, 474, 482).

155. *GO*, vol. V p. 113.

By the time he read this, Scheiner had already completed all his observations as well as two of the three letters that were to make up the *Accuratior*.[156] It is not clear, however, whether exposure to Galileo's pictures and method would have changed Scheiner's own use of pictorial evidence.

Scheiner accompanied his second, much longer set of letters with new illustrations. The new pictures, however, were not significantly different in size, number, or organization from the earlier ones. Some of them overlapped with those previously published in the *Tres epistolae* and were barely distinguishable from them (fig. 18).[157] The main improvement came from the frequency of observations, not drawing technique. For his second publication, Scheiner (possibly helped by better weather) observed more regularly and was able to avoid many of the gaps that had marred his earlier visual narratives. He also drew lines across the solar circles to mark the orientation of the ecliptic and of the spots' path across the Sun. Despite these improvements, however, Scheiner did not seem impressed by his own pictures, nor did he seem concerned about their flaws:

> All these observations, [made] as often as the weather allowed (and that was almost always when I observed) are the most accurate possible, though they are perhaps not so accurately drawn on the paper because of the failing of my hand.[158]

The illustrations included in the last letter of the *Accuratior* were of the usual size and quality (fig. 19). While apparently unconcerned with the quality of his illustrations, however, Scheiner was very keen to report that they showed a perfect match between his observations and those conducted on the same days at the Collegio Romano.[159] Perhaps he considered the endorsement of a fellow Jesuit as more weighty than the evidential narrative developed by his own inscriptions.

While it might have been tactically counterproductive for Scheiner to

156. Welser acknowledged receipt of Galileo's first letter on June 1, 1612 (*GO*, vol. XI, pp. 303–4).

157. The overlap concerns December 10, 11, 12, 13, and 14 (*GO*, vol. V, pp. 33, 47). The slight difference between the two editions may be due to the engraver's different rendering of the same drawings on two different occasions.

158. *GO*, vol. V, p. 48. Similar cautionary claims occur elsewhere in the text: "Because if the drawing on paper of their shadows does not agree to a hair, it is to be attributed to my eyes and hand," and "I tried to transfer the shapes to paper faithfully" (*GO*, vol. V, pp. 53, 49).

159. *GO*, vol. V, pp. 62–63.

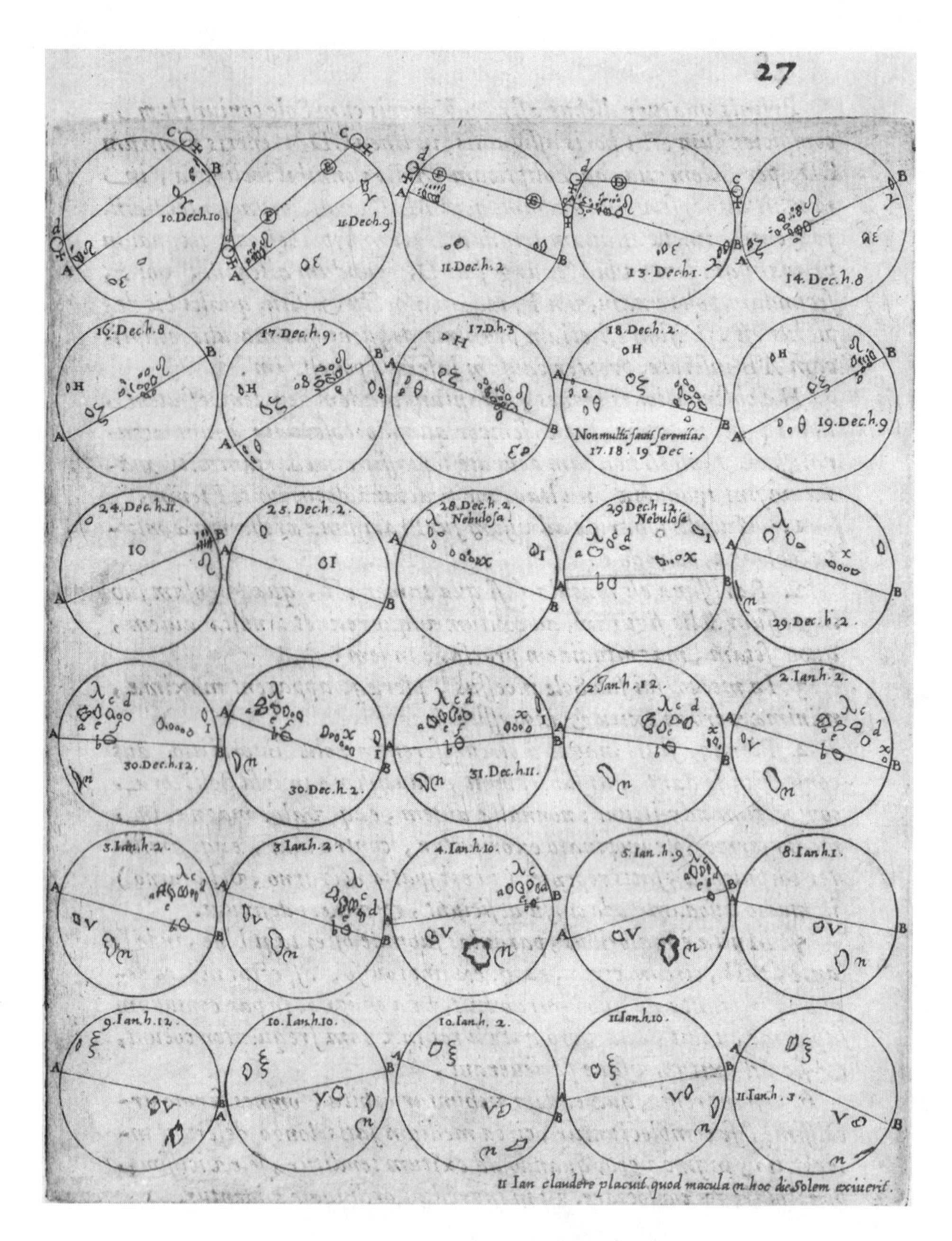

FIGURE 18. Sunspots illustrations from Scheiner's *Accuratior disquisitio* (1612), as reprinted in the Roman 1613 edition. Compare to fig. 16. (Reproduction courtesy of the William Andrews Clark Memorial Library, UCLA.)

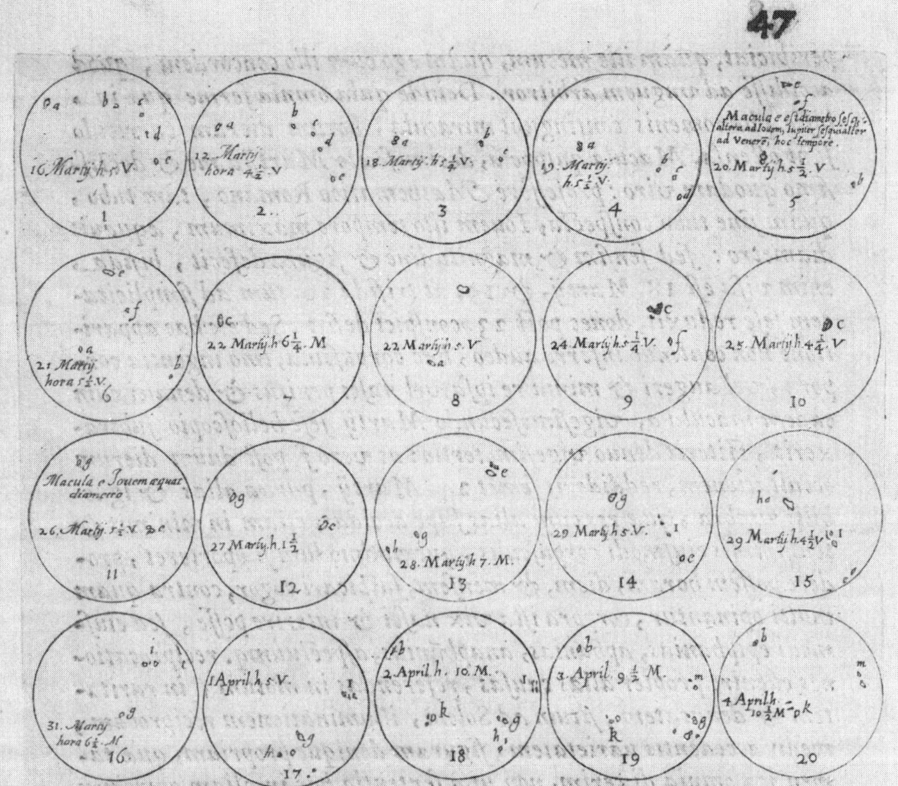

Christophorus Gruenberger Soc. Iesu, insignis Mathematicus, eas videre cœpit 2. Februarij, in festo B. Virginis Purificationis. Sed & Paulus Gulden. itidem Romæ eiusdem Soc. Mathematicus nobilis, à 18. Martij vsque ad 22. eiusdem in Sole maculas obseruauit. Quarum obseruationum maculæ, quia animaduersiones dignas comprehendunt, sunt altius repetendæ. Et quia omnes absolutæ sunt per foramen inuersionis, idcirco tenendum illarum figuram & situm atque amplitudinem talem esse, qualis sufficiat ad multa inde concludenda, à die igitur 16. mensis Martij vsque ad 4. Aprilis isti fuerunt Solis aspectus.

Has obseruationes apponere necessarium visum est, vt & tu videas, quàm censorem minimè timeam, cum vix ambigam horum dierum animaduersiones ab alijs factas, & Paulus Gulden perspi-

FIGURE 19. Final sunspots illustrations from Scheiner's *Accuratior disquisitio* (1612), as reprinted in the Roman 1613 edition. (Reproduction courtesy of the William Andrews Clark Memorial Library, UCLA.)

produce detailed pictures, there may have been other considerations behind his lukewarm interest in visual representations. Although he had not seen a description of Galileo's projection system, Scheiner too reported several experiments involving the projection of images from the telescope against a flat white surface:

> If during the day you place a tube [telescope], which is positioned before you in a window of your room, before a nearby white wall or hold a sheet of very white paper up to it, you will still observe all these appearances as before.[160]

And in another passage:

> This [effect] is evident when you transmit the Sun through a similar lens onto a smooth wall or reflect it onto a wall from a similar lens, for the entire image of the Sun will undulate with these tracks.[161]

Strikingly, the images Scheiner was studying on walls or sheets of paper were not of sunspots but of flaws in the lenses (bubbles in the first case, swirls in the second). He used the projection system not to make pictures of sunspots, but to map out how the optical artifacts produced by the telescope looked, and then to demonstrate that sunspots were clearly distinct from those artifacts. What characterizes the *Accuratior* is not the detailed visual mapping of sunspots (which Scheiner was technically equipped to produce), but a truly relentless analysis of optical effects in the atmosphere, in the eye, and in the telescope that could be used by critics to dismiss the reality of his discovery. Evidently, the "more accurate" in the title of his second text did not refer to the pictures but to his philosophical arguments.

This points to the significantly different (if overlapping) audiences that Scheiner and Galileo sought to address. In both of his texts, Scheiner seemed much more concerned than Galileo with responding to possible philosophical objections to his use of the telescope, and described the painstaking procedures he followed to prove that the spots were not optical artifacts. The physical existence of the spots was as important to him as their categorization as solar satellites. While Galileo continued to take the reliability of the telescope as a nonissue and did not seem to worry about people taking the sunspots to be optical artifacts, Scheiner was behaving as if he were still in 1609, when the telescope had just been introduced and

160. *GO*, vol. V, p. 59.

161. *GO*, vol. V, p. 58.

many were still skeptical about it. His apparently anachronistic behavior may have reflected Scheiner's institutional affiliation: Jesuit mathematicians needed to be concerned about what the philosophers and theologians of their order thought of their work—work they could censor. He seemed to be addressing his superiors (and other philosophically and theologically concerned people) much more directly than astronomers like Galileo or Kepler.

Scheiner's concerns with philosophers and theologians explain, I believe, the *Accuratior*'s frequent references to observations of sunspots not with the telescope but with the camera obscura.

> These spots are neither delusions of the eye nor a mockery by the tube or its lenses, since without a tube they are seen on paper.[162]

Or:

> If I now show that the solar spots are also seen without a tube, by the eye of any man, what will he oppose, whoever opposes, that this is not a fraud? Certainly neither the eye, nor the glasses, nor the air can be blamed.[163]

Scheiner's camera obscura (like the projection of telescopic images on walls or paper) was not part of a system to produce visual sequences about sunspots—sequences that could then be printed and distributed to a wide audience of nonspecialists. It was instead a "natural instrument," one that, being lens-less, could be assumed not to lie.[164]

Because several people could be admitted into the camera obscura, it was also a place where perceptual biases could be collectively checked and corrected:

162. *GO*, vol. V, p. 53. This point is repeated in the *Accuratior:* "For almost all of these observations were made not only with a tube but also with the Sun projected through an aperture onto a sheet of paper held perpendicularly, and thus the disk of the Sun, cast on the paper supplied the true location and motion of the spots, and the tube directed to the Sun supplied the shape" (*GO*, vol. V, p. 64). And elsewhere: "This is confirmed by the fact that the Sun, projected through an aperture onto a sheet of paper, also distinctly represented the shadows of the spots" (*GO*, vol. V, p. 67).

163. *GO*, vol. V, p. 61.

164. Scheiner also mentions another lens-less method of observation: "If you hold a clean mirror to the Sun and reflect the species of the Sun from the mirror onto a clean wall or sheet of paper at the required distance, you will see spots on the Sun in number, arrangement, and size in relation to each other and the Sun. And this method of observing, sought after in vain for a long time, I learned from a very good friend" (*GO*, vol. V, p. 62).

> If through a round hole of about this size—O—or a bit larger, the Sun is admitted perpendicularly onto a clean sheet of paper or some other white plane, it shows itself and all the bodies below it in proportion to the distance, position, and number that they retain among themselves and to the Sun. And I have made observations in this manner, *and to all willing I have shown,* whenever possible, spots so large, dense, and black, that they were quite apparent even through thin clouds.[165]

Scheiner's camera obscura was, quite literally, a darkened room, *a place for in-house demonstrations.* It was where he brought the people he needed to convince, people who lived close to him, like his fellow Jesuits.[166]

The same reasons that could have led Scheiner to use the camera obscura as a "conversion site" may have also discouraged him from spending much time devising a more efficient system for the pictorial display of sunspots. Given Scheiner's skills in instrument making and in drawing techniques there is little doubt that he could have developed a system similar to Galileo's—which in fact he adopted and perfected a few years later (fig. 17).[167] That he acknowledged the shortcomings of his images while apparently doing little to correct them (when he had the technical capability to do so) suggests that Scheiner did not see pictures as the best argument to convince his audience of the spots' existence.[168] He may have had a point: philosophers and theologians were not known for their reliance on pictorial evidence.

What differentiated Scheiner's and Galileo's sunspot illustrations, therefore, was not just their pictorial style or the narratives they told. They were aimed at quite different audiences. Scheiner and Galileo were not in com-

165. *GO,* vol. V, pp. 61–62 (emphasis mine).

166. Scheiner reported that the Jesuits at the Collegio Romano had also used the camera obscura for their observations, perhaps for the same reasons (*GO,* vol. V, p. 63). Michele Camerota, "Aristotelismo e nuova scienza nell'opera di Christoph Scheiner," *Galilaeana* forthcoming, discusses the serious opposition to Scheiner's claims within his own college at Ingolstadt.

167. Scheiner pictured his apparatus both in the 1630 *Rosa Ursina* and (with a much coarser image) in his *Refractiones coelestes, sive solis elliptici . . .* (Ingolstadt: Eder, 1617), p. 91. Besides building his own telescopes, Scheiner was an accomplished sundial maker and the author, later on, of *Pantographice, su ars delineandi res quaslibet . . .* (Rome: Grignani, 1631).

168. Scheiner, however, changed his mind quite drastically a few years later. By the time he published the *Rosa Ursina* in 1630 he had become a major user of pictorial evidence. But by that time he did not have to worry anymore about people not believing the existence of sunspots.

petition to produce the "best" pictures of sunspots. Rather, they made different tactical decisions about what claims to make, what kind of credit to seek, what audiences to target. Given where Scheiner stood institutionally and philosophically, he would have not benefited from a more detailed mapping of the spots' changes. Galileo not only needed detailed sequential pictures, but he also needed to *disseminate* his visual narratives because, unlike Scheiner, his audience was not primarily local or concentrated in a few institutional sites like Jesuit colleges. While Galileo mentioned placing a sheet of paper on the floor of a church to observe spots in the solar disk being projected there through a broken window pane acting like a pinhole, there is no evidence that he ever conducted public demonstrations in churches or in other large "dark rooms."[169] He let his printed images do the work.

FROM FLUID SKIES TO BLURRY SATELLITES

Scheiner had reasons to be apprehensive about what his superiors thought of his claims. While the observations he had conducted since the *Tres epistolae* had strengthened his belief in the reality of the sunspots, those same observations had made him much less certain about their permanence. Scheiner kept referring to the sunspots as stars, but he also kept redefining, with relative clarity, what he meant by that. He reiterated his belief in the incorruptibility of the Sun, but the evidence he shared with his readers pointed in the opposite direction. Unless he managed to keep the sunspots safely away from the solar surface, his new evidence about the extensive variability of sunspots could have ended up supporting the very claim he had been trying to refute all along: that there was plenty of change and corruption in the Sun.

For example, he acknowledged that the spots' shape was hardly circular to begin with, and that it appeared to become even less circular as the spots moved across the solar disk:

> Spherical spots appear rarely, while most spots are combined, oblong, and polygonal [. . .] Very rare is the spot (if it exists at all) that retains the shape

169. "E V.S. vedendo in chiesa da qualche vetro rotto e lontano cader il lume del Sole nel pavimento, vi accorra con un foglio bianco e disteso, che vi scorgerà sopra le macchie" (*GO*, vol. V, p. 137). On the use of churches as solar observatories, see John Heilbron, *The Sun in the Church* (Cambridge: Harvard University Press, 1999).

> that it shows at ingress of the Sun all the way to egress. Indeed, there are none that I know of that displays exactly the same size.[170]

He also acknowledged other features that would not seem to be reducible to the behavior of satellites:

> The perimeter of almost all spots was roughened with, as it were, whitish and blackish fibers; and most spots, wherever they appear, were diluted by a greater whiteness around the edges than in the middle of their bodies. Indeed, the shape of very many of the spots reminds the observer now of some blackish snowflake, now of some small piece of black bread, now of a balled-up mass of hair hidden in a large torch, and now of a blackish cloud.[171]

He finally remarked quite candidly on the ephemeral nature of the spots:

> [Some large spots] suddenly spring up around the middle of the Sun. Others just as large, on the contrary, suddenly decay [. . .] in the middle of their path and cease to be seen.[172]

Aware that the evidence of radical change in the proximity of the Sun was growing, Scheiner probably realized that he could no longer save the permanence of both the Sun and the solar stars. In the last page of the *Accuratior,* in fact, Scheiner made the surprising remark that "about only one thing we are still at a loss: whether these bodies are generated and perish, or whether they are eternal."[173] Galileo himself was puzzled by the seeming inconsistency of Scheiner's claims: "He says a thousand times that they are stars, but now he doubts whether or not they are generated and perish."[174]

He had another problem. In the second letter of the *Accuratior* (dated April 4, 1612) he claimed to have discovered yet another novelty in the heavens: a fifth satellite of Jupiter (which he dedicated to Welser and his family).[175] With typical concern for priority, Scheiner waited only five days from the first observation of the satellite to write up his discovery and send it to Welser. Excitement gave way to embarrassment when the new satellite

170. *GO*, vol. V, p. 48.

171. *GO*, vol. V, p. 48.

172. *GO*, vol. V, p. 48.

173. *GO*, vol. V, pp. 69–70.

174. *GO*, vol. V, p. 70, note 9.

175. *GO*, vol. V, p. 56.

disappeared.[176] About six months later, Castelli sent a few observations to Galileo, joking, "During my first observation I saw a star one could dedicate to Welser."[177]

It seems, then, that Scheiner was in trouble on three fronts as he was writing the *Accuratior:* (1) he could have lost face with astronomers for having blundered the transit of Venus and for having probably mistaken sunspots for satellites; (2) he could have embarrassed Welser both by dedicating an artifact to him and his family, and by backtracking on claims Welser had been instrumental in publishing and disseminating; and (3) he could have run afoul with Jesuit philosophers and theologians for declaring (without censors' approval) that the spots were real and satellite-like, and then to make things worse for the Society by being unable to keep the spots (whose reality he has so convincingly demonstrated) off the solar surface.

As the costs of backtracking would have been too onerous, Scheiner charged ahead, determined to prove that *all* of his claims in the *Tres epistolae* were correct. He did so by introducing even more sweeping and potentially controversial cosmological claims.[178] For example, his remark about not knowing whether the sunspots were eternal or generated and perishable does not mean that, as Galileo believed, Scheiner doubted whether the sunspots were satellites or not. It rather suggests that Scheiner was attempting to redefine what satellites and, more generally, stars and planets were about so as to save both his claims about sunspots and about his "fifth satellite" of Jupiter.

To Scheiner, the analogy between the sunspots and the satellites of Jupiter had become so strong that, toward the end of the *Accuratior,* he stated that "I said not in vain [. . .] that the theory of the sunspots and of the stars of Jupiter appeared to be the same."[179] He did not simply use the satellites of Jupiter to explain the sunspots, but he also used what he had observed

176. It is unclear what Scheiner had observed. A few modern astronomers have suggested that his "fifth satellite" was, in fact, a variable star (Joseph Ashbrook, "Christopher Scheiner's Observations of an Object near Jupiter," *Sky and Telescope* 42 [1977]: 344–45).

177. "Nella prima osservazione viddi una stella da donare al Welsero, come ho notato" (*GO*, vol. XI, p. 456). Castelli's drawing included in the body of the letter shows Jupiter surrounded by four satellites plus, on the right, further out, and well below the orbital plane of the Medicean Stars, the joke "new satellite."

178. "And I have gladly communicated this completion of my earlier work to Your Excellency so that you would know how wrongly this great phenomenon is called into doubt by some and how wrongly torn to pieces by most. For all the other things that I have shown in the first painting are correct" (*GO*, vol. V, p. 69).

179. *GO*, vol. V, p. 57.

about the behavior of sunspots to explain what he had observed (or not observed) about the satellites of Jupiter.[180] For instance, he took the fact that some sunspots did not appear to return to suggest that maybe his newly discovered fifth satellite of Jupiter would not return either: "since some [satellites] suddenly appear and others suddenly disappear in almost the same way as the shadows on the Sun."[181] Vanishing sunspots helped him defend his discovery of the vanishing satellite of Jupiter while the vanishing satellite of Jupiter helped him defend that the sunspots could be both real and disappearing at the same time. All these disappearing acts, however, were just that: *appearances* of change, not real change.

In Scheiner's hands, the phases of Venus became the paradigm of one of the ways in which real astronomical bodies appear to almost disappear, that is, of how spherical bodies like planets end up assuming very different, nonspherical (and thus much less visible) appearances. In the *Tres epistolae* Scheiner had already used the phases of Venus to explain why sunspots seemed to become thinner as they approached the limb of the solar disk, that is, how a presumably hard and round object would end up looking less so.[182] He had also suggested that the peculiar appearance of Saturn may be explained through a combination of satellites and phases.[183] The implication was that all astronomical objects, due to how light strikes them, are bound to appear irregularly shaped like the sunspots.[184]

180. Van Helden argues that poor telescopes were at the root of Scheiner's position. With the instruments he had available, Scheiner could not detect the periods of the satellites of Jupiter. Scheiner did not think there was a problem with his telescopes, but that his difficulty in tracking the satellites reflected the satellites' inconsistent behavior or appearance. Once he came to believe that even the satellites of Jupiter behaved somewhat erratically, it was easy for him to assume that sunspots could behave that way too (Albert van Helden, "Scheiner," unpublished manuscript).

181. *GO*, vol. V, p. 56.

182. *GO*, vol. V, pp. 29–31.

183. *GO*, vol. V, p. 31.

184. His statement that the dramatic changes in the sunspots' appearance "are to be referred to motion: to rarity and density, position with respect to the Sun, mutual illumination, change of the accidental medium, and finally, particular shape" (*GO*, vol. V, p. 64) was meant to apply, I believe, to all celestial bodies, not only to sunspots. He already hinted at this claim in the *Tres epistolae*: "Eandem fortassis esse rationem, quo ad sui illustrationem, aliorum astrorum" (*GO*, vol. V, p. 31). The mention of "mutual illumination" in the *Accuratior* refers to a generalized view of phases involving more than one light-emitting (or light-reflecting) body.

Scheiner introduced another analogy to cover the opposite case—that of irregularly shaped bodies appearing circular. This time it was the humble candle, not the phases of Venus, that supplied Scheiner with the paradigmatic example. He argued that although a candle's flame is far from circular, it appears like a luminous dot when observed from a distance. To Scheiner this meant that we cannot be sure that the stars are really round (implying that even the fixed stars—the emblems of permanence—may actually have contours as messy as those of sunspots).[185] In a way reminiscent of the astronomers' devices aimed at accounting for different anomalies in planetary motions, Scheiner created a toolbox of analogies through which some of the apparent irregularities of sunspots (their tendency to emerge and disappear, their noncircular appearance, their thinness) could be either explained away as optical artifacts or could be claimed to be shared by other canonical astronomical objects. Either way, he could defuse the claim that the sunspots' complex appearance showed that they were not physical objects or, worse, that they were signs of corruption. He was trying to decouple changing appearances from corruption.

A last normalizing analogy introduced by Scheiner involved the relative transparency of apparently opaque bodies. After he had openly admitted that some spots appeared quite frayed while other seemed denser and more compact, he tried to explain their less than starry appearance by attributing their apparently irregular contours to an effect of the spots' uneven opacity. The Moon was the body that, according to Scheiner, displayed a similarly uneven opacity. He described a partial eclipse of the Sun he had observed with several people, some of them using telescopes, others only their eyes. Scheiner drew two lessons from these observations: One was that sunspots were as opaque as the Moon (something he decided after observing sunspots in the section of the solar disk that was not obscured by the Moon and judging them as black as the lunar disk). The other was that the Sun shone through the Moon (though feebly). This was an astonishing claim, but Scheiner trusted his observation and that of one of his colleagues who "most firmly asserted that through a tube he saw the entire circumference of the Sun even though the Moon still occupied some portion of it."[186]

Scheiner had much to gain by believing this observation. A partially

185. *GO*, vol. V, p. 53.

186. *GO*, vol. V, p. 68. A similar claim is at p. 53.

transparent Moon allowed Scheiner to explain both the cosmologically thorny question of the Moon's secondary (or ashen) light while lending support to his and other Jesuits' attempt to explain the apparent irregularities of the lunar surface by invoking differences in the Moon's density.[187] It was also an observation that could be used to argue that sunspots were no less real because they seemed to be sometimes "transparent" and some other times dense and black:

> The Moon itself is transparent throughout, more and less according to the greater or lesser density (which is also the case with many spots, and because of which it is maintained that many have tears in them [i.e., that they appear frayed at the edges]).[188]

That at times the sunspots appeared "transparent" was not evidence of them not being real or undergoing change. Rather, they were as opaque (or as transparent) as the Moon.

Scheiner's extraordinary claims about the variable appearance of planets, stars, sunspots, and satellites were no sign of skepticism. On the contrary, by emphasizing that things were not what they seemed, Scheiner tried to explain that although sunspots became invisible or looked ephemeral, they were as real as other very real astronomical objects like stars or planets—objects that all shared, under certain conditions, the peculiar appearances of sunspots. This is also reflected in the long discussion of the many varieties of optical distortions involved in telescopic observations that fill the *Accuratior*'s third letter. Scheiner's point there was not so much to prove how unreliable the telescope was, but to analyze distortions and artifacts in order to show how the sunspots could *not* be reduced to them. While the disappearance of the spots and their apparent changes were artifacts, the spots themselves were real.

I would frame Scheiner's candid assessment of the flaws in his illustrations in this context. He probably saw his poor pictures as just another "distortion" that he admitted like all other optical distortions involved in the spots' observations, but which was ultimately irrelevant because it did not take away from the necessary reality of the spots. In sum, he started with a cosmologically orthodox assumption about the permanence of astronomical objects (not unlike those astronomers who assumed circularity

187. *GO*, vol. V, p. 68.

188. *GO*, vol. V, p. 68. The mention of sunspots "transparent" in the middle due to sunlight shining through them is at p. 50.

and uniformity as the paradigm for celestial motions) and then showed how much that permanence could appear to disappear:

> I am forced to suspect, against what many believe, that these bodies can hardly be born and perish, but rather that such appearances, disappearances, and changes back and forth of appearances result from other causes, which are to be referred to motion: to rarity and density, position with respect to the Sun, mutual illumination, change of the accidental medium, and finally, particular shape.[189]

His attempt to show that there was something very real (and cosmologically orthodox) behind the appearances of his observations and that, among other things, he deserved credit for the discovery of real but tricky objects (sunspots and one additional satellite of Jupiter) involved a remarkable redefinition of the features of most astronomical objects. Scheiner did take substantial additional risks to salvage his claims and the credit he believed he deserved for them, but he did so by articulating what he assumed to be the cosmological directions being adopted by his senior colleagues in Rome.

At the very end of the *Accuratior,* he stated that "according to the opinion of the astronomers [i.e., Clavius' group], hardness and this constitution of the heavens cannot endure, especially in the heaven of the Sun and Jupiter."[190] This passage suggests Scheiner's near endorsement of the doctrine of the fluid skies ("hardness cannot endure") and of Tycho's planetary model ("this constitution of the heavens cannot endure"). Cardinal Bellarmine, a leading Jesuit theologian, had believed in the fluid skies since 1572—a support he confirmed in a 1618 letter to one of Galileo's closest supporters, Prince Federico Cesi.[191] According to Bellarmine, the cosmos above the Moon was not made of rigid crystalline spheres, nor was it composed of Aristotle's fifth incorruptible element, the aether. More likely, it

189. *GO,* vol. V, p. 64.

190. *GO,* vol. V, p. 69.

191. Bellarmine's own cosmological views are in Ugo Baldini and George Coyne, *The Louvain Lectures of Bellarmine and the Autograph Copy of His 1616 Declaration to Galileo* (Vatican City: Specola Vaticana, 1984). See also Ugo Baldini, "L'astronomia del Cardinale Bellarmino," in Paolo Galluzzi (ed.), *Novità celesti e crisi del sapere* (Florence: Giunti, 1984), pp. 293–305, and his "Bellarmino tra vecchia e nuova scienza," in *Legem impone subactis,* pp. 305–44. Not surprisingly, Scheiner reprinted Bellarmine's letter in his final text on sunspots, the 1630 *Rosa Ursina,* pp. 783–84. On the history of the debate about the fluid skies, see Miguel Granada, *Sfere solide e cielo fluido: Momenti del dibattito cosmologico nella seconda metà del Cinquecento* (Milan: Guerini, 2002).

was composed of a fire-like substance that was *not* incorruptible, could undergo substantial change, and could be easily penetrated.

Because of these features, the fluid skies could accommodate the novae observed in 1572 and 1604, the superlunary trajectory of comets (which otherwise would have had to plow through various crystalline spheres), and the satellites of Jupiter (whose orbits might have been hard to accommodate within the crystalline sphere of their "host" planet).[192] The mathematicians at the Collegio Romano were fully familiar with this doctrine, and it is most likely that Scheiner had read some of the letters that circulated between Rome and Germany on this topic.[193]

Given the brevity of Scheiner's remarks, it is not clear how well the mutability of the fluid skies could fit his relentless attempts to save celestial incorruptibility by explaining away all instances of anomalous change as optical effects of different kinds. It could be that he viewed a celestial medium more mutable than the Aristotelian aether as yet another tool to explain puzzling optical effects without necessarily questioning the permanence of the objects behind those appearances. Or perhaps he mentioned the fluid skies doctrine at the end of his book to hedge his bets in an increasingly uncertain dispute. Having grown more concerned about whether the spots were immutable or perishable satellites (or even whether they could be kept outside of the Sun), Scheiner may have presented himself open to the fluid skies option so as to be able to change his mind about the nature of sunspots (as he did eventually) without too much collateral damage.

Be that as it may, the doctrine of the fluid skies helped Scheiner to switch from defense to offense. He did not have to struggle to explain the anom-

192. Galileo's support of the fluid skies doctrine is implicit in the texts from the sunspots dispute, including his discussion of epicycles as nongearlike. An explicit endorsement can be found in the 1632 *Dialogue* (*GO*, vol. VII, p. 146).

193. Paul Guldin (one of Clavius' students at the Collegio Romano) reported to his colleagues in Munich on February 1611 on the Roman Jesuits' corroboration of Galileo's discoveries. This letter, which Scheiner is most likely to have read, asked: "Where shall we locate these new planets? Which and how many orbs and epicycles shall we attribute to them? Do not Tycho's views encompass them? Shall we have it that *the stars move freely like the fish in the sea?*" (Guldin to Lanz, February 13, 1611, cited in Lattis, *Between Copernicus and Galileo*, p. 210). The last line is a clear reference to Bellarmine's statement that the planets "move by themselves like the birds in the air and the fish in the water" (Baldini and Coyne, *The Louvain Lectures of Bellarmine*, p. 20). On the Jesuit mathematicians' trouble with their fellow theologians concerning the publication of claims about the "fluid skies" doctrine, see Blackwell, *Galileo, Bellarmine, and the Bible*, pp. 148–53, and Lattis, *Between Copernicus and Galileo*, pp. 94–102, 211–16.

alous behavior of sunspots and of the vanishing satellite of Jupiter, but could simply declare them paradigmatic. The new objects he had discovered were not anomalous but epitomized the features of *all* known astronomical bodies once those were redefined within the fluid skies doctrine. That the "satellite of Jupiter" he had dedicated to Welser had vanished did not mean that it was an illusion but perhaps a rare, fleeting nova-like object. Scheiner, then, could hope to keep the credit for the discovery of the sunspots and the fifth satellite of Jupiter, gain further credit for having explained the secondary light of the Moon, all the while helping his Jesuit superiors develop a new theologically safe cosmology.

It was not unreasonable for Scheiner to believe that the discoveries of 1609–11 could force the Jesuits to consider replacing Ptolemy's astronomy and his crystalline spheres with a cosmology that would combine the fluid skies with Tycho's planetary model. However, it is far from clear that these options were being entertained as viable possibilities in Rome by mid-1612, when Scheiner was writing the *Accuratior.*[194] As far as we know, the mathematicians of the Collegio Romano managed to adopt Tycho and the fluid skies only in 1620 with the publication of Giovanni Biancani's *Cosmographia.*[195] Scheiner was also surprisingly ignorant of the fact that the head mathematician in Rome, Clavius, was a staunch opponent of the fluid skies (not to mention that his call for a "new constitution of the heavens" was much less revolutionary than how Scheiner represented it at the end of the *Accuratior*). As Clavius' closest associate, Christoph Grienberger, put it, "I know, as does anyone who was familiar with Clavius, that up to the end of his life he abhorred the fluidity of the heavens, and that he constantly sought arguments to explain the phenomena by ordinary means."[196]

Clavius was not the highest-ranking Jesuit to condemn the fluid skies.[197]

194. On the ambiguous cosmological position of Clavius right before his death, see Lattis, *Between Copernicus and Galileo,* pp. 180–202.

195. Baldini, *Legem impone subactis,* pp. 217–50.

196. Grienberger to Biancani, 1618, quoted in Lattis, *Between Copernicus and Galileo,* p. 201. In his 1630 *Rosa Ursina,* Scheiner came up with a very subtle interpretation of Clavius' position in order to argue that, while not a supporter of the fluid skies, Clavius was so open to alternatives that could have "saved the phenomena" that, were he still alive, he would have embraced it (Corrado Dollo, "Tanquam nodi in tabula—tanquam pisces in aqua: Le innovazioni della cosmologia nella *Rosa Ursina* di Christoph Scheiner," in Baldini, *Christoph Clavius e l'attività scientifica dei gesuiti nell'età di Galileo,* pp. 152–53).

197. In May 1611 and again in December 1613, the general of the Jesuits had written detailed letters to all heads of Jesuit colleges stressing the necessity to follow Aristotle (in philosophy) and Aquinas (in theology) in all teachings and publications (Blackwell, *Ga-*

Toward the end of 1614, Scheiner's superiors in Germany complained to General Aquaviva, questioning his philosophical orthodoxy. The general admonished Scheiner not to uphold the doctrine of the fluid skies and instructed the head of the Society's Upper German Province to "take care that Father Scheiner not put forward any new opinions about the fluidity of the heavens and the movements of the stars on the basis of some still uncertain observations."[198] It even appears that, earlier in 1614, Scheiner was summoned to Rome to defend himself from charges of philosophical unorthodoxy.[199]

---

lileo, Bellarmine, and the Bible, pp. 137–41). The doctrine of the fluid skies had been specifically singled out as non-Aristotelian (which may explain why Bellarmine never published anything about it, as suggested by Cesi in *GO*, vol. XIII, pp. 429–30). The opposition continued well after 1612. In July 1616, the general congratulated a provincial censor in Sicily for doing "very well to oppose Father Blandino's teaching on the fluidity of the heavens," and in 1631 he ordered the rector of the Jesuit College of Avignon to "make sure that this opinion [the fluid skies] is not proposed or defended in any way in the theses of our pupils" (Gorman, "A Matter of Faith?" pp. 294–95).

198. Bernhard Duhr, *Geschichte der Jesuiten in den Landern Deutscher Zunge* (Freiburg: Herdersche Verlagshandlung, 1913), vol. II, pt. 2, p. 438. The full statement reads: "This man [Adam Tanner] turned, at the end of 1614, to the general with a complaint. Aquaviva answered him on December 23, 1614, that he had admonished Father Scheiner to abandon the new opinions about the heavens. This admonition really did happen on December 13, 1614. After a decided encouragement of his studies, which Scheiner related in a letter of November 11, 1614, Aquaviva, before anything else, answers the question of Scheiner if in a new edition of his writings on sunspots he should add his name: on good grounds it would be better to let them appear under the old name of Apelles. In refuting Galileo, Aquaviva elaborates, it would seem to me more fitting to put forward the evidence for the truth, then to refute the opposing evidence without mentioning the author, and finally to draw the necessary conclusions. In this way the entire refutation will proceed with more benevolence and humility. This one thing I would like to counsel you, that, basing yourself on the solid doctrine of the ancients, you avoid new opinions about certain novelties. Please be convinced that such things displease us very much and that we will not allow their publication by our members. In particular, against the universal doctrine of the Fathers and the Schoolmen no new hypothesis about the fluidity of the heavens will be presented, or about the stars that move like fish in the sea or birds through the air. Over the same date and instruction from Aquaviva to Hartel, the Upper German Provincial, was issued: he should take care that Father Scheiner not put forward any new opinions about the fluidity of the heavens" (ibid., pp. 436–38, trans. Albert van Helden).

199. In a March 1614 letter to Guldin, Scheiner mentioned that he had been summoned to Rome because of some of his philosophical opinions (most likely the fluidity of the heavens), but there is no evidence he actually traveled to Rome to answer these charges (Ziggelaar, "Jesuit Astronomy North of the Alps," pp. 104, 124).

There was, therefore, a substantial gap between the state of the debate in Rome (both among mathematicians and between mathematicians and philosophers) and Scheiner's representation of it. It is hard to gauge whether this was the effect of ignorance, ambition, or both. Being a junior mathematician without direct contacts with Rome, Scheiner may have mistaken Guldin's 1611 letter to the German Jesuits—a student's excited report about the new discoveries and the problems and possibilities they brought up—as the cosmological manifesto of Clavius' group. The enthusiasm and pride he might have experienced for moving, in a few months, from being a rookie teacher of mathematics to being the discoverer (and domesticizer) of new, important and controversial objects probably made him feel like a major participant in the Society's growing involvement with cutting-edge astronomical research and cosmological debates.

Or perhaps he had invested so much in his claims about the spots being stars and about Jupiter having an extra satellite that, in order to save his cultural capital, he ignored what he knew about the cosmological debates in Rome and charged ahead with claims that effectively committed the Society to positions it was not ready to endorse at that time. While Scheiner's superiors were not happy about his cosmological claims and tried to curb them, the readers of his book did not know this. All the readers soon came to realize was that these two texts about sunspots attributed to a pseudonymous "Apelles" were in fact the work of a young German Jesuit. As the Society of Jesus did not take any public action against these texts, it would have been quite reasonable to take them as a sign of the Society's tacit endorsement of the fluid skies doctrine. After all, his direct German superiors had authorized him to publish, however pseudonymously.

This is not altogether different from the distance-based construction of Galileo's authority discussed in chapter 1. Information was indeed partial in both cases. Scheiner did not know (or pretended not to know) a lot about the state of the cosmological debate in Rome; his immediate superiors in Germany most likely did not inform the Roman headquarters of their decision to allow for Scheiner's pseudonymous publication; and most readers knew nothing about the general's unhappiness with Scheiner's claims. Taken together, these partial perceptions created a situation in which the credit of the Society was forcibly lent to Scheiner's texts. Consciously or not, Scheiner stuck out his superiors' necks for his claims. However messy or unpleasant this might have been in the short term, it paved the way for the eventual acceptance of the fluid skies, just as Galileo's construction of the Medici's endorsement as firmer than it actually was nudged them into making such a commitment. But even more simply, had Scheiner played by the publications rules of the Society of Jesus by going through its internal censorship process, it is very likely that there would have been no *Tres epistolae,* no *Accuratior,* and no sunspots dispute as we know it.

CONCLUSION: EXCREMENTAL REALISM

Only two years after the publication of the *Istoria e dimostrazioni intorno alle macchie solari,* Galileo wrote a series of letters on Copernicanism and the epistemological status of astronomy. They were never published during his lifetime. One of these letters contains a startling claim about sunspots being either the Sun's food or its excrement:

> I have already discovered the constant generation on the solar body of some dark substances, which appear to the eye as very black spots and then are consumed and dissolved; and I have discussed how they could perhaps be regarded as part of the nourishment (or perhaps its excrements) that some ancient philosophers thought the Sun needed for its sustenance. By constantly observing these dark substances, I have demonstrated how the solar body necessarily turns on itself, and I have also speculated how reasonable it is to believe that the motion of the planets around the Sun depends on such a thing.[200]

It is difficult to gauge how committed Galileo was (or remained) to such organicistic views of solar processes and of the Sun's role in the cosmos.[201] However, the fact that the same views appear in a passage from the manuscript of the *Istoria* that was dropped from the printed text shows that he was entertaining them as he was arguing against Scheiner.[202] It is then clear

200. Galileo to Dini, 23 March 1615, *GA,* p. 66. It is important to notice that Galileo did not claim that the Sun necessarily produced light and heat, but that it acted as a powerful distributor of those substances or fertilizing spirits, which it received from somewhere else. Whether this reflects his actual thoughts on the specific theological debate in which he introduced these views is a matter of conjecture.

201. In the 1615 "Letter to the Grand Duchess," Galileo argued that the Sun's rotation on its axis is somehow transmitted to the planets, so that they move in orbits centered on the Sun. He described this in organicistic terms: "just as all motion of an animal's limbs would cease if the motion of its heart were to cease, in the same way if the Sun's rotation stopped then all planetary revolutions would also stop" (*LGD,* p. 116).

202. There he speculated that the spots might be "food drops" delivered (in an unspecified manner) by the planets to the Sun to sustain its light- and motion-giving function: "Resterà per l'avvenire campo a i fisici di specolare circa la sustanza e la maniera del prodursi ed in brevi tempi di dissolversi moli così vaste, che di lunga mano superano, alcune

that Galileo had quickly moved beyond the empirical statement that the Sun turned on itself carrying the spots with it to a more ambitious speculation about the Sun's transmitting that motion to the other planets within a Copernican cosmos and, from there, to thinking that the Sun's role included bringing life to the cosmos:

> It seems to me that there is in nature a very spirited, tenuous, and fast substance that spreads throughout the universe, penetrates everything without difficulty, and warms up, gives life to, and renders fertile all living creatures. It also seems to me that the senses themselves show us the body of the Sun to be by far the principal receptacle of this spirit, and that from there an immense amount of light spreads throughout the universe and, together with such a calorific and penetrating spirit, gives life and fertility to all vegetable bodies.[203]

Believing the Sun to be the vital center of the cosmos, Galileo did not find it surprising that it could also display some of the attributes of life such as the consumption of food and the production of excrements.

This, I believe, points to the most fundamental difference between Scheiner's and Galileo's positions. Like the Aristotelians, Scheiner saw change in negative terms, that is, as a sign of corruption or of "reduced being." Galileo, on the other hand, treated change as a positive notion linked to movement, life, and generation. He made that very claim in his last letter to Welser, where he criticized the Aristotelians' abhorrence of celestial change. Fearing their own mortality, Galileo argued, humans idolize what they do not and cannot have (permanence) and hate what they fear (fragility).[204]

---

di loro, in grandezza e tutta l' Affrica e l'Asia e l'una e l'altra America. Intorno al qual problema io non ardirei affermar di certo cosa alcuna, e solo metterei in considerazione a gli specolativi come il cadere che fanno tutte in quella striscia del globo solare che soggiace alla parte del cielo per cui trascorrono e vagano I pianeti, e non altrove, dà qualche segno che essi pianeti ancor possino esser a parte di tale effetto [the generation of sunspots]. E quando [. . .] fosse a sì grande lampada sumministrato qualche restauramento all'espansione di tanta luce da i pianeti che intorno sè gli raggirano, certo, dovendo correrci per brevissime strade, no potrebbe arrivar in altre parti della solar superficie" (*GO*, vol. V, p. 140 n. 26). For a full discussion of this speculation and its possible relation to ancient sources, see Bucciantini, *Galileo e Keplero*, pp. 224–41.

203. Galileo to Dini, March 23, 1615, *GA*, p. 63.

204. "I suspect that our wanting to measure the universe by our own inadequate yardstick, makes us fall into strange fantasies, and that our particular hatred of death makes us hate fragility" (*GO*, vol. V, p. 235).

But their fears of mortality get in the way of conceptualizing the world, leading them to confuse change with death:

> If that which is called corruption were annihilation, the Peripatetics would have some reason for being such staunch enemies of it. But if it is nothing else than a mutation, it does not merit so much hatred. Nor does it seem reasonable to me that anyone would complain about the corruption of the egg if what results from it is a chick.[205]

Galileo's critique of the claim that an egg is corrupted when in fact its "corruption" leads to the production of a chick—a process mapped out in great detail by his friend Aquapendente—dovetails with his claim about the sunspots being signs not of the Sun's corruption but of its role as the power plant of the cosmos.[206] An incorruptible egg is a useless egg. An incorruptible Sun does not belong in a dynamic cosmos.

This suggests that Galileo's notion of the real was closely linked to movement—both physical motion and generation of living beings. While Galileo's early use of visual sequences did not reflect specific ontological commitments, it seems that by the end of the sunspots dispute he came to see the real as inherently and positively tied to periodic change, not permanence. As he put it in a letter written to Paolo Gualdo in the midst of the sunspots debate:

> Now she [nature] finally shows us with indelible characters [the sunspots] who she is and how much she dislikes idleness. Instead, she likes to work, generate, produce, and dissolve always and everywhere. These are her highest achievements.[207]

This adds yet another (confusing) twist to the distinction between realism and nominalism. Compared to Scheiner, whom he criticized for conceptualizing the sunspots according to his ontological boxes, Galileo did not appear to be a realist because he did not structure his argument around ontological assumptions. But it turns out that was only because Galileo's assumptions were so unorthodox that it would have been unwise for him to publish them. If we look beyond his published texts, Galileo ontologized change as much as Scheiner ontologized its impossibility. Similarly, Gali-

205. *GO*, vol. V, pp. 234–35.

206. Galileo presented almost identical views in the first day of the *Dialogue on the Two Chief World Systems*, in his critique of celestial incorruptibility (*GO*, vol. VII, pp. 83–85).

207. *GO*, vol. XI, p. 327.

leo's assumptions involving solar excrements were no less extravagant than Scheiner's comprehensive "fluidification" of things astronomical.

Be that as it may, Galileo's nonessentialist conceptualization of new objects was closely related to his ontology of change. His claim that the periodic features of his discoveries proved them to be nonartifactual did not need to be tied to a specific ontology, but it certainly dovetailed with his organicistic view of change and of the real. Because he ultimately linked the real to change (and life), the signature of the real was that it moved, not the linguistic labels humans might attach to it. For the same reasons, his pictorial evidence about discoveries was not made up of mimetic snapshots but of visual narratives about the movements—the "life"—of the new object. They were movies because only the movie (not the individual frames) could capture the real.

CHAPTER FOUR

The Supplemental Economy of Galileo's Book of Nature

SINCE the late medieval period, nature had been represented as a book that, like the Scripture, had signs, meanings, and secrets for the reader to interpret.[1] In 1623, Galileo turned this topos on its head, stating that the understanding of the book of nature required reading, but not interpretation. One did not need to understand the meaning of words and sentences, but only to recognize the characters through which they were composed. As he wrote in the *Assayer,*

> Philosophy is written in this grand book, the universe, which stands continually open to our gaze. But it cannot be understood unless one first learns to comprehend the language and recognize the letters in which it is composed. It is written in the language of mathematics, and its characters are triangles, circles, and other geometric figures, without which it is humanly impossible to understand a single word of it. Without these, one wanders about in a dark labyrinth.[2]

Although the understanding of nature remained exceedingly complex and laborious, it was not a hermeneutical process. The book of nature was open

1. James Bono, *The Word of God and the Languages of Man* (Madison: University of Wisconsin Press, 1995), pp. 123–98; Ernst Robert Curtius, *European Literature and the Latin Middle Ages* (New York: Harper & Row, 1963), pp. 319–26; Olaf Pedersen, *The Book of Nature* (Vatican City: Vatican Observatory, 1992), esp. pp. 42–53; Hans Blumenberg, *Die Lesbarkeit der Welt* (Frankfurt: Suhrkamp Verlag, 1981).

2. Galileo Galilei, *The Assayer,* in Stillman Drake and C. D. O'Malley, *The Controversy on the Comets of 1618* (Philadelphia: University of Pennsylvania Press, 1960), pp. 183–84.

and transparent to anyone with a specific linguistic competence: geometry.[3] To those with such skills truth would present itself, unmediated and immediately, in the things themselves. Probably the most quoted passage from Galileo's oeuvre, the image of book of nature has come to characterize his methodology and mathematical realism.[4]

But if we reconstruct the argumentative context of the book of nature and retrace its genealogy—especially to the various texts Galileo wrote between 1613 and 1616 and to his use of the "open book of the heavens" to defend Copernican astronomy from theological objections—we see that the topos, while offering an image of methodological transparency, was fraught with unavoidable contradictions. These contradictions, I argue, offer a map of the tensions inherent in the positions Galileo developed during the controversy over the relationship between astronomy and theology. The book of nature did not emerge as an abstract methodological reflection, but as a remarkably context-specific response to critics who had invoked the absolute authority of another book: the Scripture.

The book of nature was not a single topos but a constellation of closely related topoi he used from 1612 to 1641: the "grand book of the universe," the "open book of the heavens," the "book of nature," the "grand book of the world," and the "book of philosophy." What tied these images together was their role in Galileo's attempt to legitimize his brand of natural philosophy by casting nature as a material inscription of God's logos—a "text" that was simultaneously opposed to the Aristotelian corpus and complementary to the Scripture. Throughout this chapter, I use "book of nature" to refer both to Galileo's constellation of topoi as well as to the argument

3. *GC*, pp. 306–7; Bono, *The Word of God and the Languages of Man*, pp. 193–98. Bono shows that Galileo's book of nature marked a sharp break with previous characterizations of the topos.

4. Alexandre Koyré, "Galileo and Plato," *Journal of the History of Ideas* 4 (1943): 400–28; Mario De Caro, "Galileo's Mathematical Platonism," in Johannes Czermak (ed.), *Philosophy of Mathematics* (Vienna: Verlag Holder-Pichler-Tempsky, 1993), pp. 13–22. An analysis of the "book of nature" in Galileo's corpus, and its relationship to Galileo's views on the differences between mathematics and physics, human and divine knowledge, human language and mathematics is in Carla Rita Palmerino, "The Mathematical Characters of Galileo's Book of Nature," in Klaas van Berkel and Arjo Vanderjagt (eds.), *The Book of Nature in Modern Times* (Leuven: Peeters Publishers, 2005). On the relationship between Galileo's treatment of the language of nature in the *Sunspots Letters* and the Copernican texts of 1613–16 see Giorgio Stabile, "Linguaggio della natura e linguaggio della scrittura in Galilei, dalla *Istoria* sulle macchie solari alle lettere copernicane," *Nuncius* 9 (1994): 37–64.

about the complementarity of nature and the Scripture encapsulated in those topoi—his concept of the book of nature.

While not articulated in simple opposition to the claims of Galileo's adversaries, the book of nature was, in certain ways, made possible by them. Emerging in (and needing to comply with) a discursive context framed by the theologians' Scripture-based regime of truth, Galileo's book of nature did not and could not try to cast the domain of astronomy and philosophy as merely independent from theology (as most commentators have read him to do). The space he tried to develop for astronomy was not carved away from that of theology but rather constructed through the features and discursive practices of that more authoritative field. Galileo's book of nature deferred to the Scripture while differing from it.

Whenever possible, Galileo tried to turn the theologians' positions and authority into supplements for his own discourse. As a result, his arguments incorporated and multiplied several of the irresolvable tensions underlying the theologians' claims to authority—claims they rooted in their alleged ability to find God's original speech in the pages of the Scripture. Like the Scripture, Galileo's book of nature was not cast as an actual book but as the materialization of God's speech.[5] While the content of Galileo's claims was often in conflict with the exegetical positions held by the Church, the logic of his discourse (which, as shown in previous chapters, was distinctly nonessentialist) became, during this controversy, as logocentric as that of the theologians.[6]

5. Derrida sees the book of nature as an example of "natural writing" like the Platonic writing of truth in the soul, a kind of writing that, like the logos, is opposed to literal, material writing. This exposes a crucial paradox: "natural and universal writing, intelligible and nontemporal writing, is thus named by metaphor. A writing that is sensible, finite, and so on, is designated as writing in the literal sense; it is thus thought on the side of culture, technique, and artifice; a human procedure, the ruse of a being accidentally incarnated or of a finite creature. Of course, this metaphor remains enigmatic and refers to a 'literal' meaning of writing as the first metaphor. This 'literal' meaning is yet unthought by the adherents of this discourse. It is not, therefore, a matter of inverting the literal meaning, but of determining the 'literal' meaning of writing as metaphoricity itself." Jacques Derrida, *Of Grammatology* (Baltimore: Johns Hopkins University Press, 1976), p. 15.

6. My argument resonates with some of the critiques made by Lily Kay to a similar construct—the book of life—in her *Who Wrote the Book of Life?* (Stanford: Stanford University Press, 1999), and by Richard Doyle in *On Beyond Living: Rhetorical Transformations of the Life Sciences* (Stanford: Stanford University Press, 1997), esp. pp. 86–108. See also Colin Milburn, "Monsters in Eden: Darwin and Derrida," *Modern Language Notes* 118 (2003): 603–21, esp. 611–15.

CONSTRAINTS, SUPPLEMENTS, AND DEFERRALS

There is no evidence that Galileo wanted to enter a debate on the relationship between Copernicanism and the Scripture. His previous astronomical publications presented a number of observations that contradicted Ptolemaic astronomy while supporting Copernicus, but his claims remained within the bounds of mathematics and natural philosophy without trespassing into theology.[7] Shortly thereafter, however, he was forced to confront issues of scriptural exegesis by critics who questioned the religious orthodoxy of the Copernican doctrine as well as Galileo's personal piety by citing scriptural passages that, if interpreted literally, instantiated a geocentric cosmology.

Benedetto Castelli, a disciple of Galileo, was unexpectedly queried about the religious orthodoxy of Copernicanism during a meal at court in Pisa in December 1613. Known for her piety, Grand Duchess Christina took a personal interest in the issue and, prodded by an Aristotelian philosopher from the university, engaged Castelli for about two hours.[8] Castelli was proud of his performance but Galileo seemed to worry that the grand duchess might want to continue the conversation and perhaps develop doubts about Galileo's religious orthodoxy and his suitability as a recipient of Medici patronage.[9] In a matter of days, and without the opportunity to familiarize himself with the relevant theological literature, he composed and sent to Castelli a small paper on the relationship between astronomical knowl-

7. There was a partial exception. In his *Sunspots Letters,* it was the licensers' intervention that prevented Galileo from invoking the Scripture in support of his arguments about the corruptibility of the heavens. Notice, however, that in this case Galileo was not challenging theology but rather citing theology to support his astronomical claims. The deleted and revised passages of this book have been reproduced in *GO,* vol. V, pp. 138–39. Additional material is found in the correspondence (*GO,* vol. XI, pp. 428–29, 431, 458).

8. Castelli to Galileo, December 14, 1613, *GA,* pp. 47–48.

9. Already in December 1611, Galileo had been warned that some local opponents of the motion of the Earth were gathering around the archbishop of Florence to plan some action against Galileo (*GO,* vol. XI, pp. 241–42). We do not know whether the Castelli incident was the result of the archbishop's leaning on the grand duchess, but Galileo could assume that the local clergy would have used it to stoke the Medici's suspicions against Galileo's piety as well. About a year later, Galileo wrote Monsignor Dini that Monsignor Gherardini "burst out with the greatest vehemence against me, appearing deeply agitated, and saying that he was going to mention the matter at great length to their Most Serene Highnesses, since my extravagant and erroneous opinion was causing much talk in Rome" (*GA,* p. 57).

edge and scriptural exegesis, including a few tips and examples for a possible follow-up discussion with the grand duchess.[10]

This text—the "Letter to Castelli"—was not meant for publication, but Castelli let it circulate widely, apparently without Galileo's approval.[11] To make things worse, the copying corrupted the text in ways that made Galileo's claims sound more controversial than they originally were.[12] The Florentine clergy grew concerned about the popularity of the "Letter to Castelli" and turned its concerns into public accusations during a sermon delivered in December 1614 by a Dominican friar, Tommaso Caccini.[13] Soon after, a copy of the "Letter to Castelli" and then a judicial deposition accusing Galileo of suspect heresy were delivered to the Congregation of the Holy Office in Rome.[14] The letter Galileo had written to control a poten-

10. Galileo to Castelli, December 21, 1613, *GA*, pp. 49–54. Later on Galileo acknowledged that his response to Castelli was written "with a quick pen" and with limited familiarity with scriptural exegesis (*GA*, p. 55). Galileo's response dates from December 21, 1613, six days after the date of Castelli's letter. The text includes no references to theological literature. Galileo went over Castelli's discussion, adding to it here and there to strengthen it—especially a fuller discussion of his heliocentric interpretation of the scriptural passage about God's stopping the Sun's motion after hearing Joshua's prayer. This was a key example used by the grand duchess against Copernicus. Castelli had given the grand duchess three arguments in support of a Copernican reading of Joshua, saying that one of them was Galileo's. In his response to Castelli, Galileo went back to this argument about Joshua and spelled out the questions to deploy against a hypothetical opponent—possibly Boscaglia, a philosopher to the University of Pisa who guided the grand duchess in her questioning of Castelli during the initial discussion (*GA*, pp. 52–54). On April 16, 1614, Castelli wrote to Galileo about a rematch with Boscaglia at court, one Castelli was confident he had won. The subject of the debate is not mentioned, but, being described as a rematch (*seguito*), it could have been on Copernicanism and the Scripture (*GO*, vol. XII, p. 49).

11. *GA*, p. 62.

12. Galileo suspected that the letter sent to Rome may have been corrupted and, on February 16, 1615, he sent the original version of the "Letter to Castelli" to Monsignor Dini in Rome, asking him to give a copy to the Jesuit mathematician Grienberger and to Cardinal Bellarmine (*GA*, p. 55). The discrepancies between the two versions are given in *GA*, p. 331 n. 16.

13. *GA*, p. 137; *GO*, vol. XII, p. 123. On February 7, 1615, Niccolò Lorini (a Florentine Dominican) wrote to the Holy Office that Galileo's letter passes "through everybody's hands, without being stopped by any of the authorities" (*GA*, p. 135).

14. The March 20, 1615, denunciation by Caccini was preceded by a more informal complaint by another Dominican, Niccolò Lorini, on February 7, 1615. Lorini's complaint included a copy of the "Letter to Castelli." Both are translated, together with an inquisitor's report on "Letter to Castelli," in *GA*, pp. 134–41.

tial local crisis ended up fueling a much wider, more dangerous conflict. Its circulation also shifted the debate from friendly forums (Florence and its court) to a remote theater of operation (the Holy Office in Rome) where he had fewer supporters and where the theologians set the rules of the game. These rules were quite different from anything Galileo had worked with before, making the transition from Florence to Rome as dramatic as that between Padua and Florence.

As the inquisitorial process was slowly beginning to turn its secretive wheels in the spring of 1615, a respected Carmelite theologian unknown to Galileo, Paolo Antonio Foscarini, arrived in Rome with a short book in which he argued that Copernicus and the Scripture were not irreconcilable.[15] Foscarini's intervention backfired.[16] The book caught the immediate attention of the Holy Office, while Foscarini's references to Galileo as a Copernican seemed only to confirm the accusations of the Florentine clergy.[17] Written in Calabria and printed in Naples, the book also fueled

15. Paolo Antonio Foscarini, *Lettera sopra l'opinione de' Pittagorici e del Copernico . . .* (Naples: Scoriggio, 1615). A full English translation is in Richard Blackwell, *Galileo, Bellarmine, and the Bible* (Notre Dame: University of Notre Dame Press, 1991), pp. 217–51. Foscarini was traveling from Calabria to Rome to preach there during Lent and had the book printed in January during a stopover in Naples. On Foscarini's life and work, see Emanuele Boaga, "Annotazioni e documenti sulla vita e opere di Paolo Antonio Foscarini teologo 'copernicano,'" *Carmelus* 37 (1990): 173–216.

16. One of Galileo's closest associates, Prince Cesi, was pleased by Foscarini's intervention when he wrote on March 7, 1615, that "non poteva venir fuori in miglior tempo" (*GO,* vol. XII, p. 150). At the time he wrote this letter, however, Cesi was not aware that the Holy Office was initiating proceedings against Galileo. Two weeks later another friend, Giovanni Ciampoli, thought that Foscarini's book could be prohibited within a month from its arrival in Rome (*GO,* vol. XII, p. 160). Ciampoli's view was more accurate than Cesi's. Informed that the Inquisition was taking aim at his book, Foscarini wrote to Bellarmine in late March or early April to defend it. The censor's report is in Blackwell, *Galileo, Bellarmine, and the Bible,* pp. 253–54. Foscarini's original letter to Bellarmine is reproduced in Boaga, "Annotazioni e documenti sulla vita e opere di Paolo Antonio Foscarini," pp. 204–14. The impact of Foscarini's text on the proceedings against Galileo is discussed in Massimo Bucciantini, *Contro Galileo: All'origine dell'affaire* (Florence: Olschki, 1995), pp. 53–68. See also Blackwell, *Galileo, Bellarmine, and the Bible,* pp. 87–110.

17. Foscarini wrote, "As far as I know, and may it be pleasing to God, I am without doubt the first one to undertake this project [the reconciliation of Copernicus and the Scripture]. I believe that considerable appreciation will be expressed by those who are studying this issue, and especially the most learned Galileo Galilei, Mathematician to the Most Serene Grand Duke of Tuscany, by the most learned Johannes Kepler, Mathematician to the Sacred and Invincible Majesty of the Empire, and by all the illustrious and most virtuous

the Holy Office's concerns about the spread of Copernicanism across the Italian peninsula and especially among Catholic theologians.[18] By the time Galileo's most articulate defense of his exegetical stance—the "Letter to the Grand Duchess"—was completed in 1615, the Church's position on astronomy and biblical exegesis had stiffened significantly.[19] On March 5, 1616, three books were placed on the Index: Copernicus' *De revolutionibus,* Antonio Foscarini's *Letter on the Pythagorean and Copernican Opinion,* and Diego de Zuniga's *On Job,* with the warning that "all other books that teach the same be likewise prohibited."[20] Galileo's name was not mentioned in this edict because neither his letter to Castelli nor that to the grand duchess had been printed. As a result his claims did not fall under the jurisdiction of the Index—a congregation charged with book censorship.[21] The Church's decision to censor Copernican books, however, set the framework for Galileo's trial of 1633.[22]

This chain of events reflects a remarkable pattern of constraints and

---

members of the Academy of the Lynx, who universally accept this opinion [Copernicanism] (unless I am mistaken)." Blackwell, *Galileo, Bellarmine, and the Bible,* p. 223.

18. Of the three pro-Copernican books prohibited in 1616 (those of Copernicus, Zuniga, and Foscarini), two were by theologians. If the Holy Office wished to eradicate the Copernican scourge, it could have compiled a substantially longer list of Copernican texts. Kepler's 1609 *Astronomia nova,* for instance, was not mentioned. Only his 1618 *Epitome Astronomiae Copernicanae* caught the Index's attention in 1619 (Pierre-Noel Mayaud, *La condamnation des livres coperniciens et sa révocation* [Rome: Editrice Pontificia Università Gregoriana, 1997], pp. 59, 65–69). The censors highlighted as particularly pernicious the few passages where Kepler referred to the Scripture. It seems, therefore, that in 1616 the Holy Office was not primarily concerned with Copernicanism in and of itself, but with those texts that brought together Copernicus and the Scripture. That Zuniga and Foscarini were Catholic theologians made their cross-disciplinary texts all the more dangerous. The March 5, 1616, decree of the Index presents Foscarini's book as evidence of the "spreading and acceptance by many of the false Pythagorean doctrine" put forward by Copernicus and supported by Zuniga (*GA,* p. 149).

19. Blackwell, *Galileo, Bellarmine, and the Bible,* p. 98.

20. *GA,* pp. 149.

21. While the proceedings were managed by the Congregation of the Holy Office (which controlled all issues of doctrinal orthodoxy), the March 6 edict was issued by the Congregation of the Index, whose jurisdiction covered the licensing and censoring of books.

22. The interpretation of the meaning of the injunction given by Bellarmine to Galileo in 1616 (*GA,* pp. 147–48, 153) was key to the later determination as to what Galileo was allowed to say about Copernicus in his 1632 *Dialogue.* He was found to have violated those injunctions.

handicaps facing Galileo: He was no expert in biblical exegesis or theology;[23] he could not control the timing, pace, or forum of the dispute; Scripture was deemed more authoritative than any astronomical text; Galileo's disciplinary authority (as a mathematician) was much inferior to that of the theologians; and the theologians' superior (the pope) was clearly more powerful than Galileo's supportive patron (Grand Duke Cosimo II).[24] But a problem even greater than these power differentials was that, despite all the anti-Aristotelian and anti-Ptolemaic ammunition provided by his discoveries, Galileo lacked the kind of evidence the theologians might have accepted as a conclusive proof of Copernicanism.[25] This absence was perhaps the

23. Galileo told Dini on February 1615 that the controversies set in motion by his letter to Castelli "have made me look at other writings on the topic" (*GA*, p. 55). Galileo relied on his friend Benedetto Castelli (a friar-mathematician who had also received theological training) to gather the appropriate sources. On January 6, 1615, Castelli reported that: "Io sono alle mani con il Padre Predicatore de' barnabiti, affezionatissimo alla dottrina di V.S., e m'ha promesso certi passi di S.Agostino e d'altri Dottori in confermazione del sentimento dato da V.S. a Giosuè" (*GO*, vol. XII, pp. 126–27). Galileo's "Letter to Castelli" includes no references to theological literature or to any of the rulings of the Council of Trent on matters of biblical exegesis, but such references are abundant in the "Letter to the Grand Duchess" and in the contemporary "Considerations of the Copernican Opinion." Besides an improved familiarity with the theological literature, this shift indicates Galileo's new awareness that, since Trent, the Catholic Church had identified the rule of faith not only in the Scripture (as the Protestants did) but also on related exegetical traditions preserved within the Church—a point that Bellarmine was to repeat quite clearly to Foscarini in April 1615. Any new interpretation needed to square not only with the text of the Scripture but also with that of its traditional commentators (*GA*, pp. 67–68). The last part of Galileo's "Considerations on the Copernican Opinion" is a direct response to some of Bellarmine's points to Foscarini, suggesting that Galileo had obtained a copy of that letter. Galileo's care not to release the "Letter to the Grand Duchess" before an extensive debugging is evident in his letters to Dini and to Castelli (esp. *GO*, vol. XII, p. 165) and testifies to his new awareness of having initially stepped on a minefield without appropriate maps.

24. I compare the power of the Medici and the pope because, in this debate, secular authority was as relevant as the hierarchy of disciplinary authority that structured its discourse. Most likely there would have been no Galileo trial in 1633 if the pope did not have the political leverage to force the Medici to send Galileo to Rome. It is evident that in 1615–16 the grand duke (but not his mother, the grand duchess) did not hesitate to put much of his clout behind Galileo, and to try to influence the outcome of the dispute by political means. The letters between Florence and the Medici ambassador in Rome, or between the Medici and various cardinals reproduced in GO, vol. XII, testify to that.

25. His attempt to prove the Earth's motion on its axis and around the Sun through an explanation of tidal phenomena dates to this period and to this specific high-stakes context. Galileo wrote the argument as a letter to Cardinal Orsini at the beginning of 1616, that is,

most important factor in structuring Galileo's tactics. Galileo's discursive edifice was, in fact, remarkably defensive. Its tone was proactive—even aggressive—but its goals were much more modest: to be allowed to put off proving Copernican astronomy by stalling its pending condemnation.

Given the remarkable structural differences between the disputes Galileo had with Scheiner and some of his early critics and this one with the theologians, it is important to understand what kind of problems Galileo's lack of a proof for Copernicus posed in this specific context. This was not a dispute over heliocentrism and geocentrism between Galileo and a Ptolemaic astronomer. In that case Galileo would have tried to bring such a dispute to a successful closure, but nothing too unpleasant would have happened had he failed to do so. The debate would have just remained open. On the contrary, Galileo's lack of a proof for Copernicus in 1615 would have allowed the theologians to bring it to their kind of "closure" by condemning Galileo or Copernicanism, or both. A second consequence, perhaps more important than the first, was that the theologians' condemnation would have formally put an end (or, as it has actually happened, a long-lasting chill) on work leading to proving Copernicus within countries under their control.

This controversy involved a dramatic change in the power differential between Galileo and his adversaries, but also in the rules of the game and in its stakes.[26] On this occasion, Galileo did not try to achieve closure (at some point down the road) but rather to *avoid* the specific kind of closure the theologians could (and did) impose at a time of their choosing. The goal was to keep the game in play; not simply to defer the proof of Copernicanism to a later time but to secure the conditions of possibility for such a deferral—to make the deferral deferrable. In so doing, he was not seeking credit in the short term, but investing in the possibility of credit in the long term. I doubt we would have heard much of Galileo's book of nature had he been able to prove Copernicus to the theologians' satisfaction by 1615. There would have been no need, in that case, to build the intricate discursive edifice discussed in this chapter.

---

just a few weeks before the condemnation of Copernicus (*GA*, pp. 119–33) and it was then revised into the fourth day of the 1632 *Dialogue*.

26. An informative review of the workings of the tribunal of the Holy Office is in Francesco Beretta, "Le Procès de Galilée et les archives du Saint-Office: Aspects judiciaires et théologiques d'une condamnation célèbre," *Revue des sciences philosophiques et théologiques* 83 (1999): 441–90 esp. 446–54; and in his "L'Archivio della Congregazione del Sant'Uffizio: Bilancio provvisorio della storia e natura dei fondi d'antico regime," in Andrea del Col and Giovanna Paolin (eds.), *L'Inquisizione Romana: Metodologia delle fonti e storia istituzionale* (Trieste: Edizioni Università di Trieste, 1999), pp. 119–44.

THE MUNDANE ROOTS OF GENERALITY

As the debate moved to Rome and into the theologians' domain, its scope and tone changed. Galileo's "Letter to the Grand Duchess" grew five times longer than the initial essay to Castelli, but its argument, while more comprehensive and dense with references to the theological literature, was also more legalistic and defensive.[27] Although cast as a letter to a patron, its function resembled that of a legal brief.

By virtue of their position, the theologians of the Holy Office did not dispute with the authors whose work they evaluated: they simply approved or condemned their books and claims insofar as they impinged on Christian doctrine.[28] As Galileo was to learn, this was a remarkably unilateral, secretive process.[29] He tried to make it more dialogical by traveling to Rome at the end of 1615 to entreat Cardinal Bellarmine (the head of the Holy Office) into face-to-face discussions where he could "make use of the tongue instead of the pen," get a sense of his exact concerns, and address them.[30] But once there he realized that

> I cannot directly make contact and have an open discussion with those people I should negotiate with [. . .], nor can they be open in the least toward me without risking the most serious censures. Consequently, I have to proceed with great care and effort by identifying third parties who, without even knowing about the matter at hand, would be willing to act as intermediaries with those important people so that, as if by accident or upon

27. Galileo wrote the "Letter to the Grand Duchess" as if he was not defending himself but all Copernicans and even the entire discipline of mathematics from the attacks being brought forward against it. For instance, "this was done with little compassion and consideration for the injury not only to the doctrine [Copernicus'] and its followers, but also to mathematics and all mathematicians," or "in order to accomplish that objective, it would be necessary not only to prohibit Copernicus' book and the writings of the other authors who follow the same doctrine, but also ban all astronomical science completely" (*LGD*, pp. 89, 102).

28. The theologians' judgment differed from those of other disciplines in another important aspect: their reading was aimed at assessing orthodoxy, not quality. They did not, in principle, assess an argument past the point it related to theology. They did not judge (and their judgment did not reward) "good" mathematics or natural philosophy. Intellectually flawed texts could be easily approved if they did not violate Christian doctrine.

29. Galileo remarked that the Holy Office was very secretive even in comparison to other tribunals (*GO*, vol. XII, p. 231).

30. *GO*, vol. XII, p. 184.

> their request, I would find myself in a situation in which I can express and explain my concerns in detail. And when it is necessary to write my arguments down on paper, I need to have them secretly delivered to those I want to get them.[31]

Galileo knew Bellarmine personally, but Bellarmine summoned him only at the end of the proceedings to make sure Galileo understood and accepted the Holy Office's condemnation of the Copernican doctrine.[32] In the year leading to that decision, Bellarmine read the "Letter to Castelli" and perhaps the "Letter to the Grand Duchess," but all Galileo knew about Bellarmine's positions were a few remarks delivered in passing to Galileo's correspondents in Rome and a copy of his critique of Foscarini's book.[33]

Galileo had a limited sense of his adversaries' positions, but knew well that this was not the usual philosophical dispute with claims, responses, and counterresponses. All he could hope for was to have Bellarmine read one letter of his before deciding the case. While the *Nuncius* and the *Sunspots Letters* were aimed at a wide audience, the "Letter to the Grand Duchess" was a single-purpose text designed to influence the outcome of the Holy Office's proceedings by targeting only a handful of theologians, within a narrow window of opportunity, and with one single (however broadly construed) discursive shot. At the same time, Galileo needed to avoid making the purpose of the letter too explicit, as he was not supposed to know about the Holy Office's proceeding against him.[34] The broad, general scope and legalistic tone of the "Letter to the Grand Duchess," therefore, resulted from the many conflicting mundane circumstances this text was asked to address without seeming to acknowledge them. Its very title (at odds with the identity of the letter's actual addressee) epitomized Galileo's predicament: He was trying to engage Bellarmine while pretending not to.[35]

31. *GO*, vol. XII, pp. 227–28.

32. "Inquisition Minutes (February 25, 1616)" and "Special Injunction (February 26, 1616)," *GA*, pp. 147–48.

33. *GA*, pp. 58–59, 67–69; *GO*, vol. XII, pp. 129, 173.

34. Referring to the Holy Office's proceedings, he wrote as if they concerned only Copernicanism, not him personally (*LGD*, pp. 91, 98, 110).

35. It appears that Galileo initially addressed the letter to Castelli (as a follow up on his first letter) but subsequently changed its title to the final one. However, the actual addressee of the letter is Bellarmine (whose positions the letter addresses and challenges) and the other cardinals of the Holy Office (who had the power to decide over the matters being discussed in the letter).

I have the impression (based on the overall feel of the text rather than on specific passages) that the "Letter to the Grand Duchess" was written to function together with the personal discussions Galileo hoped to have with Bellarmine and his associates, not as a written proxy for them.[36] However, we do not have any evidence that Galileo had such a discussion, nor do we know if Bellarmine ever read the "Letter to the Grand Duchess." While the "Letter to Castelli" was central to the Holy Office's proceedings against Galileo, the reception of the "Letter to the Grand Duchess" is undocumented.[37] Bellarmine seemed eager to see it after hearing from intermediaries that Galileo was completing a longer essay on theology and astronomy.[38] The letter, however, did not enter the Holy Office's records, its arguments were never discussed in the 1616 proceedings, and its reading by Bellarmine (or by the grand duchess) was never mentioned in Galileo's correspondence.[39] It could be that Bellarmine read it but dismissed the

36. A hint of this is in *LGD,* p. 110. As we see from Galileo's letter quoted above (*GO,* vol. XII, pp. 227–28), Galileo did have a number of private conversations in Rome from December 1615 to February 1616, but there is no evidence that they were with the people he really wanted to reach. The letter to Cardinal Orsini on the tides (*GA,* pp. 119–33) stems from one of these meetings, as probably do the notes left unpublished by Galileo and now called "Considerations on the Copernican Opinion" (*GA,* pp. 70–86). In the above mentioned letter, in fact, Galileo referred to short texts that he wrote in Rome in this period that he had "secretly delivered" to his addressees.

37. Certainly the "Letter to Castelli" was read by Bellarmine and by other theologians of the Holy Office who found it "bad-sounding" but not openly dangerous ("Consultant's Report on the Letter to Castelli," *GA,* pp. 135–36).

38. *GA,* pp. 58–59.

39. By May 16, Galileo had not yet sent the "Letter to the Grand Duchess" to Dini to have it forwarded to Bellarmine (*GO,* vol. XII, p. 181). However, he had it read in April by friendly theologians to "debug" it before sending it to Rome. This debugging, however, did not involve copying or circulation (*GO,* vol. XII, p. 165). It seems that the letter experienced great circulation later on (as shown by the many manuscript copies still extant), but not before the February 1616 condemnation. Antonio Favaro, the editor of Galileo's *Opere,* notes that the letter "perhaps did not circulate widely, as suggested by the fact that one does not find it mentioned in the Proceedings of 1615–16, though that could have been out of consideration for the addressee" (*GO,* vol. V. p. 266 n. 3). Maurice Finocchiaro has noticed that the "Letter to the Grand Duchess" (misidentified as a letter to the grand duke) was mentioned by Inchofer in his April 17, 1633, report on Galileo's *Dialogue* (*GA,* p. 263). Inchofer's report makes it very clear that the Holy Office had not asked him to consider that letter (they may not even have known of its existence), but that he included such a discussion out of his pious zeal to demonstrate a pattern in Galileo's "mental attitude" about Copernicanism and the Scripture. Inchofer, however, does say that "if I am not deceived, here in Rome [the letter] passed through the hands of quite a

costly bargain Galileo proposed there.[40] It could also be that in the end Galileo thought it too dangerous (or just useless) to send it. It is possible that, together with his manuscript "Considerations on the Copernican Opinion," Galileo's most elaborate articulation of his positions might have been a monologue or virtual dialogue with an interlocutor he never reached.[41] This was a peculiar dispute indeed.

FROM THE ARISTOTELIAN CORPUS TO THE BOOK OF NATURE

In several texts from this period, Galileo grounded his defense of Copernicanism on the following assumptions:

1. Two truths cannot contradict each other.[42]
2. Both nature and the Scripture are authored by God.[43]
3. The domains of astronomy and theology, their interpretive protocols, and their different authority need to be understood as deriving from the

few" (*GA*, p. 263). Assuming that Inchofer is right, the letter must have not circulated widely outside of Rome if Bernegger, the publisher of the printed edition of the "Letter of the Grand Duchess" (a bilingual Latin-Italian edition produced in 1636 in Strasbourg) complained that he had not been able to find a copy in time to include it in his 1635 Latin edition of Galileo's *Dialogue* (Mayaud, *La condamnation des livres coperniciens,* p. 108, esp. n. 7). On the reception of the "Letter to the Grand Duchess" see Maurice Finocchiaro, *Retrying Galileo, 1633–1992* (Berkeley: University of California Press, 2005), pp. 72–79, and pp. 379–80 n. 56.

40. I believe that Bellarmine read and understood Galileo's argument and its implications. However, in his warnings both to Foscarini and, indirectly, to Galileo, Bellarmine stated his belief that Copernicus would never be proven. Consequently, he was not likely to be too concerned with the possible embarrassment (highlighted by Galileo) that the Church might suffer down the line for having prohibited a doctrine that was later proven true. By not perceiving that risk, Bellarmine had little motivation to embark in the radical reform of the relation between astronomy and theology (and especially between theology and Aristotelian cosmology) required by Galileo's argument. Having been a key player in the post-Tridentine reformation of the canonical text of the Catholic Bible (approved in 1592), Bellarmine was probably ill-disposed to rock the exegetical boat. Concerning Galileo's key evidence—the phases of Venus—Bellarmine certainly understood that they meant that Venus went around the Sun but, knowing about the existence of Tycho's planetary model, probably did not see them as the refutation of geocentrism Galileo took them to be.

41. This of course changes nothing about the logic of the arguments presented in these texts and their linkages to specific arguments and people Galileo was responding to.

42. *GA*, pp. 51, 52, 74, 75, 81, 96.

43. *GA*, pp. 50, 93.

specific features of the two different instantiations of the Logos read by these two disciplines.[44]

Galileo and the theologians agreed on the first two assumptions. The third claim, on the other hand, encapsulated the many contentions about disciplinary boundaries, methods, and hierarchies between astronomy and theology that Galileo was trying to destabilize. According to Galileo, the theologians believed that

> theology is the queen of all the sciences and hence must not in any way lower herself to accommodate the principles of other less dignified disciplines subordinated to her; rather, these others must submit to her as to a supreme empress and change and revise their conclusions in accordance with theological rules and decrees.[45]

By positing two different instantiations of the Logos, Galileo was trying to turn a hierarchical relationship between theology and astronomy into a parallel one: Both theology and astronomy dealt with the same truth, but one that was written in two different books, in two different languages. This is a position he maintained through the trial of 1633, teaching it to his last pupils. The most famous of them, Evangelista Torricelli, declared in his *Academic Lessons* (delivered a few months after Galileo's death) of having heard "a great mind say that God's omnipotence composed once two volumes. In one, *dixit, et facta sunt,* and this was the Universe. In the other, *dixit, et scripta sunt,* and this was Holy Scripture."[46]

Galileo's stance was simultaneously reactive and proactive. By present-

44. *LGD,* pp. 101–4.

45. *LGD,* p. 99.

46. Evangelista Torricelli, "Prefazione in lode delle matematiche" in *Opere scelte,* Lanfranco Belloni (ed.) (Turin: UTET, 1975), p. 620, translated in Palmerino, "The Mathematical Characters of Galileo's Book of Nature." Torricelli continued: "Che per leggere il gran Volume dell'Universo (cioè quel libro, ne i fogli del quale dovrebbe studiarsi la vera filosofia scritta da Dio) sieno necessarie le Matematiche, quelli se ne accorgerà, il quale con pensieri magnanimi aspirerà alla gran scienza delle parti integranti [. . .] e i soli caratteri con i quali si legge il gran Manuscritto della filosofia divina nel libro dell'universo, non sono altro che quelle, misere figurette che vedete ne i Geometrici elementi" (Torricelli, "Prefazione in lode delle matematiche," 620–21). Torricelli, delivering these lines some time in 1642, clearly linked the two elements of the book of nature that appeared in different texts of Galileo's: the complementarity between the book of nature and the Scripture (in the letters to Castelli and to the grand duchess) and the fact that the book of nature is written in geometrical characters (in the *Assayer*).

ing astronomy and theology as disciplines dealing with the same truth inscribed in two different but equally sacred books, Galileo tried to cast himself as respectful of the authority of divine books, not an atheist who put scientific evidence above scriptural teachings. But as he endorsed the theologians' book-based regime of truth and made it his own, he also elevated the status of astronomy to that of a science that, like theology, dealt with divine speech—the speech that authored both nature and the Scripture. Galileo could then argue that when the reading of the two halves of creation sent theology and astronomy on a collision course, such conflicts could no longer be adjudicated by considering which discipline was the most authoritative (as their equally divine subject matters put them on an equal footing). One should instead evaluate the competing claims by considering the specific features of the two books and the exegetical options they did or did not offer to their readers. The power of solving disciplinary clashes became an attribute of the books themselves, not of their readers.

By the time the geometrical book of nature was presented in the 1623 *Assayer,* its relationship to the Scripture had already been effaced (and for good reasons),[47] thus facilitating later readings of the topos as a purely philosophical reflection on the relationship between mathematics and the physical world rather than an element of a defensive tactic developed against a specific adversary. But the genealogical link between the Scripture and the book of nature is quite clear in Galileo's 1613 "Letter to Castelli" and in his invocation of the "open book of the heavens" in the 1615 "Letter to the Grand Duchess." These two texts do not yet present nature as written in geometrical characters, but they cast it as fully transparent, that is, as something whose understanding did not require an act of interpretation. This is the defining feature of Galileo's defense of Copernicanism and of the topos of the book of nature. It was a feature that emerged precisely from his casting nature in a relation of complementarity to the Scripture during the debates of 1613–16.

Prior to those debates, Galileo did not associate the image of the book with nature. Rather, he saw the relationship between natural philosophy and the book in distinctly negative terms. In a 1611 letter to Kepler,[48] for

47. I do not find it surprising that, after the condemnation of 1616, Galileo would not want to foreground the connections between the book of nature and the Scripture, given that it was precisely the trespassing from astronomy to theology that upset the Holy Office and led it to issue the 1616 edict.

48. On Kepler's notion of the book of nature see Pedersen, *The Book of Nature,* pp. 42–46.

instance, Galileo derided the Aristotelian philosophical establishment for refusing to engage with the evidence produced by his telescope:

> What do you think of the chief philosophers of our gymnasium who, with the stubbornness of a viper, did not want to see the planets, the Moon, or the telescope, even though I offered them the opportunity a thousand times? In truth, just as he [Odysseus] closed his ears, so they closed their eyes to the light of the truth. That is monstrous but it does not astonish me, for men of this kind think that philosophy is a book, like the Aeneid or the Odyssey, and that the truth is to be sought not in the world and in nature, but in the comparison of texts (as they call it).[49]

Far from being written in the book of nature, philosophy had been wrongly reduced to a human book—Aristotle's corpus. The Peripatetics peered down on books while Galileo-style philosophers gazed up at nature. The former believed that the realm of knowledge was finite while the latter thought of it as boundless. In 1611, then, the book was still something limited, limiting, and external to the truth.

Galileo's position had not changed by May 1612, when, in a draft of the first letter on sunspots to Welser, he still cast nature as the "true and real world" in opposition to the Aristotelians' "world painted on sheets of paper."[50] Nature and the book seemed to grow closer a few months later, when Galileo introduced the image of the "book of the world" in his second letter to Welser. The shift, however, was not as drastic as it appears. As he introduced the "book of the world," Galileo also started to highlight the fundamental differences separating human books from natural books. Unlike the physically and cognitively limited books of the Aristotelians, the book of the world was "grand"—an adjective that continued to characterize the topos in its later appearances:

> Some righteous defenders of every Aristotelian minutia [. . .] have been taught and fed since the beginning of their education the opinion that philosophy is—and could not be anything else than—working Aristotle's texts over and over. Because [they believe] that one can quickly cut and paste pas-

49. Galileo to Kepler, August 19, 1610, *GO*, vol. X, p. 423, trans. in Hans Blumenberg, *The Genesis of the Copernican World* (Cambridge: MIT Press, 1987), p. 658.

50. "Ma gli ingegni vulgari timidi e servili, che altrettanto confidano [. . .] sopra l'autorità d'un altro [. . .] rivolgendo notte e giorno gli occhi intorno a un mondo dipinto sopra certe carte [Aristotle's] senza mai sollevargli a quello vero e reale, che, fabbricato dalle proprie mani di Dio, ci sta, per nostro insegnamento, sempre aperto dinanzi" (*GO*, vol. V, p. 96, note).

> sages from this corpus to come up with answers to all questions, they never want to lift their eyes from these texts, as if nature had written this grand book of the world to have it read only by Aristotle and have his eyes see for all posterity.[51]

The "grand book of the world" presented here seems almost identical to the "book of nature," but an important element is still missing: the opposition between the transparency of the book of nature and the opacity of the Aristotelian corpus (or any other form of human writing). That distinction was articulated precisely during the debates of 1613–16, when the book invoked against Galileo's claims ceased to be human (Aristotle's corpus) and became divine (the Scripture).[52] It was the theologians' invocation of the divine authority of the Book that allowed and/or prodded Galileo to up his game and present natural philosophy as referencing an equally divine and suprahuman book of nature—one that shared nothing with the all-too-human books of the Aristotelians.[53]

51. *GO*, vol. V, p. 190. The book of nature remained "grand" both in the *Assayer* and in the *Dialogue*.

52. There is a handwritten comment by Galileo on Castelli's response to one of Galileo's opponents during the dispute on buoyancy that gives the book of nature the meaning it presented in the pro-Copernican texts of 1613–15. Galileo refers to himself as "Galileo, who [is] accustomed to pore over the book of nature, where things are written in one way only" (*GO*, vol. IV, p. 248). It is difficult to date this comment with precision. It was certainly written after October 28, 1612, when Castelli sent the manuscript of the response to Galileo (*GO*, vol. XI, p. 419) and before May 1615, when Castelli's response was eventually published. Given that Galileo's third letter on sunspots to Welser—the text where he presented the "intermediate" stage of the book on nature cited in this chapter—was completed on December 1, 1612 (*GO*, vol. V, p. 239), it is likely that this remark on Castelli's manuscript was written after Galileo's reference to the book on nature in the third letter to Welser.

53. The texts from 1613–16, therefore, represent the penultimate step in Galileo's articulation of the book of nature. Nature was equated to a "grand book" (in the 1613 letter to Welser) and presented as wide open, transparent, and outside of the domain of interpretation (in the letters to Castelli and the Grand Duchess Christina). What was absent in 1615 was the geometrical character of the book of nature. That was added in 1623 without, however, changing anything about the logic of the topos (its being external to interpretation) that he continued to deploy, in different forms, both in the 1632 *Dialogue on the Two Chief World Systems* and in his very last letters (*GO*, vol. VII, p. 27: "the grand book of nature that is the proper object of philosophy"). See also *GO*, vol. VII, pp. 135, 138–39, as well as pp. 128–29, where Galileo discusses the difference between human and divine knowledge. This connects with his previous discussion of the distinction between astronomy and theology in the "Letter to the Grand Duchess." Galileo reproduced

The trajectory followed by Galileo as he went from nature as a nonbook to the book of nature through successive engagements with the Aristotelian corpus and the Scripture concerned not only the kind of writing involved in these texts but also the reading practices they enabled. In the 1611 letter to Kepler but especially in the 1613 letter to Welser, Galileo connected the Aristotelian corpus to cut-and-paste readings, accusing the Peripatetics of pretending to answer whatever question was posed to them by cobbling up scattered passages from different books of their master's corpus.[54] Such reading practices (however typical they might have been) went hand in hand with the assumption that those books contained all philosophical truths and that, therefore, one did not need to look further.

At this time Galileo cast the Aristotelians' reading practices as a symptom of their unethical philosophical manners, not as something dictated by the nature of the texts. But if we fast-forward to the 1632 *Dialogue,* we find that by then Galileo had come to see the Aristotelians' cut-and-paste jobs as a general feature of all human books and forms of signification based on alphabetic languages, no matter what the readers' intentions might have been. Simplicio, the Aristotelian character in the *Dialogue,* is made to say that:

> [Aristotle] did not write for the common people, nor did he feel obliged to spin out syllogisms by the well-known formal method. Instead, using an informal procedure, he sometimes placed the proof of a proposition among passages that seem to deal with something else. Thus you must have that whole picture and be able to combine this passage with that one and connect this text with another very far from it. There is no doubt that whoever

his 1623 description of the book of nature in a January 1641 letter to the philosopher Liceti: "I truly believe the book of philosophy to be that which stands perpetually open before our eyes, though since it is written in characters different from those of our alphabet it cannot be read by everyone; and the characters of such a book are triangles, squares, circles, spheres, cones, pyramids, and other mathematical figures, most apt for such a reading" (*GO,* vol. XVIII, p. 295, as translated in Drake, *Galileo at Work,* p. 412).

54. This exact image reemerged in the January 1641 letter to Liceti, in the sentence leading to the image of the book of nature: "having as your aim the maintaining as true of every saying of Aristotle's, and sustaining that experiences show nothing that was unknown to Aristotle, you are doing what many other Peripatetics combined would perhaps be unable to do; and if philosophy were what is contained in Aristotle's books, you would in my opinion be the greatest philosopher in the world, so well does it seem to me that you have read at hand every passage he wrote" (*GO,* vol. XVIII, p. 295, as translated in Drake, *Galileo at Work,* p. 412).

ture, it would have been quite difficult to figure out what to cut and what to paste given that nature had no sentences, paragraphs, or chapters in the human sense of the terms.[59] It should be noticed, therefore, that Galileo's book of nature was specifically geometrical, not just mathematical. Algebra would not have worked for Galileo as it would have been very difficult to cast it as a system of nonarbitrary signs.[60] Instead, he was able to claim geometry as a nonsign by assuming that his readers would be able to perceive geometrical figures inherent in material objects of that shape.[61] (In this sense, Galileo's references to the reading of the book of nature should be treated as metaphors, as one would simply "see" the book of nature.)[62]

No matter what threads of the genealogy of the book of nature we pick, we see that Galileo's conceptualization of nature as a book evolved along a specific axis. Starting with ad hominem cracks about the limitations of the Aristotelian corpus and the expediency of its readers' interpretations, Galileo eventually put forward considerations about the structural semiological features of representation in general (independent of the agendas of specific readers and writers) and the essential differences that set the book of nature apart from other books. This trend of increasing generality did not reflect, I believe, an endogenous development of Galileo's philosophical interests but rather the increasing authority of his opponents and of the books they read and wielded.

The most powerful opponents—the theologians—provided Galileo with his most powerful discursive option. By fashioning the book of nature within the logocentric economy of the Scripture, Galileo not only managed to represent astronomy as a sister discipline to theology, but he simultaneously cast astronomy above Aristotelian philosophy. In Derridean terms, Galileo framed Aristotle's books as the product of writing as human tech-

59. Galileo often remarked (in the *Sunspots Letters,* in the "Letter to the Grand Duchess," and in the *Dialogue*) that nature was not constituted in a manner that would facilitate its understanding by humans, and that there was no relationship between human language and the structure of nature. So the idea of the book of nature having sentences or pages run against his own argument.

60. It would have been interesting to know whether Galileo thought of algebraic signs as an alphabet and, if so, how he conceived possible translations between geometry and algebra such as an algebraic equation and its diagram.

61. Of course, the whole edifice of the book of nature hinges on the erasure of the problems underlying this step, and the perception of geometry not as a system of representation but as inherent in the things themselves.

62. The book of nature is associated with reading at *LGD,* p. 103, and, much later, in the 1641 letter to Liceti (*GO,* vol. XVIII, p. 295).

> has this skill will be able to draw from his books the demonstrations
> knowable things, since they contain everything.[55]

Sagredo, the free-thinking critic, responds sarcastically that, if Si were right, one did not need Aristotle's texts to find the truth beca could easily form strings of words that would "explain all of the a men and the secrets of nature" by applying the same cut-and-paste to the texts of Ovid or Virgil.[56] Actually, Sagredo continues, one d even need to bother with Ovid or Virgil because truth-containing str words could be produced even more simply by working on a "much booklet in which all the sciences are contained: the alphabet."[57] Li and pieces of the Aristotelian corpus, vowels and consonants co strung together to form "words of truth" precisely because they have lation (mimetic or otherwise) to the truth.[58] The letters of the alp or fragments from Aristotle could be turned into strings of signs things—that is, representations—precisely because they were not things. The Aristotelians' cut-and-paste readings he had criticized ear as philosophical corner-cutting were presented, years later, as proof th human alphabetic texts were just bodies of signs, not of truth.

The book of nature, by contrast, did not support cut-and-paste reac because truth was just there in each of its geometrical figures—figures were natural configurations, not arbitrarily recombinable characters those of the alphabet. Even if one wanted to cut and paste the book of

55. *GO*, vol. VII, p. 134, as translated in Maurice Finocchiaro, *Galileo on the World S tems* (Berkeley: University of California Press, 1997), pp. 120–21.

56. *GO*, vol. VII, pp. 134–35.

57. *GO*, vol. VII, p. 135.

58. Sagredo extended the argument to pictorial representations as well. By mixing pigments in the appropriate order and quantity (the same procedure one follows to make words out of letters), Sagredo argued that painters can picture plants, buildings, birds, a fishes. It is crucial, however, for these pigments to be distinct from the things they represented because otherwise the painter could hardly represent anything: "It is necessary tha none of the things to be drawn nor any of the parts of them be actually among the colors, which can serve to represent everything, for if they were, for example, feathers, they woul not serve to depict anything but birds and bunches of feathers" (*GO*, vol. VII, p. 135, as translated in Finocchiaro, *Galileo on the World Systems,* p. 121). While this is clever, I think that Sagredo/Galileo's attempt to move from alphabetical to pictorial representation is only partly tenable. The relationship between colors and things depicted is not as arbitrary as between letters and concepts or things. In the case of pictorial representation the relationship is more indexical (in Peirce's sense).

nique, but presented both the Scripture and the book of nature as instances of "natural writing"—writing that was "immediately united to the voice and to breath" and whose nature "is not grammatological but pneumatological."[63] In the case of the book of nature and the Scripture, the terms "book" and "script" were used metaphorically.[64]

SELLING THE BOOK OF NATURE TO THE THEOLOGIANS

Galileo did not have the disciplinary authority to force the theologians to accept the complementary relationship between natural philosophy and theology inscribed in his book of nature. He hoped, however, that his discoveries (especially of the phases of Venus) could make the theologians perceive the book of nature as a solution to their problems.

The phases of Venus did not prove Copernicus, but, Galileo intimated, they did "clearly confute the Ptolemaic system"—the very astronomy the theologians relied on for their scriptural exegesis.[65] Galileo's discovery was confirmed in 1611 by the Jesuit mathematicians of the Collegio Romano—a report the Holy Office could not ignore since it had been requested by Cardinal Bellarmine himself.[66] The Jesuits had fully considered the cosmological implications of this discovery and, Galileo assumed, they had come to realize that Ptolemaic astronomy was no longer viable.[67] As he put it in

63. Derrida, *Of Grammatology*, p. 17.

64. Ibid., p. 15.

65. "The same Venus appears sometimes round and sometimes armed with very sharp horns and many other observable phenomena which can in no way be adapted to the Ptolemaic system" and "They [the critics] hear how I confirm this [Copernicus'] view not only by refuting Ptolemy's and Aristotle's arguments, but also by producing many for the other side, especially some pertaining to physical effects whose causes perhaps cannot be determined in any other way, and other astronomical ones dependent on many features of the new celestial discoveries; these discoveries clearly confute the Ptolemaic system, and they agree admirably with this other position [Copernicus] and confirm it." *LGD*, pp. 103, 88–89. See also p. 80.

66. Bellarmine to the mathematicians of the Collegio Romano (April 19, 1611), and mathematicians of the Collegio Romano to Bellarmine (April 24, 1611), in *GO*, vol. XI, pp. 87–88, 92–93. The report of the Jesuit mathematicians was never impugned since it was made available in April 1611.

67. On February 16, 1615, Galileo asked Monsignor Dini in Rome to deliver a copy of the "Letter to Castelli" to Father Grienberger (one of the underwriters of the corroboration of Galileo's telescopic discoveries at the Collegio Romano). He also asked him to try to enlist the Jesuits into Galileo's defense (*GA*, pp. 55, 58) and, through them, to have a

the "Letter to the Grand Duchess": "I could name other mathematicians who, influenced by my recent discoveries, [have] admitted the necessity of changing the previous conception of the constitution of the world, since it can no longer stand up in any way."[68] In the margins of the manuscript he identified those "other mathematicians" as Clavius.[69]

The theologians, Galileo was intimating, were in trouble because the refutation of Ptolemaic astronomy provided by the phases of Venus had effectively falsified their literal reading of the Scripture—a reading that was framed by geocentric assumptions. As he argued in the "Considerations on the Copernican Opinion," they were in no position to assert that Copernicus was false simply because it did not fit the (now refuted) geocentric cosmology that appeared to be inscribed in the Scripture. If the theologians wanted to condemn Copernicus, they needed first to prove that it was false. A proposition cannot be true and heretical at the same time.[70] And given that, according to Galileo, the refutation of Ptolemy through the phases of Venus meant that the theologians were in no position to use Scripture-based arguments against Copernicus, they (or any other critic) ought to refute Copernicus through astronomical arguments only.

His suggestion, I argue, was that the theologians could extract themselves from such a tight spot simply by agreeing that nature and the Scrip-

copy of the "Letter to Castelli" delivered to Bellarmine. A few weeks later, he was told that Bellarmine had relayed to Dini that he was going to discuss these matters with Grienberger (*GA*, p. 59). Bellarmine also mentioned to Dini a private discussion he had on these topics with Galileo (and the only time Galileo had been to Rome in the previous years was in the spring of 1611) (*GA*, p. 58). It is most likely, therefore, that Bellarmine understood the implications of the phases of Venus, and that Galileo did not need to say more about it in the "Letter to the Grand Duchess." Although Grienberger was probably sympathetic to Galileo's predicament, he tried to dissuade him from entering in scriptural matters prior to proving Copernicus (*GA*, p. 59).

68. *LGD*, p. 102.

69. *GO*, vol. V, p. 328.

70. "If it is inconceivable that a proposition should be declared heretical when one thinks it may be true, it should be futile for someone to try to bring about the condemnation of the earth's motion and Sun's rest unless he first shows it to be impossible and false" (*LGD*, p. 114). And, in the "Considerations" at pp. 81–82: "Whoever wants to use the authority of the same passages of Scripture to confute and prove false the same proposition would commit the error called 'begging the question.' For, the true meaning of Scripture being in doubt in the light of the arguments, one cannot take it as clear and certain in order to refute the same proposition; instead one must cripple the argument and find the fallacies with the help of other reasons and experiences and more certain observations." Similar remarks are at pp. 56, 83, 111.

ture were two different but equally true books to be read following different protocols. But he forgot to mention that his proposal would have required the theologians not only to accept a drastic redefinition of the authority of their discipline (down to their power to censor the publication of astronomy books), but also to undo Aquinas' canonical synthesis between Aristotelian (geocentric) natural philosophy and Christian theology—the mainstay of Church doctrine since the late medieval period. He also forgot to mention that another geocentric model whose existence he glaringly ignored—Tycho Brahe's—was in no way refuted by the phases of Venus—the evidence he was using to entice the theologians to accept the book of nature. (I will discuss Galileo's stunning erasure of Tycho in a moment).

The book of nature, then, was presented to the theologians as part of a two-book deal to establish a logical-looking, face-saving truce between astronomy and theology and to relieve them of the heavy burden of having to refute Copernicus to recover their disciplinary authority. It was a radical and costly proposal the theologians could have perhaps found acceptable only by acknowledging the state of emergency Galileo claimed they were in. But as the historical record reminds us, the theologians did not seem to realize they were in such a bad shape (or that they needed Galileo's help to regain their authority) when they condemned Copernicus in 1616.[71]

KEEPING THE BOOK GOING, METAPHORICALLY SPEAKING

Wanting to simultaneously use and circumvent the authority of his superiors, whose regime of truth rested on God's word as embodied in a sacred book, Galileo's first step was to argue that, like the Scripture, nature was a God-created book too. But the logic of his discourse also required nature not to be like the Scripture. It was only by claiming that the book of nature was made up of things-in-themselves that Galileo could hope to prevent the theologians from placing the authority of the Scripture above that of astronomers. As a result, Galileo's book of nature could be neither a book (an instance of human writing) nor a Book (a divinely inspired text that, like the Scripture, still allowed for some degree of interpretation).[72] In the end,

71. They probably saw the phases of Venus as refuting only Ptolemy's model of the orbit of Venus, not geocentrism in general.

72. I think, therefore, that Derrida's view of Galileo's book of nature as an example of "natural writing" (like that instantiated by the Scripture) is not completely accurate (Der-

Galileo's book of nature stretched the metaphor of the book so thin that it started to fall apart at its many seams. In the process, Galileo's topos showed itself to be more of a logocentric construct than the Scripture itself in that it claimed an immediate coexistence with the logos—one that was not mediated by any kind of alphabetic writing.

Galileo's topos rested on a fundamental metaphor, the use of "book" to refer not to an artifact made of material inscriptions, but to something immediately joined to God's logos. This is precisely the metaphorical use of "book" as "natural writing" one finds in the Scripture and in the relationship between Holy Word and Holy Writ. Galileo, however, took the next step, that is, he cast the book of nature not only as a metaphorical book like the Scripture, but as a book that, unlike the Scripture, did not allow for interpretation.

The theologians' construal of the Scripture as natural writing was inherently problematic but it was also conceptually simple as it involved one single dichotomous opposition between speech and writing. Galileo, on the other hand, construed his book of nature as operating between two dichotomies: The book of nature was *neither* a book written by humans (like the Aristotelian corpus) *nor* a metaphorical book (like the Scripture). The management of this more complex set of dichotomies required additional metaphors on top of that of the book as natural writing. And with the metaphors came the aporias.

The book of nature allowed for no interpretation whatsoever because, while immediately united with the logos (like the Scripture) it was not made up of actual words. Galileo made this point in the *Assayer*, writing that "its characters are triangles, circles, and other geometric figures," but already around 1612 he was jotting down lines about "the book of nature, where *things* are written in one way only."[73] The book of nature did not contain alphabetic letters but things in the geometrical shapes God gave them at creation. This means that when he referred to those shapes as "characters" in the very same sentence in which he stated the nonalphabetic nature of the book of nature (or when he referred to things being "written" in the book of nature) he was obviously using both "characters" and "written" as metaphors. He was also using metaphors in the *Assayer* when writing of

---

rida, *Of Grammatology*, p. 16). Galileo's "book" operates at one further level of metaphoricity above that of the Scripture.

73. *GO*, vol. IV, p. 248 (emphasis mine). The geometrical characters emerge in the 1623 *Assayer* (Drake and O'Malley, *The Controversy on the Comets of 1618*, pp. 183–84).

"letters," "words," and the "language" of the book of nature while referring to things like triangles and circles that his logic placed in precise opposition to letters, words, language, etc.[74] More metaphors were deployed in the "Letter to the Grand Duchess," where he wrote about "words" being "read" in "the open book of the heavens"—a book that contained "pages"—while his argument required that the book of the heavens contained no words or pages and could not be read but only seen or perceived.[75] (Keep in mind that, unlike Galileo, the theologians needed none of these metaphors. The Scripture had actual pages, words, and characters that could be actually read.)

If we follow Galileo's discourse, we see that its logic would have prevented him from saying that God *wrote* the book of nature. His image of the book of nature cast God not as a writer but as an architect who literally gave geometrical shape to his creation. This tension is inscribed in the fact that when Galileo says that "philosophy is written" and "things are written" in the book of nature, or mentions "very lofty words written" in the book of the heavens, he never says that God actually wrote the book of nature. He did say, instead, that the Scripture came from "the dictation of the Holy Spirit."[76] The only time he attached a writer to the book of nature, it was not God but nature itself.[77]

If the increasingly intricate and fragile construction of the book of nature as a "book" came to resemble an unstable heap of metaphors, the simultaneously close kinship and radical difference between the book of nature and the Scripture was perhaps more difficult (and certainly more dangerous) to manage. For instance, the dichotomy Galileo needed to maintain between the book of nature as a noninterpretable book and the Scripture as an interpretable text could not be marked as an opposition between a positive and a negative term—the kind of marking he could and did apply to the opposition between the (good) book of nature and the (bad) all-too-human books of the Aristotelians. Unlike Plato who, as discussed by Derrida, could use writing as the supplement for presence by casting it as a poor, dead copy of live speech, Galileo could not say that the Scripture was a poor copy of God's speech and that the book of nature was the good one.

74. Drake and O'Malley, *The Controversy on the Comets of 1618*, pp. 183–84.

75. *LGD*, p. 103. The problem with "reading" is a fortiori in the passage in the *Assayer*.

76. Galileo, "Letter to Castelli," *GA*, p. 50; *LGD*, p. 93.

77. "As if nature had written this grand book of the world to have it read only by Aristotle . . ." (*GO*, vol. V, p. 190).

As Galileo's discourse was grafted on the theologians', he simply could not kill (and probably not even irritate) his host. Galileo's sustained attempt to cast the relationship between the Scripture and the book of nature as one of complementarity reflects that predicament.

TRUTH IN THE EYES OF THE ACCIDENTAL BEHOLDER

Galileo was told that, after reading his "Letter to Castelli," Bellarmine had reiterated his commitment to a nominalist view of astronomical knowledge:

> His Most Illustrious Lordship [Bellarmine] says [. . .] the worst that could happen to the book [Copernicus'] is to have a note added to the effect that its doctrine is put forward in order to save the appearances, in the manner of those who have put forth epicycles but do not really believe in them, or something similar. And so you could in any case speak of these things with such a qualification.[78]

Bellarmine repeated the same position in a letter to Foscarini.[79] Nominalist views of the astronomers' knowledge undermined Galileo's tactics, and in fact he singled them out as one of the main obstacles facing him and the other Copernicans in 1615.[80] Actually, such an obstacle had been in place

78. Dini to Galileo, March 7, 1615, in *GA*, p. 58. A similar report is in *GO*, vol. XII, p. 160.

79. Bellarmine to Foscarini, April 12, 1615, *GA*, pp. 67–68: "It seems to me that Your Paternity [Foscarini] and Mr. Galileo are proceeding prudently by limiting yourselves to speaking suppositionally and not absolutely, as I have always believed that Copernicus spoke. For there is no danger in saying that, by assuming the Earth moves and the Sun stands still, one saves all the appearances better . . . and that is sufficient for the mathematician [astronomer]." Dini forwarded a copy of this letter to Galileo on April 18 (*GO*, vol. XII, p. 173).

80. Galileo, "Considerations on the Copernican Opinion," *GA*, p. 70: "In order to remove . . . the occasion to deviate from the most correct judgment about the resolution of the pending controversy, I shall try to do away with two ideas. These are notions which I believe some are attempting to impress on the minds of those persons who are charged with the deliberations and, if I am not mistaken, they are concepts far from the truth. [. . .] The [second] idea which they try to spread is the following: although the contrary assumption [heliocentrism] has been used by Copernicus and other astronomers, they did this in a suppositional manner and insofar as it can account more conveniently for the appearances of celestial motions and facilitate astronomical calculations and computations, and it is not the same case that the same persons who assumed it believed it to be true de facto and in nature; so the conclusion is that one can safely proceed to condemn it." The same point is made in a May 1615 letter to Dini (*GO*, vol. XII, pp. 184–85).

long before the debate as it reflected the theologians' traditional position about the status of the astronomers' claims.[81]

Trying to overcome such obstacles, Galileo added a crucial twist to the parallel between nature and the Scripture as two divinely created books. The "Letter to Castelli," the "Letter to the Grand Duchess," and the "Considerations on the Copernican Opinion" argued that in natural matters (but not in matters of morals and faith) the conclusive evidence and necessary proofs produced by natural philosophy could not be refuted by the theologians.[82] This followed from the fact that while both the Scripture and nature were equally true, there were essential differences between what one could read in them:

> For the Holy Scripture and nature derive equally from the Godhead, the former as the dictation of the Holy Spirit and the latter as the most obedient executrix of God's orders; moreover, to accommodate the understanding of the common people it is appropriate for Scripture to say many things that are different (in appearance and in regard to the literal meaning of the words) from the absolute truth; on the other hand, nature is inexorable and immutable, never violates the terms of the laws imposed upon her, and does not care whether or not her recondite reasons and ways of operating are disclosed to human understanding; but not every scriptural assertion is bound to obligations as severe as every natural phenomenon [. . .] And so it seems that a natural phenomenon which is placed before our eyes by sensory experience or proved by necessary demonstrations should not be called into

81. I do not present nominalism, realism, instrumentalism, or conventionalism as actors' categories, but as shorthand designators of a range of distinctions used to subordinate the epistemological status of mixed mathematics to that of philosophy, metaphysics, and theology. Important qualifications to the use of notions like nominalism, instrumentalism, or realism in relation to sixteenth-century astronomy are in Peter Barker and Bernard Goldstein, "Realism and Instrumentalism in Sixteenth-Century Astronomy: A Reappraisal," *Perspectives on Science* 6 (1998): 232–58. The scope of their analysis, however, is mostly restricted to the astronomers' discussions of their models and of the limits of their knowledge claims; they do not discuss the different brand of nominalism used by the theologians like Bellarmine to relativize the astronomers' claims.

82. *GA*, pp. 50, 81, 93–94, 101. Bellarmine agreed to this principle: "I say that if there were a true demonstration that the Sun is at the center of the world and the Earth in the third heaven, and that the Sun does not circle the Earth but the Earth circles the Sun, then one would have to proceed with great care in explaining the Scriptures that appear contrary, and say rather that we do not understand them than that what is demonstrated is false. But I will not believe that there is such a demonstration, until it is shown to me" (Bellarmine to Foscarini, April 12, 1615, in *GA*, p. 68). The last sentence, however, indicates that Bellarmine considered the possibility of mathematics directing theology little more than a mere hypothesis.

> question, let alone condemned, on account of scriptural passages whose words appear to have a different meaning.[83]

Galileo's argument about the differences between the two halves of God's creation relies, ultimately, on their audiences or, rather, on the fact that one book (the Scripture) has an audience while the other (nature) does not. The Scripture was written with a goal and an addressee in mind, nature was not. God, being infinitely good, had his speech written down by the prophets so that humans could reach salvation. Nature, on the other hand, was not created to guide us to heaven. As Galileo put it, quoting Cardinal Baronio: "The intention of the Holy Spirit is to teach us how to go to heaven and not how heaven goes."[84] While the Scripture had a message, nature had laws.

Because we know that the Scripture was written for an audience (and this follows from our certainty of God's infinite goodness that led him to give us a book through which we may attain salvation), it follows that such a book was written in a language that must allow for interpretation so that its message can be made clear to its intended recipients. This position was not reducible to the so-called principle of accommodation—the doctrine that the Scripture was written by God-inspired prophets in a form that could be understood by an unsophisticated audience. Galileo's argument was more radical: It depended on just the *existence* of an audience, not on its *intellectual sophistication*. The very fact that the Scripture had a message for an addressee implied that it required interpretation.

The fact that nature is "inexorable" follows from the existence of a parallel book that needs to be interpreted. As he stated in the "Letter to Castelli,"

> [N]ature is inexorable and immutable, and she does not care at all whether or not her recondite reasons and modes of operation are revealed to human understanding, and so she never transgresses the terms of the laws imposed on her.[85]

The transparency of nature and the fixity of its laws, therefore, is not presented as a methodological assumption (as people tend to read the image of the book of nature in the *Assayer*) but as a consequence of a divine choice. God chose to create two books: one opaque and full of teachings

83. *LGD*, p. 93 (emphasis mine). A very similar version in found in the "Letter to Castelli," *GA*, pp. 50–51.

84. *LGD*, p. 96. At pp. 94–95 Galileo made the bolder claim that Scripture deliberately refrained from teaching us about astronomy.

85. Galileo, "Letter to Castelli," in *GA*, p. 50.

for its human readers, and the other transparent and full of laws. If the interpretability of the Scripture derives from God's goodness, the rigidity of nature's laws derives from his power.[86] Far from being cast as an attack on theology, Galileo's whole argument hinged on the existence of God.

Besides the numerous nonbooklike features discussed previously (an infinite number of pages that are all open at the same time, all the time, and with no words, no characters, no sentence structure, and no meaning), there were other more contradictory ways in which the book of nature had to be a nonbook in order to perform the discursive task assigned to it by the logic of Galileo's argument. While a book is something to be read, Galileo argued that the book of nature's special status vis-à-vis the Scripture derived precisely from the fact that, since the beginning of time, nature was never meant to be read. And yet his central claim was that he could read nature, and read it right.

Galileo fashioned himself as the reader whom God had not planned to exist, but whose existence he had not explicitly forbidden either. The way Galileo had construed the book of nature in relation to the Scripture did not allow him to assume a less dangerous position. His ability to read the truth in the book of nature was inherently tied to his quasi-sacrilegious predicament.[87] It is precisely because he was not expected to exist as a reader that he could read the truth in the book of nature.

INEXORABILITY VERSUS INSPIRATION

The complementary opposition between the Scripture and nature was played out over and over, producing more aporias but at the same time providing Galileo with key discursive resources. For instance, since the Scripture was meant to carry a divine message conveyed by the Holy Spirit and written down by inspired prophets, the correct decoding of such a message should take place in an equally inspired context. But as Galileo put it, "[W]e cannot assert with certainty that all interpreters speak with divine inspira-

86. It should be noticed that the meaning of "nature's laws" is quite distinct from what we now call "natural laws." The former are orders imposed by God on nature; the latter are laws that are inherent in nature, and may not be of divine origin.

87. This same tension reappears in the *Dialogue on the Two Chief World Systems* in the discussion of how nature does not adjust itself to human understanding and that knowledge of the physical world is difficult and labor-intensive rather than natural (*GO*, vol. VII, p. 289). That discussion should be seen as linked to Galileo's argument about the similarities and differences between divine and human knowledge, and humans' ability to attain true knowledge comparable (in *intensio*, not *extensio*) to God's (*GO*, vol. VII, pp. 128–30).

tion since if this were so then there would be no disagreement among them about the meaning of the same passages." [88] As a result of these difficulties, one should rely on "wise interpreters," that is, on theologians authorized by the Church.[89] By contrast, because nature offers neither a message nor a path to salvation (and thus there is no coding and decoding in its creation and reading) natural philosophers do not need to be divinely inspired to read the book of nature, and read it right. One cannot be simultaneously a criminal and a prophet or a theologian. Nature, on the other hand, does not care about the moral qualifications or even the religious beliefs of the humans who might read it because it does not care about being read and understood to begin with.

This difference may also explain Galileo's characterization of the book of nature as being wide open for everyone to see. While in 1611 and 1613 he criticized the Aristotelians' practice of searching for answers not by gazing out at nature but by looking down at a finite corpus of texts (thus casting the Aristotelians as close-minded rather than nature as open), nature became an explicitly open book as soon as Galileo began to confront the theologians and the Scripture. At first, "closed" and "open" encapsulated the opposed epistemological dispositions of the old and new philosophers, but since the "Letter to the Grand Duchess," the adjective "open" referred to a feature of the book of nature: "by divine grace [God's glory and greatness] are read in the open book of the heavens." [90]

The private reading of the Bible in post-Tridentine Italy was a highly regulated practice available only to the clergy and a few selected individuals.[91] Its vernacular translations, opposed since Trent, were completely prohibited (and indeed destroyed) after the promulgation of Clement VIII's Index

88. "Letter to Castelli" (*GA*, p. 51) and *LGD*, p. 96. A similar point is at p. 97.

89. *LGD*, pp. 98, 92.

90. *LGD*, p. 103.

91. "The laity's role was to listen, absorb and appropriate the message that an authorized voice had delivered to them. One did not need direct access to the sacred texts to advance on the road to holiness. Thus Catholic reservations concerning solitary and reading print matter had a carefully argued theological and ecclesiological basis." Dominique Julia, "Reading and the Counter-Reformation," in Guglielmo Cavallo and Roger Chartier (eds.), *A History of Reading in the West* (Amherst: University of Massachusetts Press, 1999), p. 239. The choice of the Latin Bible (Vulgate) as the standard text contributed to a further reduction of the readership. Usually, members of the elite received individual counsel from spiritual advisors who also instructed on how to approach and read the Bible (ibid., p. 258). Lower-class religious instruction tended to focus more on the New Testament and other religious texts, but not on the Old Testament.

in 1596.[92] The Scripture, therefore, was a text closed to the vast majority of readers. Unlike the Aristotelian corpus, which did not contain the truth but was effectively rendered a "closed" book by the myopic approach of its readers, access to the Scripture needed to be closely controlled because it did contain the truth, a truth that could be dangerously misinterpreted by unskilled or impious readers. The book of nature, on the other hand, was open (and open to everybody irrespective of their theological training or piety) because it was meaning-free and thus not at risk of yielding to dangerous interpretations by unsuitable readers. It was open because the other divine book was practically closed. Or, to put it differently, the book of nature was not pried open by the curiosity of the Galileo-style readers, but was already open by virtue of not carrying any message. It was created wide open because nobody needed to read it.

The book of nature allowed Galileo to make a virtue of necessity. When he entered the debate, he lacked the social and disciplinary resources to present the astronomers' cognitive authority as superior to that of the theologians, or to claim at least that the theologians could not speak authoritatively about astronomical matters. The way God had created nature and the Scripture allowed Galileo to claim that the hermeneutical authority of the knower did not matter in astronomy, while at the same time confirming the theologians' authority on theological matters and their recently tightened control over the Scripture.

In 1623 Galileo treated the transparency of the book of nature as a fact. But in 1613 he was still trying to find an argument for why nature was transparent and could thus provide a condition of possibility for the certainty of the astronomer's knowledge. That the astronomer's credibility was rooted neither in moral or institutional authority nor in divine inspiration but in the transparency of the book of nature was only half of Galileo's argument. The rest was that such a transparency resulted from God having sent the teachings necessary to achieve salvation through another book (the Scripture) that, because of its message, had to be controlled and read only by specially qualified people sanctioned by the Church. As the interpretability of the Scripture constituted (as supplement) the transparency of the book of nature, it was the inspired status of prophets and theologians that constituted uninspired natural philosophers as qualified readers of the open book of nature.

92. Gigliola Fragnito, *La Bibbia al rogo: La censura ecclesiastica e i volgarizzamenti della Scrittura (1471–1605)* (Bologna: Il Mulino, 1997), pp. 173–98. The confiscation and destruction of extant copies is discussed at pp. 275–330.

INEXORABILITY OF NATURE AND THE PURSUIT OF NOVELTY

The construct of the openness of the book of nature provided Galileo with some arguments for destigmatizing a feature of astronomy and natural philosophy deemed suspect by many theologians and Aristotelians: the pursuit of novelty. Impious dispositions were thought to drive people to seek novelties instead of embracing well-established doctrines. But because the book of nature was already open, the philosopher was not one who pried nature for novelties, but simply found them right in front of his eyes. Accordingly, what could have been morally suspicious was not the philosopher's drive to discovery but the opposite drive that made some people feel that the book of nature, created open, ought to be closed:

> Who wants the human mind put to death? Who is going to claim that everything in the world which is observable and knowable has already been seen and discovered? [. . .] Nor should it be considered rash to be dissatisfied with opinions which are almost universally accepted.[93]

The interpretability of the Scripture and the inexorability of nature helped to give moral legitimation to the pursuit of novelty. Through the Scripture, God was trying to send a limited message—salvation—to very limited humans.[94] Nature, by contrast, had many laws precisely because it had no teachings to convey. Galileo, therefore, could cast the incremental nature of philosophical knowledge as a consequence of the fact that the laws that God had imposed on nature were indefinitely many (as opposed to the limited range of salvation-oriented teachings conveyed by the Scripture). And as these laws were not evident all at once (because they were not meant to be evident to begin with) their uncovering could require a potentially infinite amount of time.

Astronomy taught "how the glory and greatness of the supreme God are marvelously seen in all His works and by divine grace are read in the open book of the heavens."[95] However, one should not think that

> the reading of the very lofty words written on those pages is completed by merely seeing the Sun and the stars give off light, rise, and set, which is as far as the eyes of animals and common people reach. On the contrary, those

93. *LGD*, pp. 96–97. A similar line is in the "Letter to Castelli," in *GA*, p. 51.

94. *LGD*, pp. 92–95, 106.

95. *LGD*, p. 103.

> pages contain such profound mysteries and such sublime concepts that the vigils, labors, and studies of hundreds of the sharpest minds in uninterrupted investigation for thousands of years have not yet completely fathomed them.[96]

Scriptural meanings were "generational" while nature's laws were both eternal and too numerous to be uncovered within a lifetime.[97] If one understood this, it should not have come as a surprise that Galileo had not yet discovered a proof for Copernicus. The temporal limitation of human life and its primary focus on salvation were, therefore, supplements for the eternity and infinite number of nature's laws (as well as potential justifications for Galileo's inability to prove Copernicus).

Through its supplemental relationship to the Scripture, the book of nature allowed Galileo to uphold a notion of truth—the truth God inscribed in the book of nature—as something transparent and self-evident. At the same time, other features of the topos allowed him to say that the actual finding of such a truth was bound to be deferred, possibly forever—a claim that could be read as an admission that the self-evidence of nature is, in fact, not that evident. More generally, it is not at all clear how the book of nature could simultaneously support a view of knowledge as progressive (because of the potentially infinite levels of evidence contained in the book) and of knowledge as absolutely and permanently true (because of the transparency and inexorability of nature). In a telling passage, Galileo argued that

> because of many new observations and because of many scholars' contributions to its study, one is discovering daily that Copernicus' position is truer and truer and his doctrine firmer and firmer.[98]

Obviously, he believed that astronomical knowledge was converging toward the truth, but it is not clear how such a convergence could coexist with the book of nature as Galileo articulated it. He did not suggest that nature was an infinite book that one read page after page, building up knowledge as one went. Such an image would present each chapter as a fixed entity, but Galileo upheld a notion of knowledge that was both progressive and revisable. His statement that "those pages contain such profound mys-

96. *LGD*, p. 103.

97. On the contrast between the limited time allowed to humans to learn God's teachings of salvation and the necessary length of astronomical inquiry, see *LGD*, pp. 94–95.

98. *LGD*, p. 103.

teries and such sublime concepts that the vigils, labors, and studies of hundreds of the sharpest minds in uninterrupted investigation for thousands of years have not yet completely fathomed them" conjures a completely different image of reading—one that involves thousands of years of going back and forth over a few pages rather than reading on ad infinitum.

But how can the image of the page with a finite number of perfectly unambiguous characters be made to sustain the image of a reading that stretches over thousands of years, searching for an infinite number of laws in each page (over an infinite number of pages), and recasting the significance of each line in the light of how one has reread the previous line, or of a new character found between two old ones? Galileo's book of nature provides a very evocative image for those who think of knowledge as already achieved—as a well-organized, unambiguous map of a terrain that has been fully measured and triangulated. The book of nature conveys an image of totality, a magisterial image of knowledge like that of the encyclopedia or, even better, the Scripture—a book whose characters were all already known, without the possibility of adding new ones. Galileo put forward the book of nature because he needed an appropriate topos to counter the magisterial image of the theologians' knowledge. But the unresolvable tensions that were generated within the topos as he articulated it show that, in the end, the book of nature had become Galileo's pharmakon—simultaneously a cure and a poison.[99]

KEEPING THE DICHOTOMIES STRAIGHT

Nominalism (or similar notions like instrumentalism and conventionalism) did not just represent a view of the specific limits of human knowledge about nature, but were also a symptom of the philosophers' and theologians' attempt to keep astronomers in a subordinate position by denying them the disciplinary authority to make physical, philosophical claims.[100]

99. Jacques Derrida, "Plato's Pharmacy," in *Dissemination,* trans. Barbara Johnson (Chicago: University of Chicago Press, 1981), pp. 63–172, esp. pp. 71–72, 95–98.

100. In the texts related to this dispute, the nominalist position is sometimes defined, by Bellarmine and Galileo, as *ex suppositione* (*GA,* pp. 67, 70) or, by Cesi, as *ex hypothesi* (*GO,* vol. XII, p. 190). Galileo's realist reading of Copernicus is equally stark: heliocentrism is true *in rei natura* (*GO,* vol. XII, p. 184). Although modern philosophical notions like instrumentalism, conventionalism, nominalism, and the many varieties of realism may not be perfectly applicable to this earlier scenario, I feel that the actors' view of their opposing views was very clear, perhaps too clear. The clarity of that opposition suggests that

Bellarmine reiterated this in 1615, claiming (incorrectly) that Copernicus himself thought of his claims as hypothetical.[101]

The nominalist view of astronomy was precisely what Galileo was opposing through the book of nature, where geometrical characters were not models but truth. But unable to buttress his realist stance by proving Copernicus, he could only try to show that the opposite position—Bellarmine's nominalism—was untenable, thus creating the conditions of possibility for the theologians' acceptance of the book of nature. That move, however, introduced more important lacunae in Galileo's logic. He started by assuming that his refutation of Ptolemy through empirical evidence (the phases of Venus) amounted to a refutation of astronomical nominalism because it broke down the theologians' symmetrical treatment of Ptolemy and Copernicus as hypotheses.[102]

> It is true that it is not the same to show that one can save the appearances with the Earth's motion and the Sun's stability, and to demonstrate that these hypotheses are really true in nature. But it is equally true, or even more so, that one cannot account for such appearances with the other commonly accepted system [Ptolemy]. The latter is undoubtedly false, while it is clear that the former [Copernicus], which can account for them, may be true.[103]

He seemed to agree that it would be legitimate to be a nominalist about hypotheses if they are precisely that: hypotheses, that is, claims that are both

---

"ex suppositione" and "in rei natura" were not only philosophical distinctions but also "fighting words" lobbed back and forth by the opposing parties.

101. Bellarmine to Foscarini, April 12, 1615, in *GA*, p. 67. Bellarmine, like many other theologians, believed that Copernicus himself presented his claims as hypothetical. Bellarmine's argument appeared to be based on a short anonymous preface appended to Copernicus' *De revolutionibus* that in fact cast the book's arguments in nominalistic terms. That preface, however, was the work of a Lutheran theologian—Andreas Osiander—not Copernicus (who most probably never knew of its existence). Kepler was the first to realize that the preface was not by Copernicus. Galileo too argued from textual evidence (and largely in response to Bellarmine) that the preface was not authentic, as it contradicted the explicitly realist position taken by Copernicus in the text (Galileo, "Considerations on the Copernican Opinion," in *GA*, pp. 78–79).

102. Because the theologians tended to have a nominalist view of astronomy in general, they treated both Ptolemy and Copernicus (and later Tycho) as hypothetical models developed to "save the appearances." This position is represented in the letter from Bellarmine to Foscarini, April 12, 1615, in *GA*, p. 68.

103. Galileo, "Considerations on the Copernican Opinion," in *GA*, p. 85.

unproven and unrefuted. But one could no longer hold a nominalist position about a given hypothesis (Ptolemy's) if that hypothesis had been refuted. A hypothesis refuted through empirical evidence was no longer a hypothesis but a physically false claim. As a result, one could not continue to treat the alternative hypothesis that still stood unrefuted as a mere computational model. Galileo seems to suggest that the unrefuted half of a pair of mutually exclusive hypotheses was transformed into a positive physical claim by having its opposing hypothesis physically refuted.[104]

Galileo tried to refute Bellarmine's nominalism by asserting that the distinction between computational models and physical reality was hardly sustainable in cosmology. Because there are only two conceivable scenarios—the Sun goes around the Earth or the Earth goes around the Sun—the difference between models and reality is meaningless so far as the relative motion of the Sun and the Earth is concerned.[105] Under these circumstances, Ptolemy stands for geocentric cosmology and Copernicus stands for heliocentric cosmology:

> Note carefully that, since we are dealing with the motion or stability of the earth or of the sun, we are in a dilemma of contradictory propositions (one of which has to be true), and we cannot in any way resort to saying that perhaps it is neither this way nor that way. Now, if the earth's stability and the sun's motion are de facto physically true and the contrary position is absurd, how can one reasonably say that the false view [Copernicus] agrees better than the true one with the phenomena [phases of Venus]?[106]

104. If Galileo could demonstrate that Ptolemy and Copernicus exhausted all possible cosmological options, the refutation of Ptolemy could be construed as a proof for Copernicus. But the existence of Tycho's model precluded that possibility. That, I believe, explains why Galileo did not present his refutation of Ptolemy through the phases of Venus as a proof of Copernicanism and, at the same time, did not mention Tycho.

105. In a letter to Dini and in the unpublished "Considerations on the Copernican Opinion" Galileo drew the line, very emphatically, between two kinds of hypotheses used by the astronomers. One concerned the overall cosmological structure, the other the specific devices one may develop to account for the orbits of specific planets. In his view, given that there could be only two possible hypotheses of the first kind (geocentrism or heliocentrism), one of them was bound to be physically true and therefore not hypothetical. The second kind of hypotheses allowed for many more options, preventing one from saying exactly which one was the true one. According to Galileo, Bellarmine-style, broad-stroke nominalism was predicated on not understanding this key distinction and thus assuming that the status of cosmological hypotheses was the same as those about epicycles, eccentrics, etc. (*GA*, pp. 60–62, 70–78).

106. Galileo, "Considerations on the Copernican Opinion," in *GA*, p. 75. The same point is repeated in the same paragraph: "But, given two positions, one of which must be true

In different ways, these two statements try to argue that Copernicus is physically true because Ptolemy has been shown to be physically false. According to the first, if Ptolemy is physically false, then geocentrism itself is false, which then means that heliocentrism (and Copernicus as the only possible embodiment of heliocentrism) must be true. The second quote tries to reinforce that point through something of a reductio ad absurdum argument: If one takes Copernicus' hypothesis to be the physically false one, how can one say that the false hypothesis matches the phenomena (phases of Venus) much better that Ptolemy (the allegedly true one)?

Galileo's play of dichotomies (real/false, physical/fictional) reflects, I believe, a clear attempt to move away from a nominalist framework to a discourse where he could impose the book of nature on the theologians. To get there, Galileo tried to shift the discussion from a framework structured around a pair of hypotheses to one informed by absolute oppositions—logocentric dichotomies through which he could put supplements to work. As mere hypotheses, geocentrism and heliocentrism were neither good nor bad copies of the truth because they were not copies to begin with.[107] But once they ceased to be mere hypotheses, they became "good" or "bad" representations of the cosmos. Of course Galileo wished he could have shown that Copernicanism was not just a "good copy" of the cosmos but its very structure. He wished to find Copernicanism written in the book of nature as presence, not representation. However, unable to prove Copernicus with physical arguments, Galileo tried at least to use a refuted Ptolemy as the "bad copy" of presence thereby casting Copernicus as the "good" mimesis of presence.

It was only within a realist framework that a refuted geocentrism could function as a supplement producing an effect of presence of both mathematical realism and heliocentrism. In the context in which he operated and with the handicaps he was confronting, Galileo needed Ptolemy—a Ptolemy that was both endorsed by the theologians (so as to be authoritative) and physically refuted (so as to be usable as a supplement). He needed

---

and the other false, to say that the false one agrees better with the effects of nature is really something that surprises my imagination," and in *LGD,* p. 110.

107. Andreas Osiander, the theologian who added an anonymous preface to Copernicus' *De revolutionibus,* saw both Ptolemy and Copernicus as hypotheses and, as such, equally probable (or improbable): "Therefore alongside the ancient hypotheses [Ptolemy's], which are no more probable, let us permit these new hypotheses [Copernicus'] also to become known." Nicholas Copernicus, *Complete Works,* trans. Edward Rosen (Baltimore: Johns Hopkins University Press, 1978), vol. I, p. xvi.

Ptolemaic astronomy to be simultaneously authoritative and dead—dead wrong. It is probably for this reason that he challenged the theologians to refute Copernicus, hoping that their failure to do so could have provided him with another supplement.[108]

But if this was the logic of Galileo's argument, why did he not say that the physical refutation of Ptolemy constituted a physical proof of Copernicus? Why did he limit himself to saying only that Copernicus was becoming "firmer and firmer"?[109] Why did he only ask for a deferral, saying that Copernicus should not be condemned as heretical because, unlike Ptolemy, it had not been proven false yet? The problem Galileo was trying to hide behind the oscillations of his discourse was that between Copernicus and Ptolemy there was not a strict dichotomy as there was, instead, between the book of nature and the Scripture.

Contrary to Galileo's rhetorically confident statement that "we cannot in any way resort to saying that perhaps it is neither this way nor that way," Ptolemy and Copernicus did not exhaust all the range of possible cosmologies. Since 1588, there was a very well-known alternative to Copernicus

108. This was not presented as a challenge from a disciplinary underling but as an affirmation of the theologians' power to prosecute heretical claims. Because of theology's unique legal authority—an authority that allowed the theologians not only to declare the falsehood contained in certain people's texts and minds, but also to arrange for the punishment of their bodies—Galileo implied that their pronouncements had to be bound to particularly stringent standards: "whoever wants to condemn [Copernicanism] *judicially* must first demonstrate it to be physically false by collecting the reasons against it" (*LGD*, p. 81, emphasis mine). Condemning it as heretical without proving its falsehood first would have violated the codes of responsibility of their discipline. At the same time, because he knew that it would have been virtually impossible for the theologians to refute Copernicus with astronomical arguments, he could count on their failure (the failure of the most authoritative speakers on cosmological matters) as a way to create an effect of truth around a still unrefuted Copernicus. Even if the heliocentric cosmos instantiated the logos, Galileo was unable to prove it, to put it forward as presence. All he could do was to create effects of that presence through supplements that, in this case, took the shape of negative arguments—the untenability of other competing cosmologies or the theologians' failure to disprove Copernicus. The theologians' authority (when it failed to live up to its billing) functioned as a supplement for Galileo's authority as an astronomer.

109. The same tension is evident in his private "Considerations on the Copernican Opinion": "Now if they [Copernicans] are not more than ninety percent right, they may be dismissed; but if all that is produced by philosophers and astronomers on the opposite side is shown to be mostly false and wholly inconsequential, then the other side should not be disparaged, nor deemed paradoxical, so as to think that it could never be clearly proved" (*GA*, p. 85).

and Ptolemy, and it was called Tycho.[110] Sadly for Galileo, Tycho's hybrid planetary model could easily account for the phases of Venus while keeping the Earth at the center of the cosmos. As the dichotomy between Ptolemy and Copernicus was far from being tight, Galileo could not use one term as supplement for the other. At the same time, any mention of Tycho would have sent his card castle tumbling down. The result was a discourse that could cast Copernicus only as "firmer and firmer" (rather as plainly true) without being able to acknowledge the reason for such a waffling.[111]

Competent readers must have been flabbergasted to find no mention of Tycho in the "Letter to the Grand Duchess."[112] Nor, for that matter, was Tycho mentioned in any of the texts Galileo wrote in the context of this dispute. His famous 1632 *Dialogue on the Two Chief World Systems*—the book that triggered the final trial of 1633—claimed that there were two (not three) world systems and that these were Copernicus and Ptolemy (not Tycho).[113] Galileo's erasure of Tycho was as stunning as it was mandatory. His planetary model took the wind out of Galileo's alleged refutation of geocentrism, but, even more insidiously, it indicated that the "great book of the heavens" (as it could be read at that time) had more than one reading (Tycho's being one of them). The existence of the Tychonic system and the fact that Galileo could not refute it the way he could refute Ptolemy showed that the book of nature was just a book, not nature itself. Tycho could undermine not only heliocentrism but the very dichotomous metaphysics of truth through which Galileo was trying to constitute his brand

110. Tycho made Venus and Mercury orbit the Sun (which in turn was orbiting the Earth). Because the motions of Venus in relation to the Sun were the same in both Copernicus and Tycho, the phases of Venus could be accounted for equally well in either system.

111. Note that, in the statement quoted above, Galileo defines the opposition between geocentrism and heliocentrism as mutually exclusive by focusing only on the relative motions of the Sun and Earth, not all the other planets. This is clearly aimed at making Tycho's disappear, as his system is geocentric according to the definition given by Galileo. But Tycho is not refuted by the evidence (the phases of Venus) given by Galileo to refute Ptolemy because that evidence concerns the relative motions of the Sun and Venus (not the Earth). That is, Galileo uses different taxonomies when he sets up the heliocentric-geocentric dichotomy and then refutes one of its halves. This is a contradiction that, given Tycho's existence, he cannot avoid.

112. In some cases, as with the grand duchess herself, the readers of this nontechnical work might not have known about Tycho's model.

113. The mutual exclusivity of Copernicus and Ptolemy in repeated in the *Dialogue*, sometimes with the same turns of phrase employed in the 1615–16 texts (*GO*, vol. VII, pp. 156, 383).

of philosophical realism.[114] The logocentrism of Galileo's book of nature required Tycho's erasure.[115]

FROM OBSTACLES AND RESOURCES TO SUPPLEMENTS

Looking back at this controversy, we can safely say that Galileo confronted a remarkable number of difficulties: he could not prove Copernicus; did not have sufficient sociodisciplinary authority to oppose the theologians; did not have sufficient time to produce pro-Copernican evidence; could control neither the pace of the debate (a debate he did not initiate) nor the forums in which it would be adjudicated; could not rely on indefinitely patient and trusting patrons; did not have a notion of proof the theologians could recognize as such, and more generally, a legitimate metaphysics of truth on which to ground such a notion. This list, as we have seen, could be expanded.

If we perceive Galileo as having entered into scriptural exegetical debates without the resources he needed, it would then not come as a surprise that he was unable to avoid the theologians' first condemnation of Copernicus in 1616 and, years later, of himself. But a perspective framed by the dichotomous notions of "resource" and "obstacle" poses a series of unanswerable psychological questions about Galileo's motives for entering what would appear to be a hopelessly difficult debate: Was Galileo right or wrong in assessing the situation and his chances of success? Was he blinded by a deep commitment to Copernicanism or by his overgrown ego? Did he not receive enough reliable intelligence from his Roman supporters? Did he overestimate the power of Medici support? Answering these questions requires some way of demarcating between "enough" or "not enough" re-

114. If Galileo gave up on claims of absolute truth (such as those belonging to the theologians' regime of truth), he would have slid into the nonrealist position the theologians wished to keep the astronomers in. Again, he was in a position in which, although he had access only to probable truths, he could not present them as such, but could only create "truth effects" about an absolute truth he could only defer.

115. At the same time, the scriptural logocentrism of the theologians (and their nominalist view of astronomy) saw in Tycho the perfect tool to "save the appearances" in both astronomy and theology. The Jesuits endorsed Tycho's model in 1620. It was a patch that lasted until the nineteenth century. The Congregation of the Holy Office decided to remove books treating the motion of the Earth and the immobility of the Sun from the Index only in 1833 (Mayaud, *La condemnation des livres coperniciens et sa révocation*, pp. 271–80).

sources and "too many" or "not too many" obstacles. But these demarcations are problematic because we can infer them only after the facts, based on the closure of the dispute. The same could be said about the categories of "obstacle" and "resource" themselves. Their contours become clear only with hindsight.

A different perspective emerges if we look at the same range of historical evidence but suspend the positive and negative connotations of "resources" and "obstacles," that is, if we do not treat them as presences and absences. For instance, if we think of the articulation of Galileo's and the theologians' discourse as structured by the logic of the supplement, then what we might have read as Galileo's lack of resources and abundance of constraints (or his possible misperceptions about his own predicament) cease to appear as causes of the condemnation of 1616 and emerge, instead, as the conditions of possibility for the articulation of his discourse.

That Galileo could not deliver presence but could only effect it supplementally was not an anomaly but rather the rule of any logocentric discourse (including that of the theologians). We may say, then, that Galileo did not get in trouble with the Church because he entered into exegetical debates without the resources necessary to prove Copernicus, but that he articulated his alternative exegetical approach precisely because he could not produce the kind of proof that could have brought the controversy to a closure. Lacking that proof, he engaged the theologians on exegetical grounds to prevent them from bringing the controversy to a closure on their terms. Conversely, the theologians censored Copernicus in 1616 not because, unlike Galileo, they could access the truth and prove him wrong. Their Scripture-based metaphysics of truth was no less unstable than Galileo's and it could have been further destabilized by his exegetical proposal. In this sense, the theologians' condemnation of Copernicus was a defensive act—as defensive as the logic of Galileo's book of nature.

EPILOGUE

Unintended Differences

I WANT to comment briefly on certain patterns that have emerged in the previous chapters—patterns that cut horizontally through the main themes of the book: economies of credit, instruments, visuality, and print. I highlight them not so much as a way to bring this book to a close, but rather to point to those aspects of my argument that remain a work in progress.

Taken as a whole, these patterns relate to unstable scenarios of knowledge production where unforeseeable things happen, information is limited, the actors' tactics are steadily scrambled, and the deferred nature of knowledge claims is evidenced. But these are also situations in which new claims are produced precisely because of those unsettled conditions. Historians and sociologists of science tend to explain the closure of scientific controversies as the effect of positive, causal, social resources such as the number and clout of one's allies, the alignment of the order of knowledge and the social order, the tacit knowledge about instruments, the financial resources, political connections, and personal credibility of the actors, etc. Some of the scenarios discussed in this book, on the other hand, point to ways in which knowledge emerges from partiality, gaps, and differences—things that neither function like causes of the closure of controversies nor fall on either side of the divide between resources and constraints, between presence and absence.

Some of these scenarios are associated with radical surprises triggered by new discoveries: novae, the roughness of the Moon, the satellites of Jupiter, the sunspots, or the rings of Saturn—the kind of things Clavius called "monsters." But they also include surprises that resulted not from the unexpected behavior of nature but of human agents. I am thinking of such heterogeneous events as the development of the telescope by a Dutch glassmaker (an invention that changed astronomy forever); the grand duchess' impromptu question to Castelli over Copernicanism (that was to lead, eventually, to the condemnation of 1616); the unforeseen pro-Copernican intervention of Foscarini (that instead ended up energizing the Inquisition's concern with Galileo); Castelli's leak of Galileo's 1613 letter (which was to catch the Inquisition's fatal attention); or even the small changes introduced by the hand copying of the letter (changes that affected the Inquisition's response to the letter).

These actions were not only surprising to the people for whom they had important consequences but, more importantly, they were not intended to affect them the way they eventually did. They are quite different from, say, Capra's unexpected appropriation of Galileo's work on the compass—an action that had Galileo as its target. Although most actions by social actors are intentional, some of them may be experienced as simply agential by people who were not the targets of the actor's intentionality. Affect and effect can be quite distinct. In this sense, the consequences of the actions of social actors may share in the perceived impersonality of natural events. What results from these actions are differences that are as epistemically productive as those that traditional models of scientific change have attributed to nature, like the ability to surprise us with anomalies and emergent objects.[1] We often emphasize, quite correctly, how big a surprise for Galileo the observation of the roughness of the lunar surface was, but we do not seem to realize that the invention of the telescope was at least as unforeseen, and had even bigger consequences for him (and for astronomy) than the topographical features of the Moon. The same considerations apply not only to the unexpected introduction of new instruments but also to other kinds of difference-making events like the grand duchess' impromptu questioning of Castelli on Copernicanism.

I am particularly interested in these kinds of consequences—what, following Clavius, I'd call "social monsters"—because they may provide a

1. These events match part of Andrew Pickering's notion of temporally emergent constraints as introduced in his "The Mangle of Practice: Agency and Emergence in the Sociology of Science," *American Journal of Sociology* 99 (1993): 559–89. More specifically, the events I am talking about fit the "temporal emergence" bit, though I would not classify them as constraints. They may eventually assume the role of constraints, but only for some practitioners trying to do certain things in a certain place at a certain time. For the same reasons, they cannot be treated as "resources" either. They are pharmakon-like.

useful corrective to Derrida's exclusive association of the play of difference with writing and inscriptions. This association has been reiterated, with important changes, in Hans-Jörg Rheinberger's notion of "experimental system" where the production of differences—new things and new claims—is attributed to the system's production of material inscriptions, a production that stems from the system's own iterative operations. While I do not disagree with Derrida's and Rheinberger's claims (as far as they go), it seems to me that epistemically productive differences can emerge from the play of unexpected social actions as much as from the play inherent in writing or in the operations of inscription apparatuses. It really does not matter if monsters are social or natural, or if the differences they enable can be easily recognized as inscriptions or if instead they look more like changes or displacements in the actors' landscape. What matters is difference in whatever shape or form it may come.

A second cluster of examples concerns the productive effects of limited information about people, institutions, instruments, or nature. Such partial information has to do with the actors' location in space (at a distance from each other) and with the pace and timing of their communications, not with the noise that inevitably affects any information channel. Examples of these effects are: the construction of Galileo's authority based on the limited information people had about him and his work; the construction of the Royal Society as an authoritative node of the republic of letters by remote correspondents who overestimated it because of limited information about its mundane reality; Kepler's writing the *Dissertatio* also as a result of not having first-hand information about the Medicean Stars or about the relationship between Galileo and the Medici; and Scheiner's putting forward innovative cosmological stances during the sunspots dispute by knowing (or pretending to know) little about the state of the cosmological debate at the Collegio Romano. More specifically, the cases of Galileo and Scheiner indicate that productive effects can be had not only by aligning one's position or claim to those of a more powerful, legitimizing entity (as in a patron-client relationship, or in a Latourian "lining up" of allies, or in the identity of solutions to problem of knowledge to those of social order as proposed by SSK). Here we see that positive results derive from *avoiding* such alignments, by taking a position that is not exactly what one's patron or institution or allies would have liked or expected, while simultaneously creating a situation in which they are likely to feel compelled to endorse it anyway. Quite literally, the result is produced not by coincidence or alignment but by difference.

The acceptance of the telescopic discoveries of 1610 shows that the fact that neither Galileo nor his critics and competitors knew how the telescope worked did not impair the acceptance of Galileo's discoveries. Actually, when coupled with his withholding of information about how to build telescopes, it helped Galileo develop a monopoly over early observational astronomy. Blackboxing, it seems, was unnecessary for this kind of instrument in this kind of context. Similarly, the apparent inaccuracy of the illustrations of the Moon in the *Sidereus nuncius* did not weaken Galileo's claims. Those pictures were related as much to how the Moon looked as to Aristotle's textual statements about what it could not look like—suggesting that the very notion of referent may not fall exclusively in the category of material presence (casting an image in a mimetic relation with the referent) but also of intertextuality (in the sense that these images refer, through differences, to texts other than the ones that contain them).

The discussion of the sunspots dispute adds to the list of results produced not by positive presences, but rather from differences. Galileo's refusal to address questions about the physical nature of sunspots did not have the effect of reducing his arguments about sunspots to the realm of descriptive or nominalist discourse or to diminish his credit for the discovery. Rather he was able to make physical claims about sunspots without showing what they were simply by deploying a double negative, that is, by showing that sunspots were not artifacts. More generally, this episode questions the very notion of "representation" by showing that the effectiveness of Galileo's images cannot be explained through their ability to stand in for a preestablished physical signified, but rather through their productive management of patterns of differences.

A third pattern concerns deferral. The early distance-based construction of Galileo's authority suggests that authority itself emerges not as a thing or positive entity but as the effect of a specific flow and exchanges of partial information. The play of partial information allows for a certain delay in the delivery of evidence of authority, but it also suggests a more general point, namely, that deferral is part of the predicament of knowledge. The controversy over Copernicanism and Scripture provides a further example. The entire dispute could be read as a sustained attempt on Galileo's part to defer the condemnation of Copernicanism so as to be able to defer its proof. His remarkable discursive edifice was based as much on what he did not have (proofs, authority) as on what he did have (the phases of Venus, Medici support, etc.). And while the theologians "won" this dispute, it was not because their knowledge, unlike Galileo's, was not affected by deferral.

Both parties, in one way or another, tried to hinge their notion of truth in some presence (grounded either in the book of nature or in the Scripture) that neither of them could deliver. The theologians' censorship of Copernicus surely exemplified their power, but it was simultaneously an attempt to shield their inherently unstable claims about Scripture as presence. Deferral comes with logocentrism—an economy of truth both Galileo and the theologians shared, even if perhaps with different degrees of commitment.

This leads me to my last point about supplementation. I have highlighted the remarkable imbalance between Galileo's constraints and resources during the controversy with the theologians, and his ongoing efforts to transform a defensive position into a proactive one by turning limitations into resources. The aura of drama surrounding this episode (and the hefty doses of intentionality manifested in the actions of the two opposing parties) makes it difficult to stay focused on the systematic, less personal aspects of the discourse produced during this controversy. But if we set aside the pathos of the story, we may notice that the supplemental features of Galileo's discourse (while amplified by the stringent time frame imposed by the judicial protocols of the Holy Office) were by no means exceptional. Not only did the theologians cast writing as a supplement for the irreproducible presence of God's word inscribed in the Scripture, but Galileo in turn cast the Scripture (as well as the Aristotelian corpus) as supplements to the book of nature. A further example of supplementation can be found in the way Aristotle's strict veto on celestial corruptibility helped Galileo refute that same philosophical assumption without having the disciplinary authority to sustain that claim.

While Derrida has limited himself to analyzing the role of the supplement within individual texts produced by individual authors who did not write them in the context of real-time controversies, it seems to me that supplementation can be seen at work in intertextual settings, across texts that reference and challenge each other, written by disputing authors (as in the case of Galileo and the theologians or the Aristotelians).[2] It is also at work in pictorial inscriptions, not only in written ones (as in the relation between Galileo's images of the Moon and Aristotle's veto on celestial corruptibility). Furthermore, supplementation is not limited to inscriptions (as in the cases discussed by Derrida and Rheinberger), but it can be easily traced to embodied actors and even their institutions (as in the case of Ga-

2. The only exceptions I can find to Derrida's approach are controversies that he entered himself, such as the one with John Searle.

lileo's treating the theologians and their authority as supplements for the book of nature). I am not saying that we should follow supplementation from "texts" to "contexts," but rather that the text/context distinction could be seen as a distinction without a difference as far as the workings of supplementation are concerned.

These few observations do not amount to a positive methodological proposal. They do, however, point to the problems of those interpretations of scientific practices that cast knowledge claims as inherently weak and unstable and then proceed to account for their stabilization through the deployment of positive social resources. (My *Galileo, Courtier,* I admit, fits this mold too.) Perhaps wanting to assume the explanatory position of the so-called social sciences, many of us have adopted, in the past, a metaphysics of presence—of a presence that is usually not instantiated in nature but in social structure and resources, as well as in the social qualities of the practitioners.[3] Key attempts at reconfiguring this asymmetry, like Latour and Callon's actor-network theory, have questioned the nature/society divide, but have done so from within a discourse of positivity.[4] Claims may no longer move from naturally unstable to socially stabilized (or from local to nonlocal), but they still go from weak to strong. Something more positive or present is still added to something less positive or less present.

3. Hans-Jörg Rheinberger, "Experimental Systems, Graphematic Spaces," in Timothy Lenoir (ed.), *Inscribing Science: Scientific Texts and the Materiality of Communication* (Stanford: Stanford University Press, 1998), pp. 285–303; and Mario Biagioli, "From Difference to Blackboxing: French Theory versus Science Studies' Metaphysics of Presence," in Sande Cohen and Sylvere Lotringer (eds.), *French Theory in America* (New York: Routledge, 2001), pp. 271–87. Michel Callon, "Society in the Making: The Study of Technology as a Tool for Sociological Analysis," in Wiebe Bijker, Thomas Hughes, and Trevor Pinch (eds.), *The Social Construction of Technological Systems* (Cambridge: MIT Press, 1987), pp. 83–103; Bruno Latour, "The Politics of Explanation: An Alternative," in Steve Woolgar (ed.), *Knowledge and Reflexivity* (London: Sage, 1988), pp. 155–76; Bruno Latour, "One More Turn after the Social Turn . . . " in Biagioli, *The Science Studies Reader,* pp. 276–89; Pickering, "The Mangle of Practice." A spirited defense of SSK's asymmetrical stance about nature and society is in Harry Collins and Steven Yearley, "Epistemological Chicken" in Pickering, *Science as Practice and Culture,* pp. 301–26.

4. Michel Callon and Bruno Latour, "Don't Throw the Baby Out with the Bath School!" in Andrew Pickering (ed.), *Science as Practice and Culture* (Chicago: University of Chicago Press, 1992), pp. 343–68. Among the other authors who have tried to move past the society/nature divide are Donna Haraway (since "A Cyborg Manifesto" in her *Simians, Cyborgs, and Women* [New York: Routledge, 1991], pp. 149–81), Andrew Pickering (since "The Mangle of Practice"), and Hans-Jörg Rheinberger (see next footnote).

Galileo's Instruments of Credit, on the other hand, has tried to draw attention to specific scenarios where the opposition between presence and absence falls apart, and yet both credit and knowledge are produced. Plenty of both.[5]

5. I am here aligning myself with Hans-Jörg Rheinberger's critique of the logocentrism of science studies as articulated in his "Experimental Systems: Historiality, Narration, and Deconstruction", in Mario Biagioli (ed.), *The Science Studies Reader* (New York: Routledge, 1999), pp. 417–29; "From Microsomes to Ribosomes: 'Strategies' of 'Representation,'" *Journal of the History of Biology* 28 (1995): 49–89; and *Toward a History of Epistemic Things* (Stanford: Stanford University Press, 1997); as well as Colin Nazhone Milburn, "Monsters in Eden: Darwin and Derrida," *Modern Language Notes* 118 (2003): 603–21.

Acknowledgments

FOR someone who obsesses, as I do, over the complexities of the author function, acknowledging how much this book owes to friends and colleagues is a happy exercise in authorial deconstruction. What makes me less happy, however, is realizing how difficult, and probably impossible, it is to retrieve and acknowledge all I have received along the way. If I fail to mention something or someone, please read it as a sign of my decaying memory, not of my ingratitude.

Bernard Cohen, Sande Cohen, Paul David, Peter Galison, Michael Gordin, Adrian Johns, Phil Mirowski, Kriss Ravetto-Biagioli, Simon Schaffer, and an anonymous reviewer have read the entire manuscript at different stages of its completion and offered cogent criticism. Ongoing conversations with Kriss Ravetto-Biagioli have left yet another layer of traces on it.

Various incarnations of chapter one benefited from feedback from Hal Cook, Moti Feingold, Ofer Gal, Sharon Gallagher, the Grinch (a.k.a John Heilbron), Steve Harris, Roger Hart, Skuta Helgason, Michael Hunter, Myles Jackson, Carrie Jones, and Liz Lee. Chapter two received equal help from Jim Bennett, Allan Brandt, Reine Daston, Rivka Feldhay, Alan Franklin, Philippe Hamou, Nick Jardine, Matt Jones, Katy Park, Jürgen Renn, Randy Starn, Noel Swerdlow, Sherry Turkle, Al van Helden, and Norton Wise. Chapter three benefited from readings by Owen Gingerich, John Gorman, John Heilbron, Lisa Pon, Claus Zittel, and especially Al van Helden. Chapter four had a particularly long gestation during which it was improved by comments and suggestions from Rüdiger Campe, Arnold Davidson, Maurice Finocchiaro, Nick Jardine, Matt Jones, Lily Kay, Liz Lee, Colin Milburn, Katy Park, Andy Pickering, Hans-Jörg Rheinberger, and Randy Starn. My graduate students and the members of the Early Science Working Group at Harvard have also read and critiqued several of these chapters over the years.

Jean François Gauvin, Liz Lee, Elly Truitt, Elizabeth Yale, and especially Kristina Stewart have helped me greatly with the research and preparation of this manuscript. Lucia Prauscello helped me not to get lost in translation, while Alan Thomas and Catherine Rice at the University of Chicago Press prevented me from getting lost in never-ending revisions. Support for this project came from a grant from the National Endowment for the Humanities and fellowships from the Dibner Institute for the History of Science and Technology and the John Simon Guggenheim Foundation. Additional support was made available by the Dean of the Faculty of Arts and Sciences at Harvard, Jeremy Knowles.

Earlier versions of some of these chapters were published in *Science in Context* 13 (2000), number 3/4; in Wolfgang Detel and Claus Zittel (eds.), *Wissensideale und Wissenskulturen in der frühen Neuzeit* (Berlin: Akademie Verlag, 2002); and in *Modern Language Notes* 118 (2003), (German Issue). Material from those previous versions is included in this book with permission from the publishers.

Special thanks go to Galileo, who has given me plenty to think and write about and have fun with—not to mention a legal and socially respectable way to pay the bills. Spanning almost as long as my familiarity with Galileo's work is my friendship with Al van Helden. A mere acknowledgment would not convey how much chapters two and three owe to our conversations and to the work in progress and translations he has shared with me over the last ten years. Had I been better at long-distance collaborations, chapter three should have been part of a jointly authored book. I regret that it has not worked out that way and that, quite likely, the readers will like the new book on sunspots that Al and Eileen Reeves are finishing better than my chapter.

After thanking so many, I need to name those whom I cannot thank anymore. Three close friends, Susan Abrams, Bernard Cohen, and Lily Kay, passed away as this project was being completed. The impossibility of sharing this book reminds me of how much I have been missing them.

I dedicate this book to my beloved dudes, Gabriel and Luka, looking forward to so many more of their endless mutations. And, Kriss, thank you for everything. Davvero.

Castiglione della Pescaia, August 2004

References

Ackerman, James. "Early Renaissance 'Naturalism' and Scientific Illustration." In *The Natural Sciences and the Arts,* edited by Allan Ellenius, 1–17. Stockholm: Almqvist, 1985.

Adams, C. W. "A Note on Galileo's Determination of the Height of Lunar Mountains." *Isis* 17 (1932): 427–29.

Adelmann, Howard. *The Embryological Treatises of Hieronymus Fabricius ab Aquapendente.* 2 vols. Ithaca: Cornell University Press, 1942.

Alpers, Svetlana. *The Art of Describing: Dutch Art in the Seventeenth Century.* Chicago: University of Chicago Press, 1983.

Ambassades du Roy de Siam envoyé à l'Excellence du Prince Maurice, arrivé a la Haye le 10. Septemb. 1608. The Hague, 1608.

Aquapendente, Hieronymus Fabricius. *De ovi et pulli tractatus accuratissimus.* Padua: Benci, 1621.

———. *De venarum ostiolis.* Padua: Pasquati, 1603.

Ashbrook, Joseph. "Christopher Scheiner's Observations of an Object near Jupiter." *Sky and Telescope* 42 (1977): 344–45.

Ashworth, William. "Natural History and the Emblematic Worldview." In *Reappraisals of the Scientific Revolution,* edited by David Lindberg and Robert Westman, 303–32. Cambridge: Cambridge University Press, 1990.

Bagehot, Walter. *Lombard Street: A Description of the Money Market.* London: Murray, 1919.

Baigrie, Brian. *Picturing Knowledge: Historical and Philosophical Problems Concerning the Use of Art in Science.* Toronto: University of Toronto Press, 1996.

Baldini, Ugo. "L'astronomia del Cardinale Bellarmino." In *Novità celesti e crisi del sapere,* edited by Paolo Galluzzi, 293–305. Florence: Giunti, 1984.

———. *Legem Impone Subactis: Studi su filosofia e scienza dei gesuiti in Italia, 1540–1632.* Rome: Bulzoni, 1992.

Baldini, Ugo, and George Coyne. *The Louvain Lectures of Bellarmine and the Autograph Copy of His 1616 Declaration to Galileo.* Vatican City: Specola Vaticana, 1984.

Barker, Peter, and Bernard Goldstein. "Realism and Instrumentalism in Sixteenth-Century Astronomy: A Reappraisal." *Perspectives on Science* 6 (1998): 232–58.

Barthes, Roland. "The Brain of Einstein." In *Mythologies,* 68–71. New York: Noonday Press, 1991.

Baumgartner, Frederic. "Sunspots or Sun's Planets: Jean Tarde and the Sunspots Controversy of the Early Seventeenth Century." *Journal for the History of Astronomy* 18 (1987): 44–54.

Benjamin, Walter. "The Work of Art in the Age of Mechanical Reproduction." In *Illuminations,* 217–52. New York: Schocken, 1969.

Bennett, James. "The Mechanics' Philosophy and the Mechanical Philosophy." *History of Science* 24 (1986): 1–28.

———. "Shopping for Instruments in Paris and London." In *Merchants and Marvels,* edited by Pamela Smith and Paula Findlen, 370–95. New York: Routledge, 2002.

Beretta, Francesco. "L'Archivio della Congregazione del Sant'Ufficio: Bilancio provvisorio della storia e natura dei fondi d'antico regime." In *L'Inquisizione Romana: Metodologia delle fonti e storia istituzionale,* edited by Andrea del Col and Giovanna Paolin, 119–44. Trieste: Edizioni Università di Trieste, 1999.

———. "Le Procès de Galilée et les archives du Saint-Office: Aspects judiciaires et théologiques d'une condamnation célèbre." *Revue des sciences philosophiques et théologiques* 83 (1999): 441–90.

Berkel, K. van. "Intellectuals against Leeuwenhoek." In *Antoni van Leeuwenhoek, 1632–1723,* edited by L. C. Palm and H. A. M. Snelders, 187–209. Amsterdam: Rodopi, 1982.

Biagioli, Mario. "The Anthropology of Incommensurability." *Studies in History and Philosophy of Science* 21 (1990): 183–209.

———. "Etiquette, Interdependence, and Sociability in Seventeenth-Century Science." *Critical Inquiry* 22 (1996): 193–238.

———. "From Book Censorship to Academic Peer Review." *Emergences* 12 (2002): 11–45.

———. "From Difference to Blackboxing: French Theory versus Science Studies' Metaphysics of Presence." In *French Theory in America,* edited by Sande Cohen and Sylvere Lotringer, 271–87. New York: Routledge, 2001.

———. "Galilei vs. Capra: Of Instruments and Intellectual Property." *History of Science* forthcoming.

———. *Galileo, Courtier.* Chicago: University of Chicago Press, 1993.

———. "Knowledge, Freedom, and Brotherly Love: Homosociality and the Accademia dei Lincei, 1603–1630." *Configurations* 3 (1995): 139–66.

———. "Rights or Rewards?" In *Scientific Authorship: Credit and Intellectual Property in Science,* edited by Mario Biagioli and Peter Galison, 253–79. New York: Routledge, 2003.

———, ed. *The Science Studies Reader.* New York: Routledge, 1999.

———. "Scientific Revolution, Social Bricolage, and Etiquette." In *The Scientific Revolution in National Context,* edited by Roy Porter and Mikulas Teich, 11–54. Cambridge: Cambridge University Press, 1992.

Biagioli, Mario, and Peter Galison, eds. *Scientific Authorship: Credit and Intellectual Property in Science.* New York: Routledge, 2003.

Bijker, Wiebe, Thomas Hughes, and Trevor Pinch, eds. *The Social Construction of Technological Systems.* Cambridge: MIT Press, 1987.

Birch, Thomas. *The History of the Royal Society of London.* London: Millar, 1756–57. 4 vols.

Bisschop, W. R. *The Rise of the London Money Market.* New York: Kelley, 1968.

Blackwell, Richard. *Galileo, Bellarmine, and the Bible.* Notre Dame: University of Notre Dame Press, 1991.

Bloom, Terrie. "Borrowed Perceptions: Harriot's Maps of the Moon." *Journal for the History of Astronomy* 9 (1978): 117–22.

Blumenberg, Hans. *Die Lesbarkeit der Welt.* Frankfurt: Suhrkamp Verlag, 1981.

Boaga, Emanuele. "Annotazioni e documenti sulla vita e sulle opere di Paolo Antonio Foscarini teologo 'copernicano.'" *Carmelus* 37 (1990): 173–216.

Boas Hall, Marie. *Henry Oldenburg: Shaping the Royal Society.* Oxford: Oxford University Press, 2002.

———. *Promoting Experimental Learning: Experiment and the Royal Society, 1660–1727.* Cambridge: Cambridge University Press, 1991.

Bono, James. *The Word of God and the Languages of Man.* Madison: University of Wisconsin Press, 1995.

Bourdieu, Pierre. "Social Space and the Genesis of Groups." *Theory and Society* 14 (1985): 723–44.

———. "The Specificity of the Scientific Field and the Social Conditions for the Progress of Reason." In *The Science Studies Reader,* edited by Mario Biagioli, 31–50. New York: Routledge, 1999.

Boyle, James. *Shamans, Software, and Spleens.* Cambridge: Harvard University Press, 2003.

Brain, Robert, and Norton Wise. "Muscles and Engines: Indicator Diagrams and Helmholtz's Graphical Methods." In *The Science Studies Reader,* edited by Mario Biagioli, 51–66. New York: Routledge, 1999.

Bredekamp, Horst. "Gazing Hands and Blind Spots: Galileo as Draftsman." In *Galileo in Context,* edited by Jürgen Renn, 153–92. Cambridge: Cambridge University Press, 2001.

Bucciantini, Massimo. Contro Galileo: Alle origini dell'affaire. Florence: Olschki, 1995.

———. Galileo e Keplero: Filosofia, cosmologia e teologia nell'Età della Controriforma. Turin: Einaudi, 2003.

Bury, Michael. *The Print in Italy, 1550–1620.* London: British Museum Press, 2001.

Cajori, Florian. "History of Determinations of the Heights of Mountains." *Isis* 12 (1929): 482–514.

Callon, Michel. "Society in the Making: The Study of Technology as a Tool for Sociological Analysis." In *The Social Construction of Technological Systems,* edited by W. Bijker, T. Hughes, and T. Pinch, 83–103. Cambridge: MIT Press, 1987.

———. "Struggles and Negotiations to Define What Is Problematic and What Is Not: The Sociology of Translation." In *The Social Process of Scientific Investigation,* edited by Karin Knorr-Cetina, Roger Krohn, and Richard Whitley, 197–220. Dordrecht: Reidel, 1981.

Callon, Michel, and Bruno Latour. "Don't Throw the Baby Out with the Bath School!" In *Science as Practice and Culture,* edited by Andrew Pickering, 343–68. Chicago: University of Chicago Press, 1992.

———. "Unscrewing the Big Leviathan." In *Advances in Social Theory: Toward an Integration of Micro- and Macro-Sociologies,* edited by Karin Knorr-Cetina and Alain Cicourel, 277–303. London: Routledge, 1981.

Camerota, Michele. "Aristotelismo e nuova scienza nell'opera di Christoph Scheiner." *Galilaeana* forthcoming.

Canales, Jimena. "Photogenic Venus: The 'Cinematographic Turn' in Science and Its Alternatives." *Isis* 93 (2002): 585–613.

Capra, Baldassare. *Usus et fabrica circini cuiusdam proportionis.* Padua: Tozzi, 1607.

Carey, Daniel. "Compiling Nature's History: Travellers and Travel Narratives in the Early Royal Society." *Annals of Science* 54 (1997): 269–92.

Chalmers, Alan. *Science and Its Fabrication.* Minneapolis: University of Minnesota Press, 1990.

Chartier, Roger. "*Secretaires* for the People?" In *Correspondence,* edited by Roger Chartier, Alain Boureau, and Cécile Dauphin, 59–111. Cambridge: Polity Press, 1997.

Christianson, John Robert. *On Tycho's Island.* Cambridge: Cambridge University Press, 2000.

Clapman, John. *The Bank of England: A History.* Cambridge: Cambridge University Press, 1944.

Cohen, I. Bernard. "The Influence of Theoretical Perspective on the Interpretation of Sense Data: Tycho Brahe and the New Star of 1572, and Galileo and the Mountains on the Moon." *Annali dell'Istituto e Museo di Storia della Scienza di Firenze* 5 (1980): 3–14.

———. "What Galileo Saw: The Experience of Looking through a Telescope." In *From Galileo's "Occhialino" to Optoelectronics,* edited by P. Mazzoldi, 445–72. Padua: Cleup Editrice, 1993.

Cohen, Sande, and Sylvere Lotringer, eds. *French Theory in America.* New York: Routledge, 2001.

Collins, Harry M. *Changing Order: Replication and Induction in Scientific Practice.* London: Sage, 1985.

———. "Public Experiments and Display of Virtuosity: The Core-Set Revisited." *Social Studies of Science* 18 (1988): 725–48.

Collins, Harry M., and Steven Yearley. "Epistemological Chicken." In *Science as Practice and Culture,* edited by Andrew Pickering, 301–26. Chicago: University of Chicago Press, 1992.

Cook, Harold, and David Lux. "Closed Circles or Open Networks?: Communicating at a Distance during the Scientific Revolution." *History of Science* 36 (1998): 179–211.

Copernicus, Nicholas. *Complete Works,* translated by Edward Rosen. Vol. I. Baltimore: Johns Hopkins University Press, 1978.

Cosentino, Giuseppe. "Le matematiche nella Ratio Studiorum della Compagnia di Gesù." *Miscellanea storica ligure,* n.s., 2 (1970): 171–213.

Coulston, Christopher. "The Bank of the Republic of Letters: Johannes Hevelius and the Royal Society." Unpublished manuscript. Department of History of Science, Harvard University, 2001.

Crombie, Alistair. "Mathematics and Platonism in the Sixteenth-Century Italian Universities and in Jesuit Educational Policy." In *Prismata,* edited by Y. Maeyama and W. G. Saltzer, 63–94. Wiesbaden: Franz Steiner Verlag, 1974.

Curtius, Ernst Robert. *European Literature and the Latin Middle Ages.* New York: Harper & Row, 1963.

Daston, Lorraine, and Peter Galison. "The Image of Objectivity." *Representations* 40 (1992): 81–128.

Daxecker, Franz. *Briefe des Naturwissenschaftlers Christoph Scheiner SJ an Erzherzog Leopold V von Österreich Tirol 1620–1632*. Innsbruck: Publikationsstelle der Universität Innsbruch, 1995.

Dear, Peter. *Discipline and Experience: The Mathematical Way in the Scientific Revolution*. Chicago: University of Chicago Press, 1995.

———. "Jesuit Mathematical Science and the Reconstitution of Experience in the Early Seventeenth Century." *Studies in History and Philosophy of Science* 18 (1987): 133–75.

———. "Totius in verba: Rhetoric and Authority in the Early Royal Society." *Isis* 76 (1985): 145–61.

De Caro, Mario. "Galileo's Mathematical Platonism." In *Philosophy of Mathematics*, edited by Johannes Czermak, 13–22. Vienna: Verlag Holder-Pichler-Tempsky, 1993.

Del Col, Andrea, and Giovanna Paolin, eds. *L'Inquisizione Romana: Metodologia delle fonti e storia istituzionale*. Trieste: Edizioni Università di Trieste, 1999.

Derrida, Jacques. "Differance." In *Margins of Philosophy*, translated by Alan Bass, 1–27. Chicago: University of Chicago Press, 1982.

———. "Le facteur de la vérité." In *The Post Card*, 413–96. Chicago: University of Chicago Press, 1987.

———. *Of Grammatology*. Baltimore: Johns Hopkins University Press, 1976.

———. "Plato's Pharmacy." In *Dissemination*, translated by Barbara Johnson, 61–172. Chicago: University of Chicago Press, 1981.

———. "Signature Event Context. " In *Margins of Philosophy*, translated by Alan Bass, 307–30. Chicago: University of Chicago Press, 1982.

Dobell, Clifford. *Antony van Leeuwenhoek and His "Little Animals."* New York: Dover, 1960.

Dollo, Corrado. "Tanquam nodi in tabula—tanquam pisces in aqua. Le innovazioni della cosmologia nella *Rosa Ursina* di Christoph Scheiner." In *Christoph Clavius e l'attività scientifica dei gesuiti nell'età di Galileo*, edited by Ugo Baldini, 133–58. Rome: Bulzoni, 1995.

Donahue, William. *The Dissolution of the Celestial Spheres: 1595–1650*. New York: Arno Press, 1981.

Doyle, Richard. *On Beyond Living: Rhetorical Transformations of the Life Sciences*. Stanford: Stanford University Press, 1997.

Drake, Stillman. "Galileo and Satellites Prediction." *Journal for the History of Astronomy* 10 (1979): 75–95.

———. *Galileo at Work: His Scientific Biography*. Chicago: University of Chicago Press, 1978.

———. *Galileo Studies*. Ann Arbor: University of Michigan Press, 1970.

———. "Galileo's First Telescopic Observations." *Journal for the History of Astronomy* 7 (1976): 153–68.

———. *The Unsung Journalist and the Origin of the Telescope*. Los Angeles: Zeitlin & ver Brugge, 1976.

Drake, Stillman, and C. D. O'Malley. *The Controversy on the Comets of 1618*. Philadelphia: University of Pennsylvania Press, 1960.

Duhem, Pierre. *To Save the Phenomena*. Chicago: University of Chicago Press, 1969.

Duhr, Bernhard, *Geschichte der Jesuiten in den Landern Deutscher Zunge.* Vol. II, pt. 2. Freiburg: Herdersche Verlagshandlung, 1913.

Dupré, Sven. "Galileo's Telescopes and Celestial Light." *Journal for the History of Astronomy* 34 (2003): 369–99.

Eddy, John, Peter Gilman, and Dorothy Trotter. "Anomalous Solar Rotation in the Early 17th Century." *Science* 198 (1977): 824–29.

Edgerton, Samuel. "Galileo, Florentine 'Disegno,' and the 'Strange Spottedness' of the Moon." *Art Journal* 44 (1984): 225–32.

Evans, R. J. W. "Rantzau and Welser: Aspects of Later German Humanism." *History of European Ideas* 5 (1984): 257–72.

Fabricius, Johannes. *De maculis in sole observatis et apparente earum cum Sole conversione. . . .* Wittemberg: Typis Laurentii Seuberlichii, 1611.

Favaro, Antonio. *Amici e corrispondenti di Galileo.* Edited by Paolo Galluzzi. Florence: Salimbeni, 1983.

———. "Delle case abitate da Galileo Galilei in Padova." In *Galileo Galilei a Padova,* Vol. I, pp. 57–95. Padua: Antenore, 1968.

———. *Oppositori di Galileo, III: Cristoforo Scheiner.* Venice: Ferrari, 1919.

———. "Sulla morte di Marco Velsero e sopra alcuni particolari della vita di Galileo." *Bullettino di bibliografia e storia delle scienze matematiche e fisiche* 17 (1884): 252–70.

———. "Sulla priorità della scoperta e della osservazione delle macchie solari." *Memorie del Reale Istituto Veneto di Scienze, Lettere, ed Arti* 13 (1887): 729–90.

Feldhay, Rivka. "Producing Sunspots on an Iron Pan." In *Science, Reason, and Rhetoric,* edited by Henry Krips, J. E. McGuire, and Trevor Melia, 119–43. Pittsburgh: University of Pittsburgh Press, 1998.

Feyerabend, Paul. *Against Method: Outline of an Anarchistic Theory of Knowledge.* London: Verso, 1978.

Finocchiaro, Maurice, ed. *The Galileo Affair.* Berkeley: University of California Press, 1989.

———. *Galileo on the World Systems.* Berkeley: University of California Press, 1997.

———. *Retrying Galileo, 1633–1992.* Berkeley: University of California Press, 2005.

Fleck, Ludwik. *Genesis and Development of a Scientific Fact.* Chicago: University of Chicago Press, 1979.

Fortun, Michael. "Mediated Speculations in the Genomics Future Markets." *New Genetics and Society* 20 (2001): 139–56.

Foscarini, Paolo Antonio. *Lettera sopra l'opinione de' Pittagorici e del Copernico. . . .* Naples: Scoriggio, 1615.

Foucault, Michel. "What Is an Author?" In *Language, Counter-Memory, Practice: Selected Essays and Interviews,* edited by Donald Bouchard, 113–38. Ithaca: Cornell University Press, 1977.

Fragnito, Gigliola. *La Bibbia al rogo: La censura ecclesiastica e i volgarizzamenti della Scrittura (1471–1605).* Bologna: Il Mulino, 1997.

Gabrieli, Giuseppe. "Marco Welser Linceo augustano." *Rendiconti della Reale Accademia Nazionale dei Lincei, Classe di Scienze Morali, Storiche e Filologiche,* 6th ser., 14 (1938): 74–99.

Gaines, Jane. "Reincarnation as the Ring on Liz Taylor's Finger: Andy Warhol and the Right of Publicity." In *Identities, Politics, and Rights,* edited by Austin Sarat and Thomas Kearns, 131–48. Ann Arbor: University of Michigan Press, 1997.

Galilei, Galileo. *Dialogo sopra i due massimi sistemi del mondo, tolemaico e copernicano.* Florence: Landini, 1632.

———. *Difesa contro alle calunnie & imposture di Baldessar Capra Milanese.* Venice: Baglioni, 1607.

———. *Discorso intorno alle cose che stanno in su l'acqua, o che in quella si muovono.* Florence: Giunti, 1612.

———. *Discoveries and Opinions of Galileo.* Translated by Stillman Drake. New York: Doubleday, 1957.

———. *Istoria e dimostrazioni intorno alle macchie solari e loro accidenti: comprese in tre lettere scritte all'illustrissimo signor Marco Velseri.* Rome: Mascardi, 1613.

———. *Operations of the Geometric and Military Compass.* Translated by Stillman Drake. Washington: Smithsonian Institution Press, 1978.

———. *Operazioni del compasso geometrico e militare.* Padua: Marinelli, 1606.

———. *Le opere di Galileo Galilei.* Edited by Antonio Favaro. 20 vols. Florence: Barbera, 1890–1909.

———. *Sidereus nuncius.* Frankfurt: Paltheniano, 1610.

———. *Sidereus nuncius: or the Sidereal Messenger.* Translated with introduction, conclusion, and notes by Albert van Helden. Chicago: University of Chicago Press, 1989.

Galison, Peter. "Context and Constraints." In *Scientific Practices,* edited by Jed Buchwald, 13–41. Chicago: University of Chicago Press, 1995.

Galluzzi, Paolo. "Evangelista Torricelli: concezione della matematica e segreto degli occhiali." *Annali dell'Istituto e Museo di Storia della Scienza di Firenze* 1 (1976): 71–95.

———, ed. *Novità celesti e crisi del sapere.* Florence: Giunti, 1984.

———. "The Sepulchers of Galileo: The 'Living' Remains of a Hero of Science." In *Cambridge Companion to Galileo,* edited by Peter Machamer, 417–48. Cambridge: Cambridge University Press, 1998.

Gerulaitis, Leonardas Vytautas. *Printing and Publishing in Fifteenth-Century Venice.* Mansell: London, 1976.

Giere, Ronald. "Visual Models and Scientific Judgment." In *Picturing Knowledge: Historical and Philosophical Problems Concerning the Use of Art in Science,* edited by Brian Baigrie, 269–302. Toronto: University of Toronto Press, 1996.

Gingerich, Owen. "Dissertatio cum Profesor Righini and Sidereo Nuncio." In *Reason, Experiment, and Mysticism,* edited by Maria Luisa Righini Bonelli and William Shea, 77–88. New York: Science History Publications, 1975.

Gingerich, Owen, and Albert van Helden, "From Occhiale to Printed Page: The Making of Galileo's *Sidereus nuncius.*" *Journal for the History of Astronomy* 34 (2003): 251–67.

Gingerich, Owen, and Robert Westman. *The Wittich Connection: Conflict and Priority in Late Sixteenth-Century Cosmology.* Philadelphia: American Philosophical Society, 1988.

Goldgar, Ann. *Impolite Learning: Conduct and Community in the Republic of Letters, 1680–1750.* New Haven: Yale University Press, 1995.

Goldstein, Bernard. "Some Medieval Reports of Venus and Mercury Transits." *Centaurus* 14 (1969): 49–59.

Gorman, Michael John. "Mathematics and Modesty in the Society of Jesus: The Problems of Christoph Greimberger." In *The New Science and Jesuit Science,* edited by Mordechai Feingold, 1–120. Dordrecht: Kluwer, 2003.

———. "A Matter of Faith? Christoph Scheiner, Jesuit Censorship, and the Trial of Galileo." *Perspectives on Science* 4 (1996): 283–320.

———. "The Scientific Counter-Revolution: Mathematics, Natural Philosophy, and Experimentation in Jesuit Science, 1580–1670." Ph.D. diss., European University Institute, 1998.

Granada, Miguel. *Sfere solide e cielo fluido: Momenti del dibattito cosmologico nella seconda metà del Cinquecento.* Milan: Guerini, 2002.

Hamou, Philippe. *La mutation du visible: Microscopes et Télescopes en Angleterre de Bacon à Hooke.* Villeneuve d'Ascq: Presses Universitaires du Septentrion, 2001.

Haraway, Donna. "A Cyborg Manifesto." In *Simians, Cyborgs, and Women,* 149–81. New York: Routledge, 1991.

———. "Situated Knowledges: The Science Question in Feminism and the Privilege of Partial Perspective." In *The Science Studies Reader,* edited by Mario Biagioli, 172–88. New York: Routledge, 1999.

Harcourt, Glenn. "Andreas Vesalius and the Anatomy of Antique Sculpture." *Representations* 17 (1987): 28–61.

Harris, Steven. "Expanding the Scales of Scientific Practice through Networks of Travel, Correspondence, and Exchange." In *Cambridge History of Science,* vol. III, edited by Lorraine Daston and Katharine Park. Cambridge: Cambridge University Press, forthcoming.

———. "Long-Distance Corporations, Big Sciences, and the Geography of Knowledge." *Configurations* 6 (1998): 269–304.

Harvey, William. *Exercitatio anatomica de motu cordis at sanguinis in animalibus.* Frankfurt: Fitzer, 1628.

Heilbron, John. *Physics at the Royal Society during Newton's Presidency.* Los Angeles: Clark Library, 1983.

———. *The Sun in the Church: Cathedrals as Solar Observatories.* Cambridge: Harvard University Press, 1999.

Herr, Richard. "Solar Rotation Determined from Thomas Harriot's Sunspots Observations of 1611 to 1613." *Science* 202 (1978): 1079–81.

Hooke, Robert. *A Description of Helioscopes.* London: Martyn, 1676.

Horky, Martinus. *Brevissima peregrinatio contra nuncium sidereum.* Modena: Cassiani, 1610.

Hosie, Alexander. "The First Observations of Sun-Spots." *Nature* 20 (1879): 131–32.

Hughes, Justin. "The Personality Interest of Artists and Inventors in Intellectual Property." *Cardozo Arts and Entertainment Law Journal* 16 (1998): 81–181.

Humbert, Pierre. "Joseph Gaultier de la Vallette, astronome provençal (1564–1647)." *Revue d'histoire des sciences et de leurs applications* 1 (1948): 316.

Hunter, Michael. *Establishing the New Science: The Experience of the Early Royal Society.* Woodbridge: Boydell, 1989.

———. *The Royal Society and Its Fellows, 1660–1700: The Morphology of an Early Scientific Institution.* Oxford: British Society for the History of Science, 1994.

———. *Science and Society in Restoration England.* Cambridge: Cambridge University Press, 1992.

Hutchinson, Keith. "Sunspots, Galileo, and the Orbit of the Earth." *Isis* 81 (1990): 68–74.

Iliffe, Rob. "Butter for Parsnips: Authorship, Audience, and Incomprehensibility." In *Scientific Authorship,* edited by Mario Biagioli and Peter Galison, 33–66. New York: Routledge, 2003.

———. "Foreign Bodies: Travel, Empire, and the Early Royal Society of London. Part 1. Englishmen on Tour." *Canadian Journal of History* 33 (1998): 358–85.

———. "Foreign Bodies: Travel, Empire, and the Early Royal Society of London. Part 2. The Land of Experimental Knowledge." *Canadian Journal of History* 34 (1999): 24–50.

———. "'In the Warehouse': Privacy, Property, and Priority in the Early Royal Society." *History of Science* 30 (1992): 29–68.

International Committee of Medical Journal Editors (ICMJE). "Uniform Requirements for Manuscripts Submitted to Biomedical Journals." *JAMA* 227 (1997): 928.

Ivins, William. *Prints and Visual Communication.* Cambridge: MIT Press, 1953.

———. "What about the 'Fabrica' of Vesalius?" In *Three Vesalian Essays,* edited by S. W. Lambert, 45–99. New York: McMillan, 1952.

Jardine, Lisa. *Erasmus, Man of Letters: The Construction of Charisma in Print.* Princeton: Princeton University Press, 1993.

Jardine, Nicholas. *The Birth of the History and Philosophy of Science.* Cambridge: Cambridge University Press, 1984.

———. "The Forging of Modern Realism: Clavius and Kepler against the Sceptics." *Studies in History and Philosophy of Science* 10 (1979): 141–73.

Johns, Adrian. *The Nature of the Book: Print and Knowledge in the Making.* Chicago: University of Chicago Press, 1998.

Julia, Dominique. "Reading and the Counter-Reformation." In *A History of Reading in the West,* edited by Guglielmo Cavallo and Roger Chartier, 238–68. Amherst: University of Massachusetts Press, 1999.

Kay, Lily. *Who Wrote the Book of Life?* Stanford: Stanford University Press, 1999.

Kemp, Martin. "Temples of the Body and Temples of the Cosmos." In *Picturing Knowledge: Historical and Philosophical Problems Concerning the Use of Art in Science,* edited by Brian Baigrie, 40–85. Toronto: University of Toronto Press, 1996.

Kepler, Johannes. *Conversation with the Sidereal Messenger.* Translated by Edward Rosen. New York: Johnson, 1965.

———. *Dioptrice.* Augsburg: Franci, 1611.

———. *Dissertatio cum Nuncio Sidereo.* Prague: Sedesan, 1610.

———. *Gesammelte Werke.* Edited by Max Caspar and Franz Hammer. 20 vols. Munich: Beck'sche Verlagsbuchhandlung, 1937–.

———. *Narratio de observatis a se quatuor Iovis satellitibus erronibus.* Frankfurt: Palthenius, 1611.

King, Nicholas. "Narrative and the Effacement of the Visual in the *De motu cordis.*" Unpublished manuscript. Department of History of Science, Harvard University, 1996.

Knorr-Cetina, Karin, and Alain Cicourel, eds. *Advances in Social Theory: Toward an Integration of Micro- and Macro-Sociologies.* London: Routledge, 1981.

Knorr-Cetina, Karin, Roger Krohn, and Richard Whitley, eds. *The Social Process of Scientific Investigation.* Dordrecht: Reidel, 1981.

Koyré, Alexandre. "Galileo and Plato." *Journal of the History of Ideas* 4 (1943): 400–428.

Latour, Bruno. "One More Turn After the Social Turn . . ." in *The Science Studies Reader*, edited by Mario Biagioli, 276–89. New York: Routledge, 1999.

———. "The Politics of Explanation: An Alternative." In *Knowledge and Reflexivity*, edited by Steve Woolgar, 155–76. London: Sage, 1988.

———. *Science in Action*. Cambridge: Harvard University Press, 1987.

———. "Visualization and Cognition: Thinking with Eyes and Hands." *Knowledge and Society* 6 (1986): 1–40.

Latour, Bruno, and Steve Woolgar. *Laboratory Life: The Construction of Scientific Facts*. Princeton: Princeton University Press, 1986.

Lattis, James. *Between Copernicus and Galileo*. Chicago: University of Chicago Press, 1994.

Law, John. "On the Methods of Long-Distance Control: Vessels, Navigation, and the Portuguese Route to India." *Sociological Review Monographs* 32 (1986): 234–63.

Leeuwenhoek, Antoni van. *The Collected Letters of Antoni van Leeuwenhoek*. Edited by G. van Rijnberk, A. Schierbeek, J. J. Swart, J. Heniger, and L. C. Palm. 12 vols. Amsterdam: Sweets & Zeitlinger, 1939–89.

Lenoir, Timothy. *Inscribing Science*. Stanford: Stanford University Press, 1998.

Lindberg, David, and Robert Westman, eds. *Reappraisals of the Scientific Revolution*. Cambridge: Cambridge University Press, 1990.

Long, Pamela. *Openness, Secrecy, Authorship: Technical Arts and the Culture of Knowledge from Antiquity to the Renaissance*. Baltimore: Johns Hopkins University Press, 2001.

Lynch, Michael. "Discipline and the Material Form of Images: An Analysis of Scientific Visibility." *Social Studies of Science* 15 (1985): 37–66.

———. "Representation Is Overrated: Some Critical Remarks about the Use of the Concept of Representation in Science Studies." *Configurations* 1 (1994): 137–49.

Machamer, Peter K. "Feyerabend and Galileo: The Interaction of Theories, and the Reinterpretation of Experience." *Studies in History and Philosophy of Science* 4 (1973): 1–46.

MacKenzie, Donald. *Inventing Accuracy: A Historical Sociology of Nuclear Missile Guidance*. Cambridge: MIT Press, 1990.

MacLeod, Christine. *Inventing the Industrial Revolution*. Cambridge: Cambridge University Press, 1988.

Mahoney, Michael. "Diagrams and Dynamics: Mathematical Perspectives on Edgerton's Thesis." In *Science and the Arts in the Renaissance*, edited by John Shirley and David Hoeniger, 198–220. Washington: Folger Books, 1985.

Mayaud, Pierre-Noel. *La condamnation des livres coperniciens et sa révocation*. Rome: Editrice Pontificia Università Gregoriana, 1997.

Meeus, Jean. 1964. "Galileo's First Records of Jupiter's Satellites." *Sky and Telescope* 27, no. 2 (1964): 105–6.

Merton, Robert. "Priorities in Scientific Discoveries." In *The Sociology of Science: Theoretical and Empirical Investigations*. Chicago: University of Chicago Press, 1973.

Micanzo, Fulgenzio. *Vita del Padre Paolo*. In *Istoria del Concilio Tridentino*, edited by Paolo Sarpi, II: 1273–1413. Turin: Einaudi, 1974.

Milburn, Colin. "Monsters in Eden: Darwin and Derrida." *Modern Language Notes* 118 (2003): 603–21.

———. "Nanotechnology in the Age of Posthuman Engineering: Science Fiction as Science." *Configurations* 10 (2002): 261–95.

Milgrom, Paul, and Nancy Stokey. "Information, Trade, and Common Knowledge." *Journal of Economic Theory* 26 (1982): 17–27.

Mosley, Adam, Nicholas Jardine, and Karin Tybjerg. "Epistolary Culture, Editorial Practices, and the Propriety of Tycho's *Astronomical Letters*." *Journal for the History of Astronomy* 34 (2003): 419–51.

North, John. "Thomas Harriot and the First Telescopic Observations of Sunspots." In *Thomas Harriot: Renaissance Scientist,* edited by John Shirley, 129–57. Oxford: Clarendon Press, 1974.

Oldenburg, Henry. *The Correspondence of Henry Oldenburg.* 13 vols. Edited by Rupert Hall and Marie Boas Hall. Madison: University of Wisconsin Press; London: Mansell; London: Taylor and Francis, 1965–86.

Palm, L. C. "Leeuwenhoek and Other Dutch Correspondents of the Royal Society." *Notes and Records of the Royal Society of London* 43 (1989): 191–207.

Palm, L. C., and H. A. M. Snelders, eds. *Antoni van Leeuwenhoek,* 1632–1723. Amsterdam: Rodopi, 1982.

Palmerino, Carla Rita. "The Mathematical Characters of Galileo's Book of Nature." In *The Book of Nature in Modern Times,* edited by Klaas van Berkel and Arjo Vanderjagt, vol. II, pp. 27–45. Leuven: Peeters Publishers, 2005.

Pedersen, Olaf. *The Book of Nature.* Vatican City: Vatican Observatory, 1992.

Peters, J. S. "The Bank, the Press, and the 'Return to Nature.'" In *Early Modern Conception of Property,* edited by John Brewer and Susan Staves, 365–88. London: Routledge, 1996.

Pickering, Andrew. "The Mangle of Practice: Agency and Emergence in the Sociology of Science." *American Journal of Sociology* 99 (1993): 559–89.

———, ed. *Science as Practice and Culture.* Chicago: University of Chicago Press, 1992.

Pinch, Trevor. *Confronting Nature: The Sociology of Solar-Neutrino Detection.* Dordrecht: Reidel, 1986.

Porter, Roy, and Mikulas Teich, eds. *The Scientific Revolution in National Context.* Cambridge: Cambridge University Press, 1992.

Powell, Ellis. *The Evolution of the Money Market.* London: Financial News, 1915.

Prickard, A. O. "The 'Mundus Jovialis' of Simon Marius." *Observatory* 39 (1916): 367–503.

Priuli, Antonio. "Dalla Cronica di Antonio Priuli." In Galileo, *Opere,* Vol. XIX, pp. 587–88.

Quarrell, W. H., and Margaret Mare, eds. and trans. *London in 1710: From the Travels of Zacharias Conrad von Uffenbach.* London: Faber & Faber, 1934.

Reeves, Eileen. *Painting the Heavens: Art and Science in the Age of Galileo.* Princeton: Princeton University Press, 1997.

Renn, Jürgen, ed. *Galileo in Context.* Cambridge: Cambridge University Press, 2001.

Rennie, Drummond, et al. "When Authorship Fails." *JAMA* 278 (1997): 580.

Rheinberger, Hans-Jörg. "Experimental Systems, Graphematic Spaces." In *Inscribing Science,* edited by Timothy Lenoir, 285–303, 425–29. Stanford: Stanford University Press, 1998.

———. "Experimental Systems: Historiality, Narration, and Deconstruction." In *The Science Studies Reader,* edited by Mario Biagioli, 417–29. New York: Routledge, 1999.

———. "From Microsomes to Ribosomes: 'Strategies' of 'Representation.'" *Journal of the History of Biology* 28 (1995): 49–89.

———. *Toward a History of Epistemic Things.* Stanford: Stanford University Press, 1997.

Righini, Guglielmo. "New Light on Galileo's Lunar Observations." In *Reason, Experiment, and Mysticism,* edited by Maria Luisa Righini Bonelli and William Shea, 59–76. New York: Science History Publications, 1975.

Righini Bonelli, Maria Luisa. "Le posizioni relative di Galileo e dello Scheiner nelle scoperte delle macchie solari nelle pubblicazioni edite entro il 1612." *Physis* 12 (1970): 405–10.

Righini Bonelli, Maria Luisa, and William Shea. *Reason, Experiment, and Mysticism.* New York: Science History Publications, 1975.

Roche, John. "Harriot, Galileo, and Jupiter's Satellites." *Archives internationales d'histoire des sciences* 32 (1982): 9–51.

Roffeni, Giovanni. *Epistola apologetica contra caecum peregrinationem cuiusdam furiosi Martini.* Bologna: Rossi, 1611.

Rosen, Edward. "Did Galileo Claim He Invented the Telescope?" *Proceedings of the American Philosophical Society* 98 (1954): 304–12.

Rossini, Giuseppe, ed. *Lettere e documenti riguardanti Evangelista Torricelli.* Faenza: Lega, 1956.

Royal Society of London. *The Record of the Royal Society.* London: Royal Society, 1940.

Rudwick, Martin. "The Emergence of a Visual Language for Geological Science, 1760–1840." *History of Science* 14 (1976): 149–95.

Sarat, Austin, and Thomas Kearns, eds. *Identities, Politics, and Rights.* Ann Arbor: University of Michigan Press, 1997.

Sarton, George. "Early Observations of the Sunspots?" *Isis* 37 (1947): 69–71.

Schaffer, Simon. "Glass Works: Newton's Prism and the Uses of Experiment." In *The Uses of Experiment,* edited by David Gooding, Trevor Pinch, and Simon Schaffer, 67–104. Cambridge: Cambridge University Press, 1989.

———. "The Leviathan of Parsontown: Literary Technology and Scientific Representation." In *Inscribing Science,* edited by Timothy Lenoir, 182–222. Stanford: Stanford University Press, 1998.

Scheiner, Christoph. *De maculis solaribus et stellis circa Iovem errantibus, accuratior disquisitio ad Marcum Welserum . . .* Augsburg: Ad insigne pinus, 1612.

———. *Pantographice, su ars delineandi res quaslibet. . . .* Rome: Grignani, 1631.

———. *Refractiones coelestes sive solis elliptici phaenomenon illustratum.* Ingolstadt: Eder, 1617.

———. *Rosa Ursina sive Sol ex admirando facularum & macularum.* Bracciano: Phaeum, 1630.

———. *Tres epistolae de maculis solaribus scriptae ad Marcum Welserum.* Augsburg: Ad insigne pinus, 1612.

Schove, Justin. "Sunspots and Aurorae." *Journal of the British Astronomical Association* 58 (1948): 178–90.

———. "Sunspots, Aurorae, and Blood Rain." *Isis* 42 (1951): 133–38.

Shapin, Steven. "Here and Everywhere: Sociology of Scientific Knowledge." *Annual Review of Sociology* 21 (1995): 289–321.

———. "Pump and Circumstance: Robert Boyle's Literary Technology." *Social Studies of Science* 14 (1984): 481–520.

———. *A Social History of Truth*. Chicago: University of Chicago Press, 1995.

Shapin, Steven, and Simon Schaffer. *Leviathan and the Air Pump*. Princeton: Princeton University Press, 1985.

Shea, William. *Galileo's Intellectual Revolution*. New York: Science History Publications, 1977.

Sirtori, Girolamo. *Telescopium, sive ars perficiendi*. Frankfurt: Iacobi, 1618.

Sizi, Francesco. 1611. *Dianoia astronomica, optica, physica*. Venice: Bertani, 1611.

Smith, Mark. "Galileo's Proof of the Earth's Motion from the Movement of Sunspots." *Isis* 76 (1985): 543–51.

Smith, Pamela, and Paula Findlen, eds. *Merchants and Marvels*. New York: Routledge, 2002.

Snyder, Joel. "Visualization and Visibility." In *Picturing Science, Producing Art*, edited by Caroline Jones and Peter Galison, 379–97. New York: Routledge, 1998.

Sprat, Thomas. *A History of the Royal Society*. London: Martyn, 1667.

Stabile, Giorgio. "Linguaggio della natura e linguaggio della scrittura in Galilei, dalla *Istoria* sulle macchie solari alle lettere copernicane." *Nuncius* 9 (1994): 37–64.

Starn, Randolph, and Loren Partridge. *Arts of Power: Three Halls of State in Italy, 1300–1600*. Berkeley: University of California Press, 1992.

Stieglitz, Joseph. "The Contributions of Information to Twentieth Century Economics." *Quarterly Journal of Economics* 114 (2000): 1441–78.

Thoren, Victor. *The Lord of Uraniborg*. Cambridge: Cambridge University Press, 1990.

Topper, David. "Galileo, Sunspots, and the Motions of the Earth." *Isis* 90 (1999): 757–67.

Torricelli, Evangelista. "Prefazione in lode delle matematiche." In *Opere scelte*, edited by Lanfranco Belloni, 615–26. Turin: UTET, 1975.

Traweek, Sharon. *Beamtimes and Lifetimes: The World of Particle Physicists*. Cambridge: Harvard University Press, 1988.

Turkle, Sherry. *Life on the Screen: Identity in the Age of the Internet*. New York: Simon & Schuster, 1995.

Van Helden, Albert. "The Accademia del Cimento and Saturn's Ring." *Physis* 15 (1973): 237–59.

———. "Galileo and Scheiner on Sunspots: A Case Study in the Visual Language of Astronomy." *Proceedings of the American Philosophical Society* 140 (1995): 357–95.

———. "Galileo and the Telescope." In *Novità celesti e crisi del sapere*. Edited by Paolo Galluzzi, 150–57. Florence: Giunti, 1984.

———. "Introduction." In *Sidereus nuncius: or the Sidereal Messenger*. Translated by Albert van Helden, 1–24. Chicago: University of Chicago Press, 1989.

———. "The Invention of the Telescope." *Transactions of the American Philosophical Society* 67, pt. 4 (1977): 1–67.

———. "Scheiner." Unpublished manuscript.

———. "The Telescope in the Seventeenth Century." *Isis* 65 (1974): 38–58.

———. "Telescopes and Authority from Galileo to Cassini." *Osiris* 9 (1994): 8–29.

Van Helden, Albert, and Eileen Reeves. *Galileo and Scheiner on Sunspots*. Chicago: University of Chicago Press, forthcoming.

Viviani, Vincenzio. "Racconto istorico di Vincenzio Viviani." In Galileo, *Opere*, Vol. XIX, pp. 597–632.

Voelkel, James. "Publish or Perish: Legal Contingencies and the Publication of Kepler's *Astronomia Nova.*" *Science in Context* 12 (1999): 33–59.

Von Braunmuehl, Anton. *Christoph Scheiner als Mathematiker, Physiker, und Astronom.* Bamberg: Buchnersche Verlagsbuchhandlung, 1891.

Weitzmann, Kurt. *Illustrations in Roll and Codex.* Princeton: Princeton University Press, 1970.

Westman, Robert. "The Astronomer's Role in the Sixteenth Century." *History of Science* 18 (1980): 105–47.

Whitaker, Ewen. "Galileo's Lunar Observations and the Dating of the Composition of the *Sidereus nuncius.*" *Journal for the History of Astronomy* 9 (1978): 155–69.

———. *Mapping and Naming the Moon: A History of Lunar Cartography and Nomenclature.* Cambridge: Cambridge University Press, 1999.

Wilding, Nick. *Writing the Book of Nature: Natural Philosophy and Communication in Early Modern Europe.* Ph.D. diss., European University Institute, 2000.

Winkler, Mary, and Albert van Helden. "Johannes Hevelius and the Visual Language of Astronomy." In *Renaissance and Revolution: Humanists, Scholars, Craftsmen, and Natural Philosophers in Early Modern Europe,* edited by Judith Field and Frank James, 97–116. Cambridge: Cambridge University Press, 1993.

———. "Representing the Heavens: Galileo and Visual Astronomy." *Isis* 83 (1992): 195–217.

Worp, J. A. *De Briefwisseling van Constantijn Huygens.* The Hague: Nijhoff, 1917.

Yau, K. C. C., and F. R. Stephenson. "A Revised Catalogue of Far Eastern Observations of Sunspots." *Quarterly Journal of the Royal Astronomical Society* 29 (1988): 175–97.

Ziggelaar, August. "Jesuit Astronomy North of the Alps. Four Unpublished Jesuit Letters, 1611–1620." In *Christoph Clavius e l'attività scientifica dei gesuiti nell'età di Galileo,* edited by Ugo Baldini, 101–32. Rome: Bulzoni, 1995.

Zik, Yaakov. "Galileo and Optical Aberrations." *Nuncius* 17 (2002): 455–65.

———. "Galileo and the Telescope." *Nuncius* 14 (1999): 31–67.

Zik, Yaakov, and Albert van Helden. "Between Discovery and Disclosure: Galileo and the Telescope." In *Musa Musaei: Studies on Scientific Instruments and Collections in Honour of Mara Miniati,* edited by Marco Beretta, Paolo Galluzzi, and Carlo Triarco, 173–90. Florence: Olschki, 2003.

Zucchi, Luca. "Brunfels e Fuchs: L'illustrazione botanica quale ritratto della singola pianta o immagine della specie." *Nuncius* 18 (2003): 411–65.

Index

Page numbers in italics refer to figures.